Hydraulic Conductivity: Engineering Essentials

Hydraulic Conductivity: Engineering Essentials

Editor: William Sobol

MURPHY & MOORE
www.murphy-moorepublishing.com

www.murphy-moorepublishing.com

MURPHY & MOORE

Cataloging-in-publication Data

Hydraulic conductivity : engineering essentials / edited by William Sobol.
p. cm.
Includes bibliographical references and index.
ISBN 978-1-63987-700-3
1. Hydraulic engineering. 2. Hydraulics. 3. Hydrodynamics. I. Sobol, William.
TC153 .H93 2023
627--dc23

Murphy & Moore Publishing
1 Rockefeller Plaza,
New York City,
NY 10020, USA

ISBN 978-1-63987-700-3

Contents

Preface

Hydraulic conductivity of materials such as plants, soil and rocks is defined as the ability of the fluid to pass through pores, spaces or fractures. Saturated hydraulic conductivity is the ratio of volume flux to hydraulic gradient yielding a quantitative measure of a saturated soil's ability to transmit water when subjected to a hydraulic gradient. The level of hydraulic conductivity relies on many factors such as the intrinsic permeability of the material, the degree of saturation, and the density and viscosity of the fluid. There are two approaches that are used for the determination of hydraulic conductivity. In the first approach, the empirical approach, hydraulic conductivity is associated with soil properties such as pore size and particle size (grain size) distributions, and soil texture. In the second approach, the experimental approach, the hydraulic conductivity is determined from hydraulic experiments using Darcy's law. This book contains some path-breaking studies related to hydraulic conductivity. The topics included herein are of utmost significance and will provide incredible insights to the readers.

This book has been the outcome of endless efforts put in by authors and researchers on various issues and topics within the field. The book is a comprehensive collection of significant researches that are addressed in a variety of chapters. It will surely enhance the knowledge of the field among readers across the globe.

It gives us an immense pleasure to thank our researchers and authors for their efforts to submit their piece of writing before the deadlines. Finally in the end, I would like to thank my family and colleagues who have been a great source of inspiration and support.

Editor

1

Unsaturated Hydraulic Conductivity of Fractal-Textured Soils

Yongfu Xu

1. Introduction

The increasing concern with groundwater pollution and contamination of soils has stimulated the development of numerous mathematical models of pollutant transport in soils. The most important approaches to model transient water and solute transport in the vadose zone are based on the Richards equation. To solve this equation, the knowledge of the soil hydraulic properties, namely, the soil-water characteristic curve (SWCC) and the unsaturated hydraulic conductivity is required. The laboratory measurements show that the value of unsaturated hydraulic conductivity varies considerably from soil to soil with different water content (Khaleel and Relyea, 1995). Indeed, it is found that the unsaturated hydraulic conductivity decreases by one to three orders of magnitude across a small pressure head range even near saturation (0~10cm pressure head), due to the effects of structural macropores (Jarvis and Messing, 1995). Of all hydraulic properties, the unsaturated hydraulic conductivity is most difficult to measure. Therefore the use of indirect methods has become more and more common to estimate the unsaturated hydraulic conductivity from more easily measured soil properties (van Genuchten et al., 1992).

Modern hydrological models require information on hydraulic conductivity and soil-water retention characteristics. All hydraulic properties, the soil-water characteristics, hydraulic conductivity and soil-water diffusivity (SWD) are closely related to the geometry of a porous media (Brooks and Corey, 1966; Burdine, 1953). Measurements of hydraulic properties are expensive, time-consuming and highly variable (Dirksen, 1991). Prediction of these properties is a viable alternative, especially when the predictive model contains a few parameters sensitive to structural conditions. Porous media (e.g. soils, rocks, etc.) are heterogeneous systems composed of numerous, different and interacting components (van Damme, 1995).

The complex nature of these porous media complicates any prediction of their hydraulic properties.

A potentially powerful method results from the pore surface models, in which the soil–water characteristic curve of unsaturated soils is interpreted as statistical measure of its equivalent pore-size distribution (PSD) (van Genuchten, 1980; Corey, 1992). The frequency of different pore radii is related to matric suction by the soil-water characteristic curve (SWCC) and the relation between matric suction and pore radius is described by the Young-Laplace equation. The relative hydraulic conductivity (RHC) of unsaturated soils can be deduced from the soil-water characteristic curve (SWCC) through simplifying assumptions on pore topology and using the Hagen-Poiseuille law as an approximation for water flow. An additional empirical parameter is introduced, which includes all uncertainties. This empirical parameter is often referred to as "tortuosity", but its physical meaning is unclear (Vogel and Roth, 1998). The choice of the analytical model for the soil-water characteristic curve (SWCC) can significantly affect the predicted function of the unsaturated hydraulic conductivity.

Fractals describe hierarchical systems and are suitable to model the heterogeneous soil structure with tortuous pore space (Rieu and Sposito,1991; Xu and Sun, 2002). Toledo et al. (1990) modeled the soil-water characteristic curve (SWCC) and unsaturated hydraulic conductivity using fractal geometry and thin-film physics. Tyler and Wheatcraft (1990) gave the unsaturated hydraulic conductivity functions based on the fractal model for the soil-water characteristic curve (SWCC) and the relative conductive models developed by Mualem (1976) and Burdine (1953). Crawford (1994) studied the influence of heterogeneity of both the solid matrix and the pore space, as well as the shape of the pore boundary, on the saturated and unsaturated hydraulic conductivity. Fuentes et al. (1996) derived an expression for unsaturated hydraulic conductivity using the fractal dimension obtained from the soil-water characteristic curve (SWCC). Hunt and Gee (2002) applied critical path analysis from percolation theory to calculate the unsaturated hydraulic conductivity of soils with fractal pore space. Xu (2004), Xu and Dong (2004) Xu et al. (2004) derived the unsaturated hydraulic conductivity using the fractal model for the pore surface and gave a simple method to determine the fractal dimension. Models of the unsaturated hydraulic conductivity incorporate fractal dimension characterizing scaling of different properties including parameters representing connectivity and tortuosity. Thus, it is encouraged to derive the functions of the soil–water characteristic curve and unsaturated hydraulic conductivity from the fractal model for the pore surface. In this chapter, the soil-water characteristic curve (SWCC) and relative hydraulic conductivity (RHC) function were derived and expressed by the effective degree of saturation based on the fractal model for the pore surface. The proposed soil–water characteristic curve and unsaturated hydraulic conductivity function were examined in detail by the published experimental data. Comparisons between the prediction using both the fractal model and the van Genuchten-Mualem (G-M) model and the measurement of the relative hydraulic conductivity (RHC) were conducted. Using the fractal model for the soil-water characteristic curve (SWCC) and relatively hydraulic conductivity (RHC), one-dimensional rainfall infiltration and slope stability due to rainfall infiltration were analysized in chapter.

2. Fractals

Some of the most common and useful examples of fractals are: the Koch curve, the Sierpinski gasket and carpet, and the Menger sponge. All of these fractals enjoy the property of self-similarity. Roughly speaking, a subset of $\Re^n$ is said to be self-similar if it is a union of a number of smaller similar copies of itself. Some of the essential notions of the theory of fractals as it applies to self-similar sets should be briefly reviewed. A previous acquaintance with at least one of the sets mentioned above is useful but not necessary. Given an arbitrary subset B of $\Re^n$, a *cover* of B is a family U of sets $U_0 \subset \Re^n$ such that:

$$B \subset \bigcup_n U_\alpha \tag{1}$$

If the family U is countable (or finite) the cover is said to be countable (or finite). It is customary in that case to indicate the members of the family with a Latin subscript, namely, U_i, where i ranges over the natural numbers. The diameter of a subset of $\Re^n$ is the least upper bound (i.e., the supremum) of the distance between pairs of points in the subset. By convention, the diameter of the empty set is zero. A δ-cover of $B \subset \Re^n$ is a (countable) cover such that, for every i, diam(U_i)≤δ, where δ is a positive real number and where diam(.) is the diameter function on subsets of $\Re^n$. The members of a δ-cover are indicated by U_i^δ. The s-dimensional Hausdorff measure $H_s(B)$ of B is defined as:

$$H^s(B) = \liminf_{s \to 0} \sum_{i=1}^{n} [\mathrm{diam}(U_i^s]^s \tag{2}$$

where the inf extends over all δ-covers. It can be shown that this definition indeed provides an outer measure for all non-negative values of s and for any subset of $\Re^n$. The Hausdorff dimension of B is defined as:

$$\dim_H B = \inf\{s \geq 0 : H^s(B) = 0\} \tag{3}$$

The Hausdorff dimension of a subset of $\Re^n$ cannot exceed n. If $\dim_H B > 0$, then it can be shown that for all values of s strictly smaller (respectively, larger) than the dimension, the s-dimensional Hausdorff measure of B is infinite (respectively, zero). In this sense, the Hausdorff dimension represents a critical value of discontinuity of the s-dimensional Hausdorff measure, thus making Eq. (3) meaningful. Of particular interest is the value of the s-dimensional Hausdorff measure for s equal to the Hausdorff dimension of the set. This is usually called the *Hausdorff measure* of the set. The Hausdorff measure of a set may turn out to have any value, including zero or infinite.

An important feature of the s-dimensional Hausdorff measure is the *scaling property*, namely, the way it changes under similarity transformations of $\Re^n$. Recall that a transformation S: $\Re^n \to \Re^n$ is called a similarity if there exists a real scale factor $\lambda>0$ such that, for all $x, y \in \Re^n$, the following equation is satisfied:

$$|S(y) - S(x)| = \lambda |y - x| \tag{4}$$

where $|.|$ denotes the Euclidean distance in $\Re^n$. It is not difficult to prove that for all subsets $B \subset \Re^n$ and for all values of s, under a similarity transformation S with scale factor λ the s-dimensional Hausdorff measure transforms according to the formula:

$$H^s(S(B)) = \lambda^s H^s(B) \tag{5}$$

Another important property of the s-dimensional Hausdorff measure is that it is preserved under Euclidean isometries (translations, rotations, reflections). In other words, congruent sets have the same s-dimensional Hausdorff measure. The behavior of the Hausdorff measure under general affine transformations, on the other hand, cannot be captured under the umbrella of a simple formula.

The properties just described of the s-dimensional Hausdorff measure can be used to obtain a straightforward evaluation of the Hausdorff dimension of self-similar fractals. Indeed, let B be the union of m copies of λ-scaled mutually congruent copies of itself (with $\lambda<1$). If these copies are disjoint Borel sets, we have:

$$H^s(B) = m\lambda^s H^s(B) \tag{6}$$

by virtue of the scaling property (5). Assume now that the Hausdorff measure of B (namely, the value of $H_s(B)$ for $s=\dim_H(B)$) is finite and positive. In that case, it follows Eq. (6) that:

$$\dim_H B = -\frac{\log m}{\log \lambda} \tag{7}$$

The fractal dimension for a pore surface can be defined in the following way. Imagine the pores being enclosed by a set of spheres of radius r, the number of spheres N necessary to do this is clearly a function of radius of the spheres. The definition of the fractal dimension D is:

$$D = \lim_{r \to 0} \left[\frac{\ln(N(A,r))}{\ln(1/r)} \right] \tag{8}$$

Equation (8) is often used to determine the fractal dimension by experiments. A typical fractal set, the middle third Cantor set may be constructed from a unit interval by a sequence of deletion operations (Fig. 1). Let E_0 be the interval [0,1], E_1 is the set obtained by deleting the middle third of E_0 so E_1 consists of the two intervals $\left[0, \frac{1}{3}\right], \left[\frac{2}{3}, 1\right]$. Deleting the middle third of E_1 gives E_2, thus E_2 comprises the four intervals $\left[0, \frac{1}{9}\right], \left[\frac{2}{9}, \frac{3}{9}\right], \left[\frac{6}{9}, \frac{7}{9}\right], \left[\frac{8}{9}, \frac{9}{9}\right]$. Proceeding in this like manner, E_i is obtained by deleting the middle third of each interval in E_{i-1} and Ei consists of 2^i intervals with length of 3^{-i}. Using Eq. (8), the fractal dimension of the middle-third Cantor set is $D=\lim_{i\to\infty}\left\{\frac{\log(2^i)}{\log(1/(3^{-i}))}\right\}=0.63$. The Cantor set is often used to model the dust distribution.

Figure 1. Middle-third Cantor set with fractal dimension D=0.63

3. Fractal model for the pore surface

Many research results show that the soil pore surface is fractal (Avnir and Jaroniec 1989; Xu and Sun 2002). To investigate the impacts of fractal scaling upon hydraulic properties of porous media, a fractal representation of a porous media is developed. A cube of size 1 by 1 is formed by initially subdividing the cube into $1/a^3$ subcubes, each of size a by a. From the original cube, N subcubes are removed and represent a pore of size a by a. The remaining $1/a^3$-N subcubes are then each divided into $1/a^3$ subcubes and N subcubes are removed from each of the original subcubes. Such a recursion algorithm results in a cube everywhere filled with pores of all sizes, with a predominance of small pores. The number of size of a^3 needed to cover the pores equal to or larger than a^3 is given by N. The Menger sponge (Fig. 2) is often used to model porous materials, such as soils. For the Menger sponge, a=1/3, N=7, and the fractal dimension is given by $D=\lim_{i\to\infty}\left\{\frac{\log(20^i)}{\log(1/(3^{-i}))}\right\}=2.73$.

When we covered a soil pore surface using balls with the same radius, the soil pore surface is covered by N_1 balls with radius r_1, the same surface is covered by N_i balls with radius r_i, and it is covered by N_j balls with radius r_j, and usually $r_1>r_i>r_j$, it is found that the surface area of the soil pore surface contents has the following relationship: $N_0r_0^2<N_ir_i^2<N_jr_j^2$. If the soil pore

Figure 2. The Menger sponge with fractal dimension D=2.73

surface is a self-similar surface, i.e., the pore surface is a fractal surface, the relationship between the number of covered balls and its radius can be written as (Mandelbrot 1982)

$$N(r) = Cr^{-D} \tag{9}$$

where C is a constant, r is the size corresponding to the pore radius, D is the fractal dimension of the pore surface. All materials have a surface fractal dimension in the whole range of physically meaningful values, i.e., from 2 to 3. The limit D=2 would correspond to a perfectly regular, smooth (Euclidean) surface, whereas D=3 would correspond to a self-similar surface so intricate that it would be space filling.

The fractal dimension of the pore surface can be determined using mercury intrusion porosimetry and adsorption isotherm. These two methods are introduced in detail as follows.

3.1. Mercury intrusion porosimetry

The d-measure is obtained from Eq. (9), and is expressed as follows:

$$M(r) = N(r)r^{d} = Cr^{d-D} \tag{10}$$

For the fractal pore surface, the relationship between the surface area with the radius less than r and the pore radius r can be obtained from Eq. (10). The surface area can written as

$$\begin{aligned} A_{\mathrm{p}}(\leq r) &= Cr^{2-D} \qquad a) \\ V_{\mathrm{p}}(\leq r) &= Cr^{3-D} \qquad b) \end{aligned} \tag{11}$$

where in 11 a) $A_p(\leq r)$ is the surface area of the soil pores with the radius smaller than r. Similarly, the total volume of the pores with the radius less than r is given by 11b), where in $V_p(\leq r)$ is the cumulative volume of the soil pores with the radius smaller than r.

Neimark (1992) gave a method to measure the surface fractal dimension of the soil pores using mercury intrusion porosimetry. The soil pore surface can be approximated by the equilibrium interface between mercury and soil particles in the close vicinity of the pore surface. According to the Young–Laplace equation, the mean radius of the equilibrium interface is expressed as follows:

$$r(p) = \frac{2\sigma \cos\alpha}{p} \tag{12}$$

where $r(p)$ is the mean radius of the equilibrium interface under pressure p, σ is surface tension. From the thermodynamic viewpoint, the equilibrium interface area can be calculated from the balance between the work of formation of the equilibrium interface and the work of mercury intrusion (Neimark 1992), i.e.,

$$A_{\mathrm{p}}(p) = \int_0^{V(p)} \frac{V(p)}{T} \mathrm{d}p \tag{13}$$

where $A_p(p)$ is the equilibrium interface area under pressure p. The surface fractal dimension can be determined theoretically from Eq. (11a). If the slope of the straight line is λ in log $r(p)$ versus log $A_p(p)$, the surface fractal dimension of soil pores is given by

$$D = 2 - \lambda \tag{14}$$

The pore-size distribution (PSD) is usually obtained using mercury intrusion porosimetry. According to Eq.(11b), the fractal dimension of the pore surface can be obtained from the slope of the regressed linear relation in the plane of log$r$ vs. logV_p. If the slope of the fitting line in the log$r$-logV_p plane is κ, the fractal dimension of the pore surface is written as

$$D = 3 - \kappa \tag{15}$$

The value of fractal dimension spans a large range from 1.0 to 3.0 (Gimenez et al., 1997). Larger fractal dimension is associated with clayey soils (van Damme, 1995).

The accumulative volume of the pores can be measured by many methods. Mercury intrusion porosimetry provides a useful and potentially valuable measurement of the pore volume for the porous medium. The ranges of equivalent pore diameter explored cover almost five orders of magnitude, from several hundred microns down to approximately 16Å (Watabe et al., 2000).

Figure 3 shows the soil pore-size distribution (PSD) obtained from mercury intrusion porosimetry (Watabe et al., 2000). Symbols S-02, S-03 and S-04 are the serial numbers of soil samples, and V_v is the total pore volume in Fig. 3. It is seen from Fig. 3 that the surface of the soil pores can be described by fractal model, and the relationship between $V_p(\leq r)$ and r can be expressed by a linear function in log-log plot. The slopes of the regressed linear relation in the plane of $\log r$ vs. $\log(V_p(\leq r)/V_v)$ are 0.34, 0.37 and 0.49, and therefore the fractal dimensions are 2.66, 2.63 and 2.51 for specimens of S-02, S-03 and S-04, respectively. The maximum radius is the radius at which the pore volume reaches the maximum value, and is defined as the intersection between the fitting line and the line of $V_p(\leq r)/V_v$=100%. The maximum radii of soil pores are 0.0125, 0.06 and 0.006 mm for specimens of S-02, S-03 and S-04, respectively. The relationship between the maximum radius R and the air-entry value ψ_e can be expressed by the Young-Laplace equation, i.e. $\psi_e=2\sigma\cos\alpha/R$. The parameters α=0 and σ=0.075kPa mm (Watabe et al., 2000), and the air-entry values are 12kPa, 2.5kPa and 25kPa corresponding to the maximum pore radius R of 0.0125, 0.06 and 0.006 mm for specimens of S-02, S-03 and S-04, respectively. The parameters obtained from the fractal model of the pore surface are listed in Table 1. In Table 1, parameters R and κ are obtained from the fractal model of the pore surface in Fig. 3.

Soil type	κ	D	δ	R (mm)	ψ_e (kPa)	Data source
S-02	0.34	2.66	-0.34	0.0125	12	Watabe et al., 2000
S-03	0.37	2.63	-0.37	0.06	2.5	
S-04	0.49	2.51	-0.49	0.006	25	
A horizon	0.22	2.78	-0.22	0.2	0.75	Vogel and Roth, 1998
B horizon	0.13	2.87	-0.13	0.25	0.6	
Toyoura sand	1.33	1.67	-1.33	0.09	1.67	Uno et al., 1998

Table 1. Parameters obtained from the fractal model of PSD

Experimental data of the pore-size distribution (PSD) for agricultural silty soils from two horizons, A and B were also taken from Vogel and Roth (1998). The pore-size distribution (PSD) was determined using the three-dimensional reconstructions of the pore space. The pore space was eroded by a spherical structuring element with a given diameter 2r and subsequently dilated by the same structuring element (Vogel and Roth, 1998). In the resulting image all pores smaller than $2r$ were removed. The cumulative volume $V_p(>r)/V_T$ was obtained using the

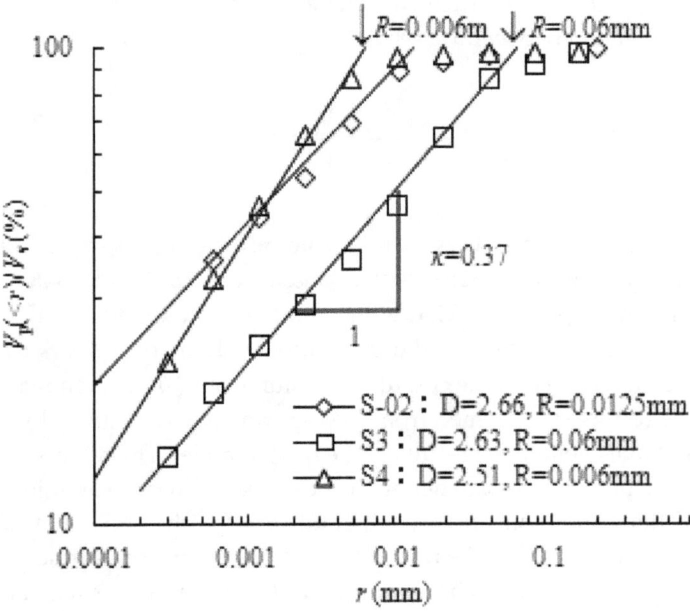

Figure 3. Fractal dimension and the maximum pore radius obtained from PSD for glacial tills (Data from Watabe et al., 2000)

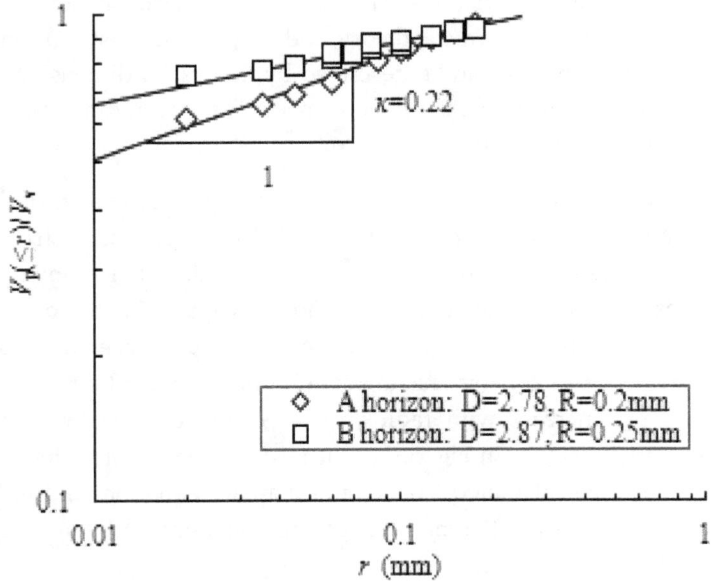

Figure 4. Fig. 4 Fractal dimension and the maximum radius of the PSD of two silty soils (Data from Vogel and Roth, 1998)

application of successively larger structuring elements by Vogel and Roth (1998). The following relationship is obtained between $V_p(>r)/V_T$ and $V_p(\le r)/V_T$.

$$\frac{V(>r)}{V_T}+\frac{V(\le r)}{V_T}=\frac{V_v}{V_T}=\theta_s \tag{16}$$

where V_T is the total volume of soil, V_v is the total volume of void, θ_s is the saturated volumetric water content. The saturated volumetric water contents are 0.502 and 0.448 for the soils at A horizon and B horizon, respectively. Hence the cumulative volume $V(\le r)/V_T$ can be obtained from the cumulative volume $V(>r)/V_T$. The cumulative volume of $V(\le r)/V_T$ can be translated into $V(\le r)/V_v$ according to Eq.(16). The cumulative volume of $V(\le r)/V_v$ is shown in Fig. 4, where the dots denote the results transformed from the experimental data given by Vogel and Roth (1998) and the solid lines denote the results of the fractal model. The relationship between the pore volume and the pore radius satisfies an approximately linear expression in log-log plot. The slopes (κ) of the regression line are 0.24 and 0.13 for the soils at A and B horizons, respectively. According to Eq. (15), the fractal dimensions of the pore surface are 2.76 and 2.87 for A and B horizons, respectively. The maximum radius is defined as the value at which the cumulative volume of pores reaches the maximum, and $V(\le r)/V_v$=100%. The maximum radii of the soil pores at the A and B horizons are 0.2cm and 0.25cm, respectively. The air-entry values can be calculated using the Young-Laplace equation, and they are 0.75kPa and 0.6kPa for the A and B horizons, respectively. The fractal dimension of the soil pore and the air-entry value of the soils at the A and B horizons are listed in Table 1. It can be seen from Figs. 3-4 that the pore surface can be modeled by fractal theory, and the pore-size distribution (PSD) satisfies with Eq. (11b). The soil structure can be described by the fractal dimension. The larger the fractal dimension, the more tortuous is the pore space. The fractal dimension is varied with the soil structure, and is different for different soils.

Uno et al. (1998) studied the relationship between the pore-size distribution (PSD) and unsaturated hydraulic properties of Toyoura sand. The pore-size distribution (PSD) of Toyoura sand was measured by the air intrusion method, and is shown in Fig. 5. The relationship between the pore volume and pore radius is approximated by a linear function in log-log plot, and satisfies with Eq. (11b). Hence the pore-size distribution (PSD) of Toyoura sand can be described by fractal model, and its fractal dimension is 1.67 obtained from Fig. 5 not increase with radius, i.e., V_p/V_v=1. The maximum pore radius of Toyoura sand is nearly 0.09 mm in Fig. 5. The surface tension (T) of water is 0.075kPa mm. The air-entry value can be calculated from Young-Laplace equation, and is 1.67kPa on the assumption that α=0. The fractal dimension of the soil pore and the air-entry value of Toyoura sand are listed in Table 1.

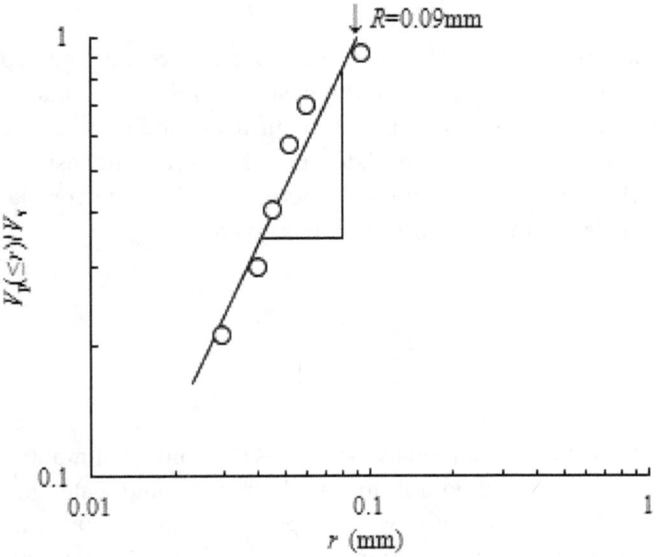

Figure 5. The fractal dimension and the maximum radius evaluated from the PSD of Toyoura sand. (Data from Uno et al., 1998.)

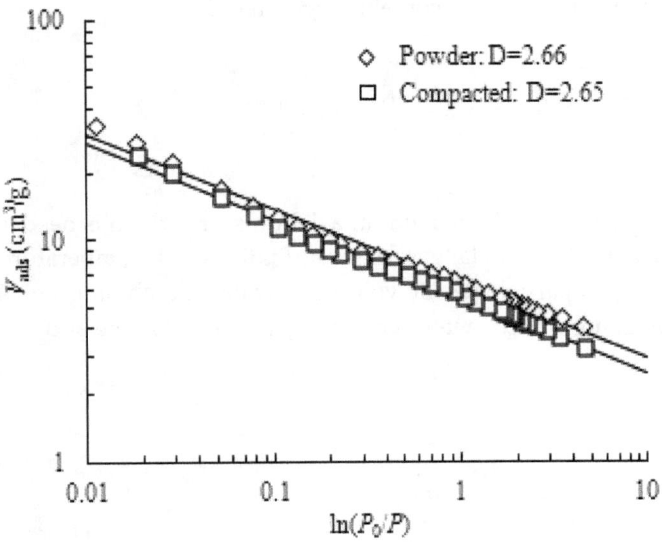

Figure 6. The surface fractal dimension of Tsukinuno bentonite from nitrogen adsorption

3.2. Adsorption isotherm

The fractal dimension of the pore surface serves to characterize the pore surface roughness or irregularities. The magnitude of the fractal dimensions is relevant to many important physicochemical processes namely adsorption, surface diffusion and catalysis. Avnir and Jaroniec (1989) proposed a convenient method to determine the fractal dimension from the nitrogen adsorption. Similarly with the adsorption isotherm equation, the correlation of the water volume absorbed by clay to the vapour pressure is given by

$$\frac{V_{\mathrm{w}}}{V_{\mathrm{m}}}=k\left[\ln\left(\frac{P_0}{P}\right)\right]^{D-3} \tag{17}$$

where V_w is the water volume absorbed by clay, V_m is the volume of montmorillonite, P is the partial water vapour pressure in equilibrium with clay at some water content w and some temperature T and P_0 is the equilibrium water vapour pressure of pure water at temperature T. The surface fractal dimension of Tsukinuno bentonite obtained from nitrogen adsorption is shown in Fig. 6. The surface fractal dimensions of bentonite powder and compacted bentonite are 2.66 and 2.65, respectively.

The adsorption isotherm also allows the calculation of swelling pressure of clay as a function of the water content (Kahr et al., 1990). Satisfactory agreement is found between the calculated and the experimental data for water content bigger than 10%.

$$p_s=-\frac{\bar{R}T}{M_w v_w}\ln\left(\frac{P_0}{P}\right) \tag{18}$$

where p_s is swelling pressure, the maximum axial pressure which is needed to maintain the original sample height, $\bar{R}$ is the molar gas constant, T is the Kelvin temperature, M_w is molecular mass of water and v_w is partial specific volume of water. Combining Eq. (17) with (18), the relationship between normalized water volume by the clay volume and swelling pressure is written as

$$\frac{V_{\mathrm{w}}}{V_{\mathrm{m}}}\propto p_s^{\,D-3} \tag{19}$$

The surface fractal dimension of clay can be obtained from Eq. (19). The experimental data of the swelling pressure and swelling deformation tests are compiled in Fig.7. It is seen that the relationship between the normalized water content and swelling pressure or vertical overburden pressure can be described by the same linear function which is expressed in Eq. (19). Therefore, Eq. (19) represents a general relationship between the water volume absorbed by clay per unit volume and vertical pressure supplied to specimen in swelling process. The surface fractal dimension of Tsukinuno bentonite is 2.63, which nearly equals to that obtained from Fig. 6. In Fig. 6, C_b is the bentonite content.

Swelling deformation tests of Wyoming bentonite were also conducted by Millins et al. (1996) and Studds et al. (1998). The relationship between the clay void ratio (e_c) and vertical pressure is shown in Fig. 8. Relationship between the clay void ratio (e_c) and normalized water volume is given by

$$e_c = \frac{1}{C_m} \frac{V_w}{V_m} \tag{20}$$

where C_m is the montmorillonite content, and C_m=75%. It is seen that the slope of loge_c-log$p$ equals to that of log(V_w/V_m)-log$p$ from Eq. (20) for C_m being constant. Thus, the surface fractal dimension of Wyoming bentonite can be obtained from the linear relationship of loge_c-log$p$. The surface fractal dimensions of Wyoming bentonite are 2.64, which obtained from the swelling deformation tests conducted by different authors.

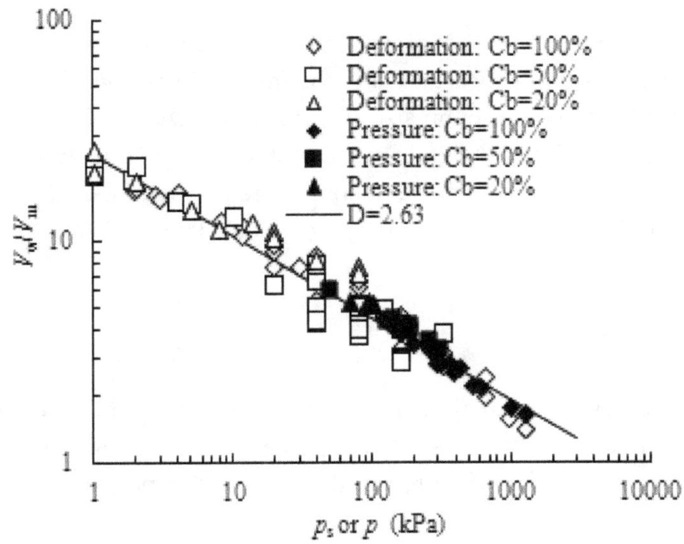

Figure 7. Relationship between absobed water volume and swelling pressure or vertical overburden pressure of Tsukinuno bentonite

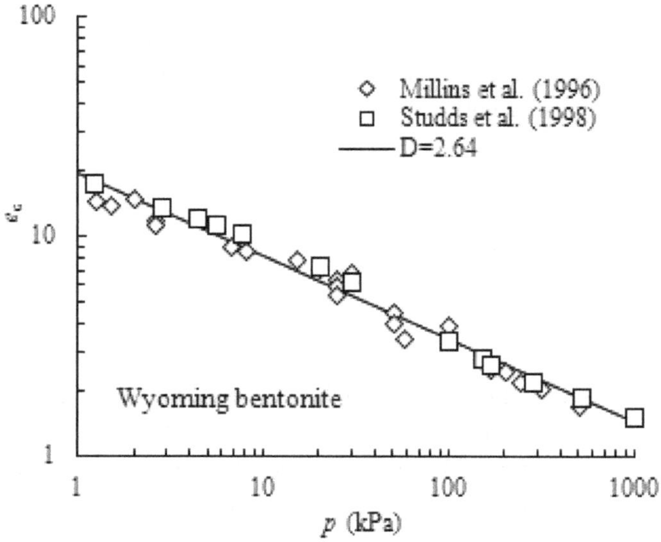

Figure 8. Relationship between clay void ratio and vertical overburden pressure of Wyoming bentonite

3.3. Correlation between the fractal dimension of PSD and that of the GSD

The pore-size distribution (PSD) is not often and easily measured experimentally. The grain-size distribution (GSD) is easily measured and the fractal dimension of the pore surface can be conveniently obtained from the grain-size distribution (GSD). The relationship between the pore-size distribution (PSD) and the grain-size distribution (GSD) can be constructed for the soils with a homogeneous fabric. Consider a pore limited by a number of grains N_g, the pore radius r is related to the radius r_g of the smallest grain as follows (Watabe et al., 2000)

$$r = f_p r_g \tag{21}$$

where f_p is a constant less than 1.0. If the particles are randomly arranged, the probability of having a grain of radius less than r_g is dP_g, the probability of having the N_g-1 grains with a radius greater than r_g is $(1-P_g)^{Ng-1}$. The probability P of having a pore with a radius less than r is given by (Watabe et al., 2000)

$$dP = N_g \left(1 - P_g\right)^{N_g - 1} dP_g \tag{22}$$

It is assumed that the grain-size distribution (GSD) satisfies the fractal model, and is given by

$$P_g = \left(r_g / R_g \right)^{3-D_g} \tag{23}$$

where D_g is the fractal dimension of the grain-size distribution (GSD), R_g is the maximum radius of soil grains. The probability P is given by (Watabe et al., 2000)

$$P = 1 - \left[1 - \left(\frac{r}{f_p R_g} \right)^{3-D_g} \right]^{N_g} \tag{24}$$

If $r/(f_g R_g) << 1$, Eq.(24) can be written as

$$P = N_g \left(\frac{r}{f_p R_g} \right)^{3-D_g} \tag{25}$$

It is seen that the pore surface has the same fractal dimension as that obtained form the grain-size distribution (GSD) from Eqs. (23) and (24). Comparisons between the pore-size distribution (PSD) and the grain-size distribution (GSD) are shown in Fig. 9, where $P=V(\leq r)/V_v$ and $P_g=M(\leq r)/M_T$, V and M are the pore volume and grain mass, respectively. It is seen from Fig. 9 that both the pore-size distribution (PSD) and the grain-size distribution (GSD) can be expressed by fractal model, and they have the nearly same fractal dimension.

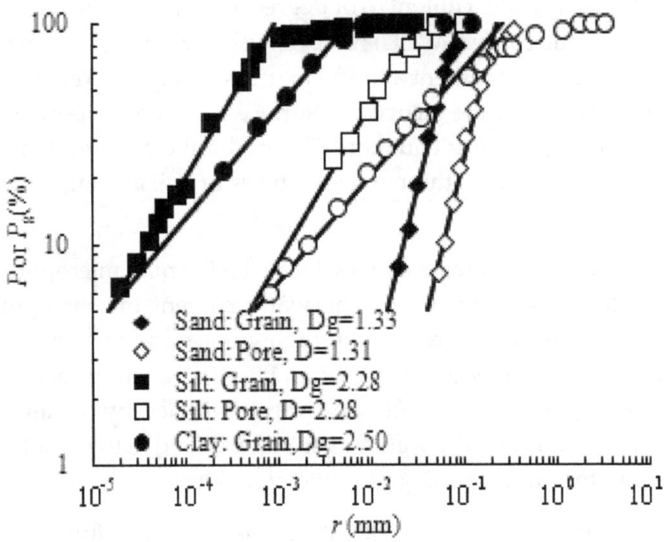

Figure 9. Correlation of the pore-size distribution (PSD) to the grain-size distribution (GSD).

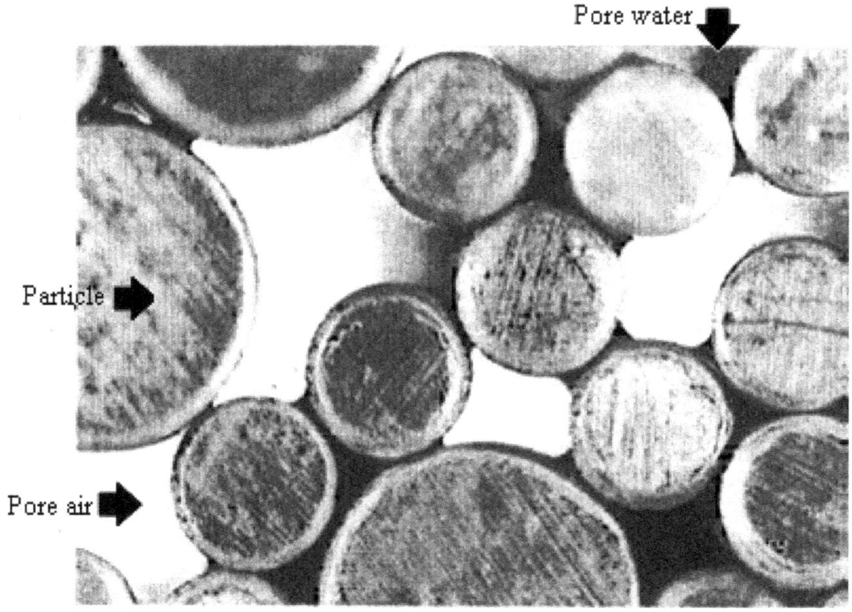

Figure 10. Model test for simulating the water distribution

4. Soil-water Characteristic Curves (SWCC)

The soil-water characteristic curve defines the relationship between water content (gravimetric water content w, volumetric water content θ) or degree of saturation (S) and matric suction (ψ). The soil water characteristic curve (SWCC) has been used extensively in the study of unsaturated soils and relates the water content and soil matric suction. The relationship between the matric suction and the radius of the incurvated surface between the pore-air and pore-water is expressed by the Young-Laplace equation. The hydraulic and mechanical properties of unsaturated soils are correlated to the microstructures of soils through the Young-Laplace equation.

For unsaturated soils, the pore water is usually localized in small micropores, and the micropores are usually fully filled with water. The water content in macropores is very little, sometimes, there is no water in macropores in highly unsaturated soils (Fig. 10). The water distribution (black domain) of model test shown in Fig. 10 proved the above assumptions. In the model test, the soil grains of different sizes were simulated by the aluminum rods with different radius (Matsuoka, 1999). The water mainly distributed in the small pore and no water exists in the large pore for unsaturated soils in the model test.

Three air-water distribution states were identified as saturated funicular state, complete pendular state, and partial pendular state, respectively. All the voids are filled with liquid in the saturated state. Air bubbles are present in the voids in the funicular state, and the liquid phase is continuous. There is no continuity in the water phase in the pendular state. Pore water within unsaturated soils can be divided into three forms (Wheeler and Karube, 1996) (Fig. 11): bulk water within those void spaces that are completely flooded, meniscus water surrounding all inter-particle contact points that are not covered by bulk water and absorbed water (which is tightly bounded to the soil particles and acts as parts of the soil skeleton). The relationship between the soil-water characteristic curve (SWCC) and the pore-water forms is shown in Fig. 11. When the volumetric water content is less than its residual value, the pore-water is tightly absorbed by soil particle and cannot move freely. This absorbed water is seen as a part of soil particles. Thus, the contribution of the filled pores with radius $r \rightarrow dr$ to the water content is given by

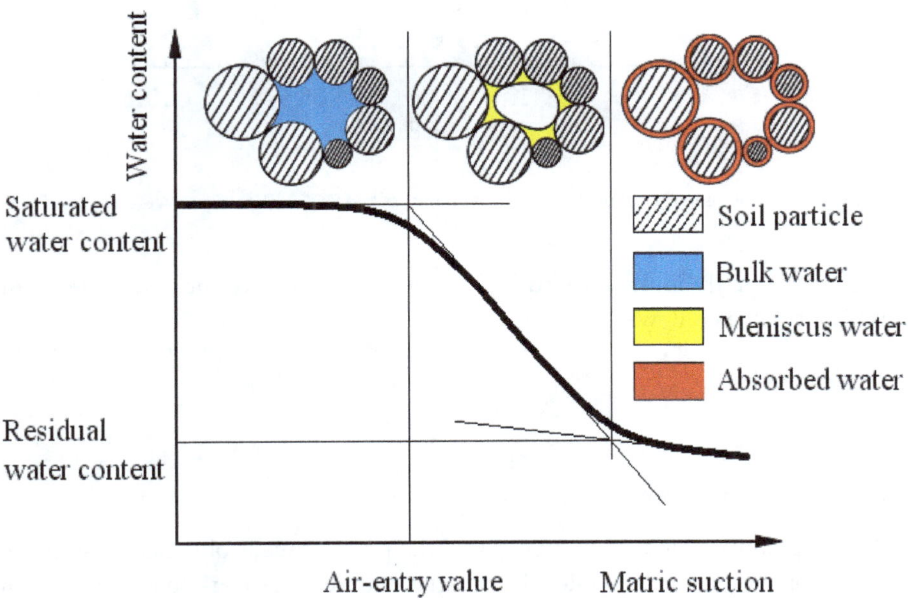

Figure 11. Relationship between the SWCCs and the pore-water forms of unsaturated soil (Matsuoka, 1999)

$$d\Lambda = \frac{N4\pi r^2 dr}{V_T} \tag{26}$$

where Λ is the relative volumetric water content, and $\Lambda = \theta - \theta_r$, θ and θ_r are the actual and residual volumetric water content, respectively. The residual volumetric water content is the volumetric water content at which the effectiveness of matric suction to cause further removal of water requires vapour migration. Substituting Eq. (9) into Eq. (26), Λ is given by

$$\Lambda = Br^{3-D} \tag{27}$$

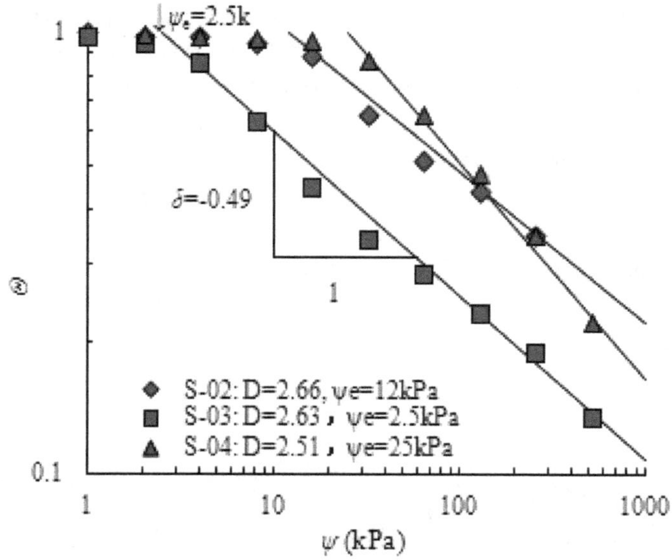

Figure 12. Comparison between the prediction and experiments of the SWCCs (Data from Watabe et al., 2000)

where $B=4\pi C/[V_T(3\text{-}D)]$. Similarly with Eq. (27), the relatively volumetric water content at saturation is written as follows

$$\Lambda_s = BR^{3-D} \tag{28}$$

where R is the maximum radius of soil pores.

The relationship between matric suction ψ and the pore radius is obtained from the Young-Laplace equation. Thus, the soil-water characteristic curve (SWCC) is obtained as follows

$$\Theta = \left(\frac{\psi}{\psi_e}\right)^{\delta} \tag{29}$$

where $\delta=D\text{-}3$, Θ is the normollized volumetric water content, and is written as follows:

$$\Theta = \frac{\theta\text{-}\theta_r}{\theta_s\text{-}\theta_r} = \frac{S-S_r}{100-S_r} \tag{30}$$

where θ, θ_r and θ_s are the volumetric water content, residual volumetric water content and saturated volumetric water content, S and S_r are the degree of saturation and residual degree of saturation. The residual volumetric water content is not always made routinely, in which case it has to be estimated by extrapolating available soil-water characteristics data towards lower water content, such as shown in Fig. 11. van Genuchten (1980) defined the residual volumetric water content as the water content for which the gradient $d\theta/d\psi$ becomes zero at high matric suction. From a practical point of view it seemed sufficient to define θ_r as the water content at some large matric suction, e.g. at the permanent wilting point ψ_e=1500kPa.

Equation (29) is the soil-water characteristic curve (SWCC) derived from the fractal model for the pore surface. The normalized volumetric water content is equal to unity at values of matric suction up to the air-entry value and equal to zero after residual saturation. The only two parameters ψ_e and D, which have obvious physical meaning, are used in Eq. (29) to express the soil-water characteristic curve (SWCC). The fractal dimension of the pore surface and the maximum pore radius can be evaluated from the pore-size distribution (PSD) measured using the mercury intrusion tests. The air-entry value can be calculated from the Young-Laplace equation using the maximum pore radius. Thus, the soil-water characteristic curve (SWCC) can be calculated from Eq. (29) using the fractal dimension and the air-entry value obtained from the mercury intrusion tests.

The soil-water characteristic curves (SWCC) were calculated from Eq. (29) using the fractal dimension of the pore surface and the air-entry value for specimens of S-02, S-03 and S-04, which listed in Table 1. The value of the normalized volumetric water content is calculated from Eq. (30). Here S_r=7.5% (Watabe et al., 2000). Experimental data show in good accord with the calculation of the soil-water characteristic curve (SWCC) in Fig. 12.

The experimental data of the soil-water characteristic curve (SWCC) measured in a multi-step outflow experiment were given by Vogel and Roth (1996) for silty soils. The saturated volumetric water contents are 0.502 and 0.448 for silty soils at the A and B horizons, respectively. The residual volumetric water content is equal 0 for both A and B horizons. Comparisons between the calculations of Eq. (29) and experimental data of the soil-water characteristic curves (SWCC) were shown in Fig. 13. The calculation of the soil-water characteristic curves (SWCCs) for silty soils was obtained from Eq. (29). The fractal dimensions of the soil pore and the air-entry values were listed in Table 1.

Uno et al. (1998) gave the experimental data of the soil water characteristic curves for Toyoura sand. Using the fractal dimension and the air-entry value obtained from the pore-size distribution (PSD), the soil-water characteristics can be simulated from Eq. (29) for Toyoura sand. The parameters obtained from the pore-size distribution (PSD) and used to predict unsaturated hydraulic conductivity are listed in Table 1. θ_s=0.425, θ_r=0. Comparison between the prediction of Eq. (29) and experimental data of the soil-water characteristics is shown in Fig. 14 for Toyoura sand.

Stingaciu et al. (2009) evaluated the feasibility of using nuclear magnetic resonance (NMR) relaxometry measurements to characterize pore size distribution and hydraulic properties in four porous samples with different texture and composition. The sandy samples FH31 and W3 were loaded in the penetrometer and packed at the same packing density as for the NMR measurements. The Mix8 and MZ samples were used as solid conglomerates. From the relaxation time distribution functions, the cumulative pore-size distribution functions were

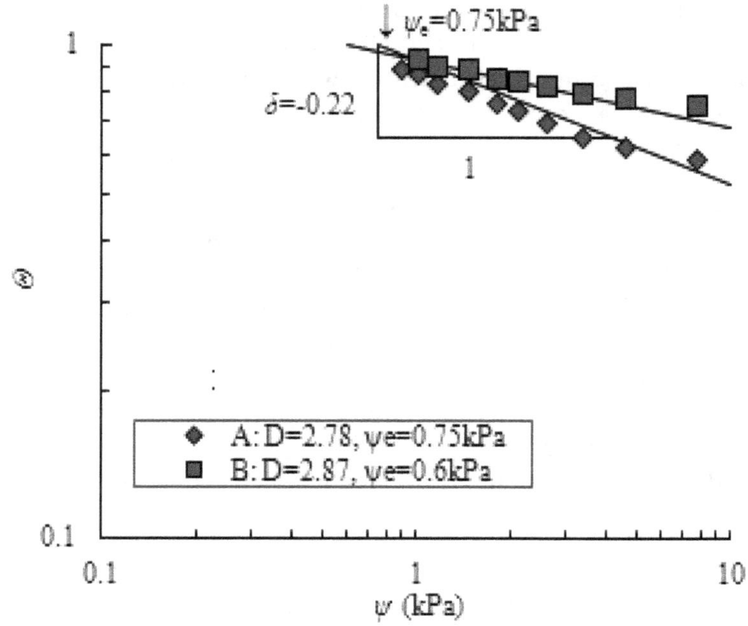

Figure 13. Comparison between prediction and experiments of SWCCs (Data from Vogel and Roth, 1996)

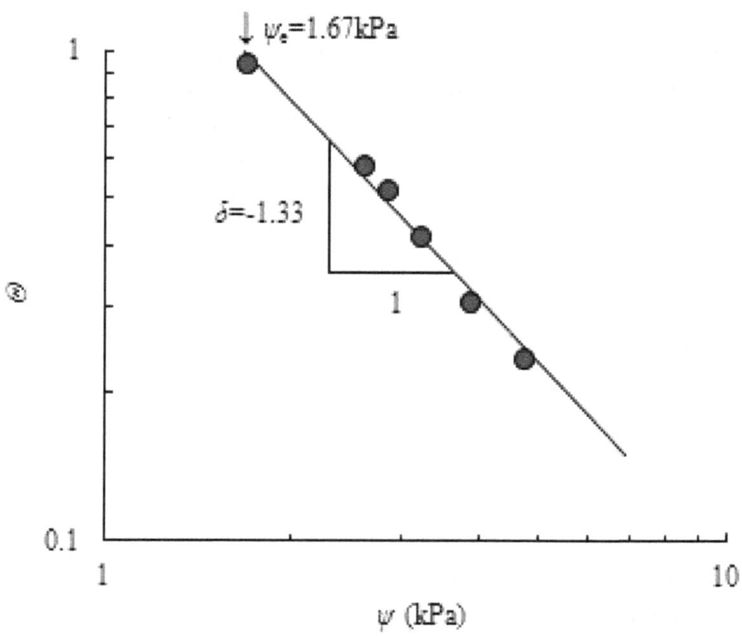

Figure 14. Comparison of the SWCCs for Toyoura sand (Data from Uno et al., 1998)

calculated with the average surface relaxivity. The normalized cumulative pore-size distributions functions were displayed in Fig. 15. The fractal dimension of pore surface and the maximum pore radius can be obtained from Fig. 15, and were listed in Table 2.

Sample	Method	Bulk density (g/cm³)	θ_s	θ_r	D	R (mm)	ψ_e (kPa)
FH31	Pressure plate	1.58	0.32	0.02	0.85	0.065	2.31
W3	MSO	1.42	0.33	0.058	1.51	0.0175	8.6
Mix8	Rosetta	1.45	0.41	0.061	1.52	0.00053	290
MZ	Pressure plate	1.60	0.44	0	2.29	0.00035	429

Table 2. Parameters for the prediction of SWCCs

The soil-water characteristic curves (SWCC) were determined using the standard sand bed, pressure cell or multistep outflow method, and were plotted in Fig. 16. SWCC of FH31 and MZ were based on pressure plate measurements. SWCC of W3 was determined using multistep outflow (MSO) and that of Mix8 using ROSETTA software (Schaap et al., 2001). The matric suction as plotted in the abscissa (Fig. 16) was transformed into pore diameter using the Young-Laplace equation. In addition, the ordinate was the normalized volumetric water content. The predictions of SWCC were conducted using Eq. (29) with the parameters listed in Table 2. The parameters were derived from the pore-size distribution (PSD) shown in Fig. 15. Comparisons of the predictions and experimental data were shown in Fig. 16. The predictions of Eq. (29) were in good accord with the experimental data.

In general, the soil-water characteristic curves (SWCC) were usually interpreted as cumulative distribution functions in comparison to pore-size distribution (PSD) functions Using the soil-water characteristic curves (SWCC), pore-size distribution can be extracted for a given porous medium on the basis of an empirical law that related the pore suction to the effective pore radius. The soil-water characteristic curves (SWCC) and the pore-size distributions (PSD) are related by correlations as: (1) the absolute values of the slopes of SWCC and PSD are equal; (2) the air-entry value of SWCC is related to the maximum pore radius of PSD through the Young-Laplace equation.

The following conclusions can be obtained from Figs. 12-16: (1) The surface of the soil pore can be described by fractal model, and the surface fractal dimension of the soil pore and the air-entry value can be evaluated from the pore-size distribution (PSD). The soil-water characteristic curves (SWCCs) can be calculated using fractal model of the soil pore. (2) The fractal dimension and air-entry value obtained from the soil pore-size distribution (PSD) are equivalent to those obtained from the soil-water characteristic curve (SWCC). Hence the fractal dimension and air-entry value can be obtained from the fitting of the soil-water characteristic curve (SWCC) if the pore-size distribution (PSD) were not measured.

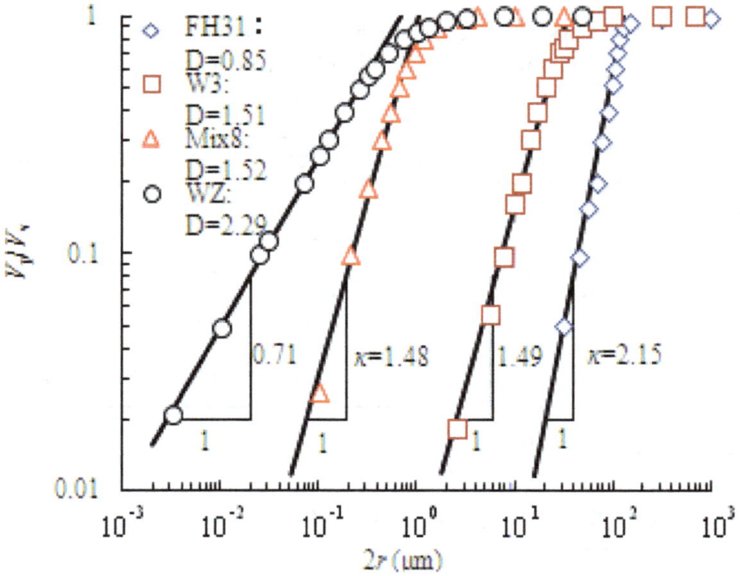

Figure 15. Cumulative pore-size distribution of for the three artificial substrates (FH31, W3, and Mix8) and the natural soil Merzenhausen (MZ), from which the fractal dimension of pore surface and the maximum pore radius were derived (Data from Stingaciu et al. 2009)

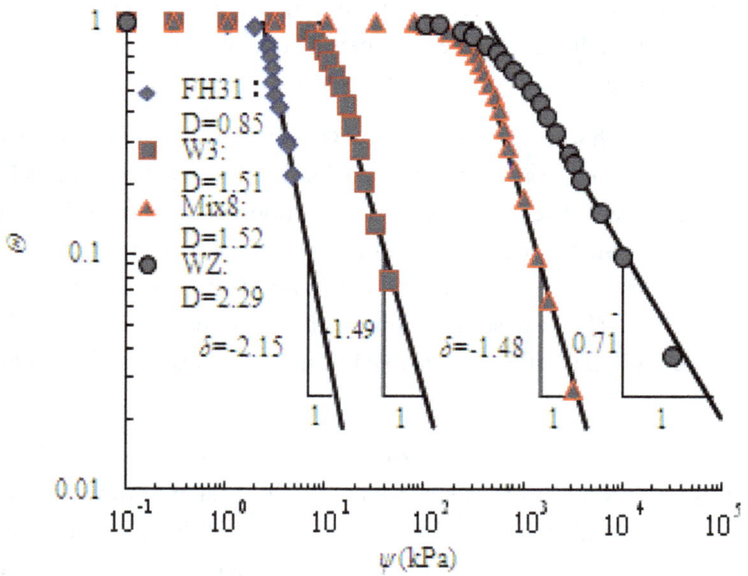

Figure 16. Comparison between the predictions of Eq. (29) and experimental data of SWCCs (Data from Stingaciu et al. 2009)

5. Unsaturated hydraulic conductivity and diffusivity

5.1. Brief review

Brutsaert (2000) made a brief review of the capillary model, which was cited as follows.

It is very difficult to determine unsaturated hydraulic conductivity through experiments. Hydraulic conductivity for unsaturated soils is a function of soil-water content θ. For this reason, over the years many attempts have been made to represent the hydraulic conductivity by simple parametric equations, in terms of other properties of the soil which are easier to determine. One of these was of the following form (Brutsaert, 2000):

$$k_{\mathrm{r}} = \Theta^{\omega} \tag{31}$$

In the past, the power form of Eq. (31) has been derived on the basis of some widely different conceptual models of the pore geometry. There have been some marked differences in the obtained values of ω, depending on the adopted model of the pore geometry.

1. *Uniform pore size models*

A simple approach was that porous medium was characterized by some equivalent uniform pore size and the variability of the pore sizes was not considered. This approach invariably resulted in a constant value of ω. The capillary tube model of Averyanov (Polubarinova-Kochina, 1952) led to ω=3.5, whereas the extension of the hydraulic radius model of Kozeny (Kozeny, 1927) to unsaturated soils by Irmay (1954) produced ω=3.

2. *Parallel models*

In parallel models, the pore system was assumed to be equivalent with a bundle of uniform capillary tubes of many different sizes. The pore-size distribution was derived from the soil-water characteristics (SWCC), which can be related to the effective pore radius r by the Yong-Laplace equation. The true mean velocity in each pore was described by a Poiseuille-like equation for creeping flow. Purcell (1949) and Gates and Tempelaar-Lietz (1950) first used this approach. Several subsequent applications of the parallel model can be written in the common form (Brutsaert, 1967):

$$k_{\mathrm{r}} = \frac{\beta\int_0^{\Theta}\psi(x)^{-2}\mathrm{d}x}{\beta_0\int_0^{1}\psi(x)^{-2}\mathrm{d}x} \tag{32}$$

where x is the dummy variable representing Θ. The variable β=$\beta(\Theta)$ is related to the tortuosity, and β_0 is its value at satiation, when Θ=1. The tortuosity concept was originally introduced by Carman (1956) as an improvement on the non-uniform hydraulic radius model of Kozeny (1927) and it can be expressed as T=$(L_e/L)^2$, in which L_e is the actual or microscopic path length of the fluid particles in the pores and L is their apparent or macroscopic path length along the Darcy stream lines. The parameter β was introduced in the parallel model by Wyllie and Spangler (1952), because without it, Eq. (32) yielded values considerably larger than their experimental data. Burdine (1953) proposed on the basis of his experimental data that $\beta/\beta_0=\Theta^2$.

A further development was performed by adopting the following soil-water characteristic curve (SWCC) (Brooks and Corey, 1966):

$$\Theta = 1 \qquad \text{for } \psi \le \psi_e$$
$$\Theta = \left(\frac{\psi}{\psi_e}\right)^b \qquad \text{for } \psi \le \psi_e \tag{33}$$

Brooks and Corey (1966) integrated (32) with Burdine's assumption for β (Burdine, 1953) to obtain Eq. (31), with an exponent

$$\omega = 3\text{-}\frac{2}{b} \tag{34}$$

It should be noted that, without Burdine's assumption for the tortuosity, the parallel model Eq. (32) applied with Eq. (33) would have yielded

$$\omega = 1\text{-}\frac{2}{b} \tag{35}$$

3. *Series-parallel models*

The theoretical construction of this model also started with a bundle of parallel pores, each with a different but uniform size. However, these pores were then cut normally to the direction of flow with two resulting faces, and finally after some random rearrangement of the tubes the faces are joined again. This way account was taken of the random variations of the pore sizes, not only in the plane normal to the direction of flow, but also along the direction of flow. The discharge rate in each single pore, which consisted of two sections in cases, was assumed to be governed by the section with the smaller diameter. The pore-size distribution was derived form of the soil-water characteristics by means of the Yong-Laplace equation. The true velocity in each pore was obtained by means of a Poiseuille-like equation.

This model was originally proposed by Childs and Collis-George (1950) in a finite-difference scheme to calculate hydraulic conductivity from experimental data of the soil-water characteristics (SWCC). It was subsequently reformulated in integral form by Brutsaert (1968), to allow the derivation of more concise analytical expressions for hydraulic conductivity. The integral form can be written as

$$k_{\mathrm{w}} = \left[\left(\frac{2T}{\gamma}\right)^2 \frac{\theta_0^{\,2}(1-S_{\mathrm{r}})^2}{G}\right] \times \left[\int_0^{\Theta}\int_0^{x} \psi(x)^{-2}\mathrm{d}y\mathrm{d}x + \int_0^{\Theta} \psi(x)^{-2}\int_0^{\Theta}\mathrm{d}y\mathrm{d}x\right] \quad (36)$$

where G is a geometrical constant. In the case of Poiseuille's equation, G=8. It was trivial to show by integration by parts that the first double integral on the right was identical to the second. Hydraulic conductivity can be expressed concisely as (Brutsaert, 2000):

$$k_{\mathrm{r}} = \frac{\int_0^{\Theta}(\Theta - x)\psi(x)^{-2}\mathrm{d}x}{\int_0^{1}(1-x)\psi(x)^{-2}\mathrm{d}x} \quad (37)$$

The capillary model similar with Eq. (37) in form was originally proposed for by Fatt and Dykstra (1951).

5.2. Fractal model for hydraulic conductivity

Soil pore systems are viewed as a collection of interconnected voids. The individual voids are considered to have two equilibrium states: either filled by water or empty. Flow within the soil pores is assumed to be described by (Burdine, 1953; Mualem, 1976)

$$v = -\frac{r^2 g}{c\eta}\frac{\mathrm{d}h}{\mathrm{d}x} \quad (38)$$

where v is the average velocity within the pore; r is the radius of the pore; g is the gravity acceleration; η is the kinematic viscosity; c is a constant depending on the pore geometry; h is the hydraulic head, x is the position axis. Let us consider a porous slab of thickness Δx isolated from a homogenous soil column by two parallel cross-sections normal to the x axis. The areal porosity at each face of the slab is the same and it is equal to the volumetric porosity. The major simplifying assumption is that the actual configuration of slab may be replaced by a set of capillary tubes with different radii, parallel to the x axis. The flux $\mathrm{d}q$ flowing through the connections between pores of radius $r_1 \rightarrow r_1+\mathrm{d}r_1$ on one side of the slab (at x) and pores of radius $r_2 \rightarrow r_2+\mathrm{d}r_2$ on the other side of the slab (at $x+\mathrm{d}x$) can be expressed by (Mualem, 1976)

$$\mathrm{d}q = \beta r_{\mathrm{e}}^{\,2}(r_1, r_2, \rho)A_{\mathrm{e}}(r_1, r_2, \rho)\frac{\mathrm{d}h}{\mathrm{d}x}\mathrm{d}r_1\mathrm{d}r_2 \quad (39)$$

where β is a constant which incorporates the fluid with the matrix properties; $r_{\mathrm{e}}(r_1, r_2, q)$ is the effective radius; $A_{\mathrm{e}}(r_1, r_2, q)$ is the effective area; ρ is the maximum radius of the water filled pores at volumetric water content θ. According to the Darcy equation, the hydraulic conductivity k_w is given by

$$k_{\mathrm{w}} = \frac{q}{\mathrm{d}h/\mathrm{d}x} = \beta\int_0^{\rho}\int_0^{\rho} r_{\mathrm{e}}^2(r_1, r_2, \rho)A_{\mathrm{e}}(r_1, r_2, \rho)\mathrm{d}r_1\mathrm{d}r_2 \tag{40}$$

The relatively hydraulic conductivity (k_r) is the ratio of the hydraulic conductivity at any volumetric water content (k_w) to the hydraulic conductivity at saturation (k_s). The relative hydraulic conductivity (RHC) is related to the radius of the soil pores, and is written as (Mualem, 1976)

$$k_r = \frac{k_w}{k_s} = \frac{\int_0^r\int_0^r r_e^2(r_1, r_2, r)A_e(r_1, r_2, r)dr_1dr_2}{\int_0^R\int_0^R r_e^2(r_1, r_2, r)A_e(r_1, r_2, r)dr_1dr_2} \tag{41}$$

where r is the maximum radius of the water filled pores at the volumetric water content θ. It is assumed that the pore distribution at the two cross-section is completely random. The probability of pores of radius $r_1 \rightarrow r_1 + \mathrm{d}r_1$ at x to encounter pores of radius $r_2 \rightarrow r_2 + \mathrm{d}r_2$ at $x + \mathrm{d}x$ is given by

$$A_e(r_1, r_2)dr_1dr_2 = f(r_1)f(r_2)dr_1dr_2 \tag{42}$$

The effective radius is assumed to be equal to the radius r reduced by a factor, which is equal to the ratio between the effective area to flow and the actual pore area. The effective radius is given by (Mualem, 1976)

$$r_e = r\frac{\int_0^r\int_0^r A_e(r_1, r_2, r)dr_1dr_2}{\int_0^R f(r_1)dr_1} = r\Lambda^{1/2} \tag{43}$$

Substituting Young-Laplace equation and Eqs.(42) and (43) into Eq.(41), the relative hydraulic conductivity (RHC) is obtained as

$$k_r = \left(\frac{\psi}{\psi_e}\right)^{\xi} \tag{44}$$

where ξ=3D-11. Substituting Eq. (29) into Eq. (44), the relative hydraulic conductivity (RHC) unction expressed by effective degree of saturation is written as follows

$$k_r = \Theta^{\xi/\delta} \tag{45}$$

Equations (44) and (45) are the relative hydraulic conductivity (RHC) function derived from the fractal model of the pore surface. The parameters in the proposed hydraulic conductivity function have an obvious physical implication, and can be determined from the pore-size distribution (PSD). Eqs. (44) and (45) offer a powerful, physical-based method to calculate the relative hydraulic conductivity (RHC). Eq. (44) is nearly similar with the Brooks and Corey equation (Brooks and Corey, 1966) in form, and the parameters in Eq. (44) have obvious physical implication. Eqs. (44) and (45) offer a powerful, physical-based method to calculate the relative hydraulic conductivity (RHC).

The expression of the soil-water diffusivity (SWD) can be derived from the unsaturated hydraulic conductivity and soil-water characteristics, and written as follows:

$$d = k_{\mathrm{w}} \left| \frac{\mathrm{d}\psi}{\mathrm{d}\theta} \right| \tag{46}$$

where d is the soil-water diffusivity (SWD), k_{w} is the saturated hydraulic conductivity. Substituting Eqs. (44) and (45) into Eq. (46), the soil-water diffusivity (SWD) is written as

$$d = k_{\mathrm{s}} \psi_{\mathrm{e}} \left(\frac{\psi}{\psi_{\mathrm{e}}} \right)^{\varsigma} \tag{47}$$

$$d = k_{\mathrm{s}} \psi_{\mathrm{e}} \Theta^{\varsigma/\delta} \tag{48}$$

where $\varsigma=2D-7$.

As given in Eqs. (29), (44), (45), (47) and (48), methods to evaluate the soil-water characteristics (SWCCs), unsaturated hydraulic conductivity (RHC) and soil-water diffusivity (SWD) are proposed using the fractal model for the pore surface. The fractal dimension and the maximum radius of soil pores can be obtained from the pore-size distribution (PSD). The air-entry value can be calculated from the Young-Laplace equation using the maximum radius of soil pores. The soil-water characteristic curves (SWCC), unsaturated hydraulic conductivity (RHC) and soil-water diffusivity (SWD) can be predicted using the fractal dimension and air-entry value, which can be determined from the pore-size distribution (PSD). According to Eq. (29), the fractal dimension of the pore surface and the air-entry value can also be evaluated from the fitted soil-water characteristic curve (SWCC). Thus, the unsaturated hydraulic conductivity (RHC) and soil-water diffusivity (SWD) can also be estimated from Eqs. (44), (45), (47) and (48) using the fractal dimension and air-entry value evaluated from the soil-water characteristic curve (SWCC).

1. Predictions of SWCC and RHC from PSD

Vogel and Roth (1998) inferred effective hydraulic properties of unsaturated soil from the structure of the pore surface. The water retention characteristic and the hydraulic conductivity were simulated by network models with 32^3=32768 nodes. The hydraulic conductivity, k_w, is determined at each step of desaturation by imposing a pressure gradient Δp across the ends of the network. Water flow q_{ij} in a bond between the nodes i and j through a horizontal surface A is described by Poiseuille's law. The hydraulic conductivity was calculated as $=qL/(A\Delta p)$, where L is the vertical length of the network. Comparison between the predictions of Eq. (44) and experimental data given by Vogel and Roth (1998) was shown in Fig. 17. The parameters used to do prediction were listed in Table 1, which were obtained from the fractal model for the pore-size distribution (PSD). The predictions were in good accord with the experimental data.

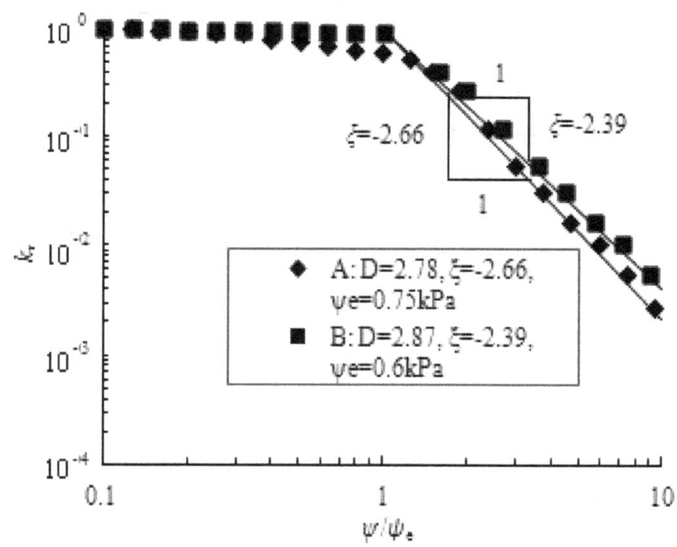

Figure 17. Comparison between the predictions of Eq. (44) and experimental data of the RHC (Data from Vogel and Roth, 1998)

Uno et al. (1998) studied the relationship between the pore-size distribution (PSD) and unsaturated hydraulic properties of Toyoura sand. The relationship between the relative hydraulic conductivity (RHC) and the normollized volumetric water content Θ can be calculated from Eq. (45) using the parameters listed in Table 1. The prediction of Eq. (45) satisfactorily agrees with the experimental data of relative hydraulic conductivity (RHC) for Toyoura sand in Fig. 18. The proposed function of relative hydraulic conductivity (RHC) (Eq. (45)) is verified by the experimental data in Fig. 18.

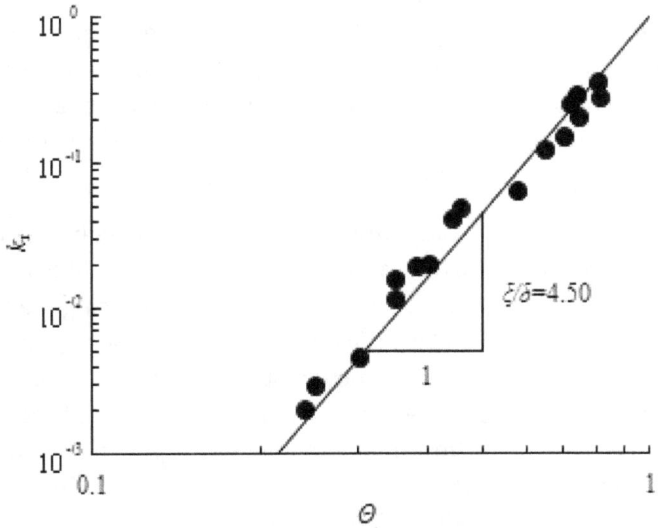

Figure 18. Comparison between predictions of Eq. (45) and experimental data of the RHC for Toyoura sand. (Data from Uno et al., 1998)

2. Prediction of RHC and SWD from SWCC

The characteristics of the pore-size distribution (PSD) were prerequisite to determine the relative hydraulic conductivity and soil-water diffusivity of unsaturated soils in Eqs. (44), (45), (47) and (48). A close relationship was established between the soil-water characteristic curve (SWCC) and the pore-size distribution by fractal model for the pore surface. The pore-size distribution (PSD) can be estimated from the soil-water characteristic curve because many soils show that the soil-water characteristic curve is equivalent to the pore-size distribution (PSD) function (Watabe et al., 2000). Thus, the fractal dimension of the pore surface and the air-entry value can be estimated from the fitting of the soil-water characteristic curve (SWCC). Thus, unsaturated hydraulic properties, such as relative hydraulic conductivity (RHC) and soil-water diffusivity (SWD) can be predicted using the fractal dimension and air-entry value obtained from the soil-water characteristic curve (SWCC).

Smettem and Kirkby (1990) offered the hydraulic properties of haploxeroll loam. Experimental data of the soil-water characteristic curve (SWCC) were shown in Fig. 19. The fractal dimension of the pore surface can be obtained from the fitted soil-water characteristic curve (SWCC). Through fitting the soil-water characteristic curve (SWCC), we obtained that δ=-0.37 for haploxeroll loam from Fig. 19. Fractal dimension of the pore surface can be calculated from Eq. (29) using the fitting value of δ. The fractal dimension (D) of the pore surface is 2.63 for haploxeroll loam. The air-entry value is 0.14kPa obtained from Fig. 19.

The hydraulic properties of unsaturated soils can be predicted using the fractal dimension of the pore-size distribution (PSD), obtained from fitted soil-water characteristic curve (SWCC). The fundamental parameters for predictions of hydraulic properties were tabulated in Table 3. The comparisons between predicted result and experimental data of the relative hydraulic conductivity (RHC) were shown in Fig. 20. The coefficient of hydraulic conductivity at saturation is 140 mmh^{-1}. Fig. 20 depicted the relationship between the relative hydraulic conductivity (RHC) and water potential for haploxeroll loam. The parameter ξ was -3.11 calculated from Eq. (44) using the fractal dimension D=2.63 for haploxeroll loam. The predictions of the relative hydraulic conductivity (RHC) calculated from the proposed function Eq. (44) are in satisfactory agreement with the experimental data.

The relationship between the soil-water diffusivity and water potential for haploxeroll loam was also shown in Fig. 20. The values of k_s and ψ_e are 140mmh^{-1} and 0.14kPa, respectively, used in Eq. (47). The parameters ς was -1.74, calculated from Eq. (47) using the fractal dimension D=2.63. The prediction of the soil-water diffusivity was the line passing through the point of (0.14, 0.64) with the slope of -1.74. The prediction of the soil-water diffusivity nearly agreed with the experimental data. The proposed function for the soil-water diffusivity was verified by the comparison between prediction and measured results. The parameters obtained from the soil-water characteristic curve (SWCC) and used to predict the relative hydraulic conductivity (RHC) and soil-water diffusivity (SWD) for haploxeroll loam were listed in Table 3.

Soil type	D	δ	ξ	ξ/δ	ς	ψ_e (kPa)
Haploxeroll loam	2.63	-0.37	-3.11	8.41	-1.74	0.14
Toyoura sand (wetting)	1.60	-1.40	-6.2	4.43	-3.8	/
Toyoura sand (drying)	1.26	-1.74	-7.22	4.15	-4.48	/
Loamy sand	2.68	-0.32	-2.96	9.25	-1.64	2.0
McGee Ranch soil	2.51(GSD)	-0.49	-3.47	7.08	-1.98	4.6

Table 3. Parameters used to predict the unsaturated hydraulic conductivity

Figure 21 shows the experimental data and the fitting of the soil-water characteristic curve (SWCC) for loamy sand. The experimental data are scaled from Simunek et al. (1999). The parameter δ was -0.32, which is obtained from the fitting of the soil-water characteristic curve (SWCC) for loamy sand from Fig. 26. Hence the fractal dimension of the pore surface was 2.68 calculated from Eq. (29) using the fitting value of δ for loamy sand. The air-entry value was given by the intersection between the line expressed by Eq. (29) and the line of Θ=1 in log-log plot. The air-entry value is 2kPa obtained from Fig. 21.

The parameters used to predict the hydraulic properties of loamy sand were listed in Table 3. Substituting D=2.68 and ψ_e=2kPa in Eq. (44), the predictions of relative hydraulic conductivity (RHC) can be obtained. The comparisons between predictions and experimental data of the relative hydraulic conductivity were shown in Fig. 22. The parameter ξ is -2.96 calculated from Eq. (44) using the fractal dimension D=2.68 for loamy sand. The predictions of the relative hydraulic conductivity were in good accord with the experimental data. The proposed function for the relative hydraulic conductivity was validated by the good agreement between the predictions and measurements.

The parameters of fractal model were first calibrated by matching the measured moisture-suction data. It should be noted that the experimental values of hydraulic conductivity were not matched by adjusting the model parameters. The calibrated SWCC model parameters were, instead, directly used to predict the relative hydraulic conductivity. The soil-water characteristic curves (SWCC) and their calibration of eight soils were shown in Fig. 23. The calibration of SWCC model parameters were listed in Table 4. As seen in these figures, the measured moisture-suction data for these soils were unavailable for the full range (0%-100%) of the degree of saturation. Predicting the suction beyond the available experimental data range was a challenging task since the pattern of variation was unknown (Ravichandran and Krishnapillai, 2011).

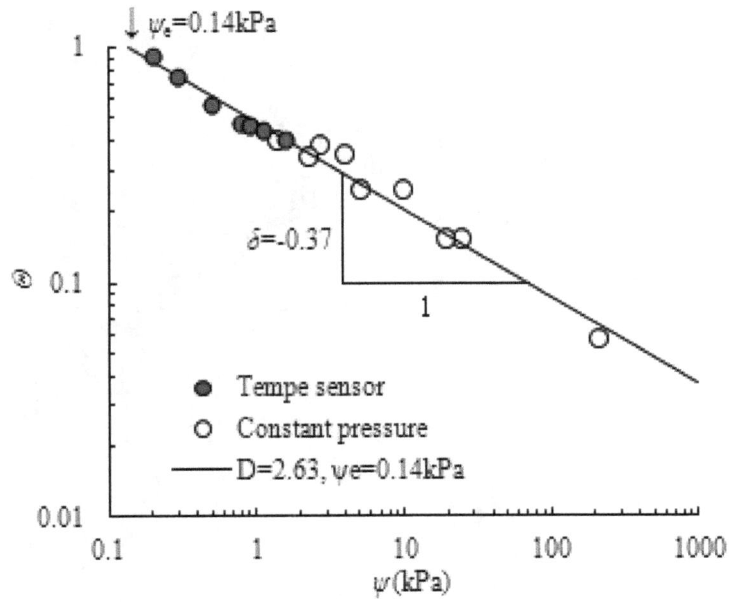

Figure 19. Fractal dimension of the pore surface and air-entry value evaluated from SWCC (Data from Smettem and Kirkby, 1990)

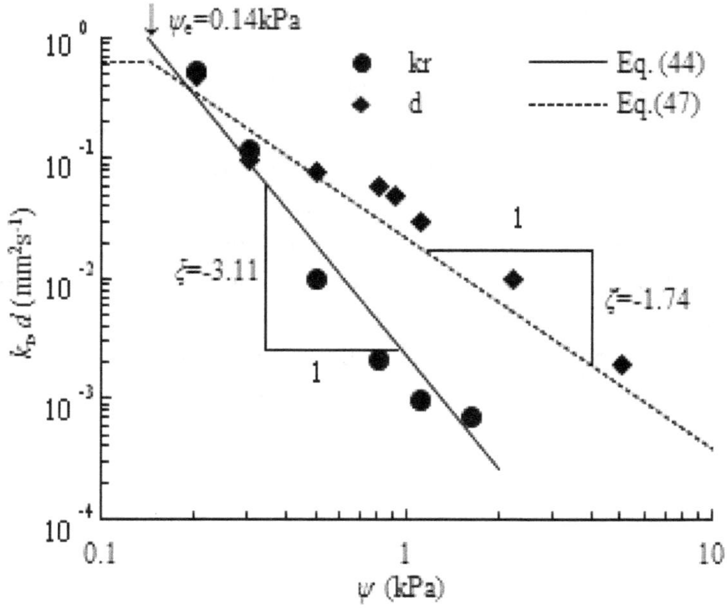

Figure 20. between predictions and experimental data of RHC and SWD for haploxeroll loam (Smettem and Kirkby, 1990)

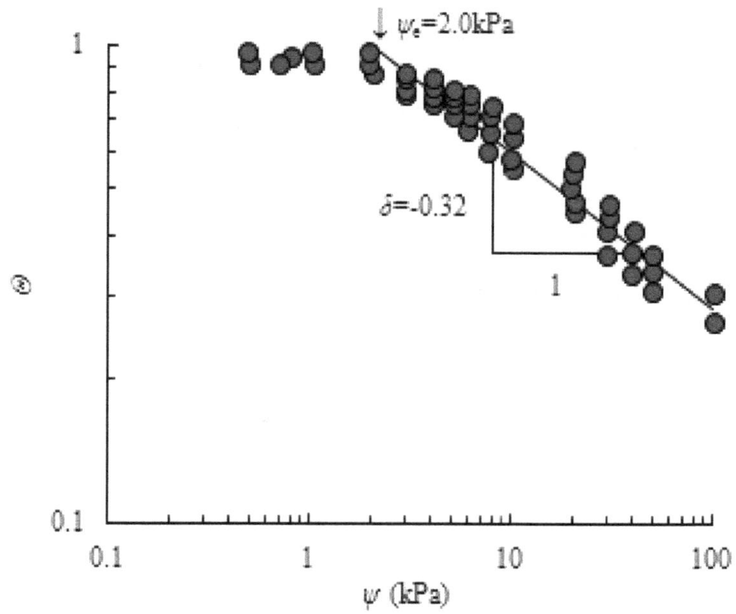

Figure 21. Fractal dimension and the air-entry value obtained from the fitting of SWCC for of loamy sand (Data from Simunek et al., 1999)

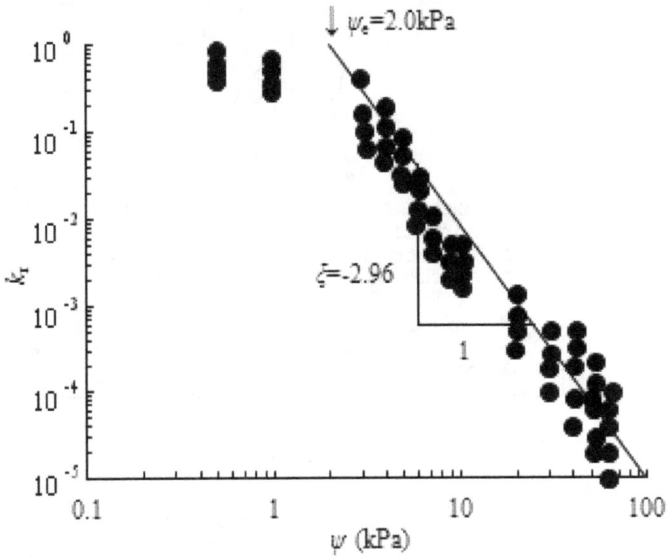

Figure 22. Comparison between measurements of RHC and prediction of Eq. (44) for loamy sand (Data from Simunek et al., 1999)

Soil	θ_s	S_r (%)	D	ψ_e (kPa)	Reference
Lakeland	0.375	30-100	1.75	2.5	Elzeftawy and Cartwright, 1981
Superstition	0.5	30-100	2.7	1.2	Richards, 1952
Columbia sandy loam	0.458	50-100	2.29	4.5	Brooks and Corey, 1964
Touchet silt loam	0.43	20-100	2.09	7	Brooks and Corey, 1964
Silt loam	0.396	50-100	2.59	15	Reisenauer, 1963
Guelph loam	0.52	45-100	2.75	3	Elrick and Bowmann, 1964
Yolo light clay	0.375	45-100	2.87	1.5	Moore, 1939
Speswhite kaolin	0.56	55-100	2.85	15	Peroni et al., 2003

Table 4. Parameters of the fractal model for eight selected soils

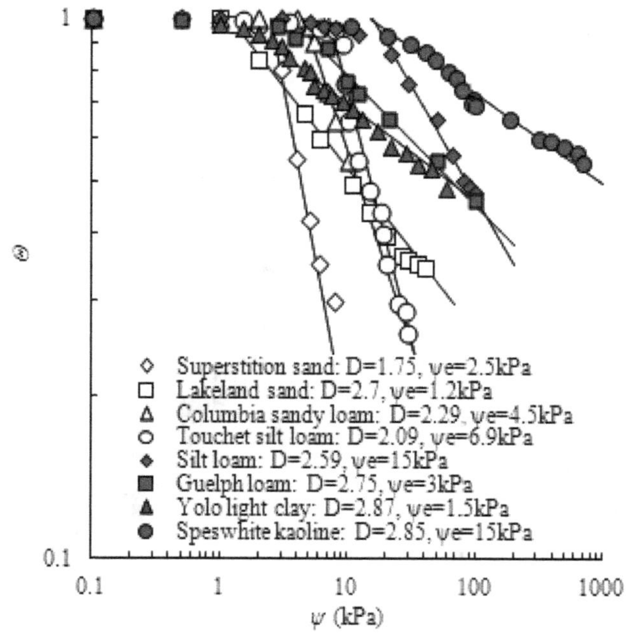

Figure 23. Fractal dimension and air-entry value obtained from the fitting of SWCC

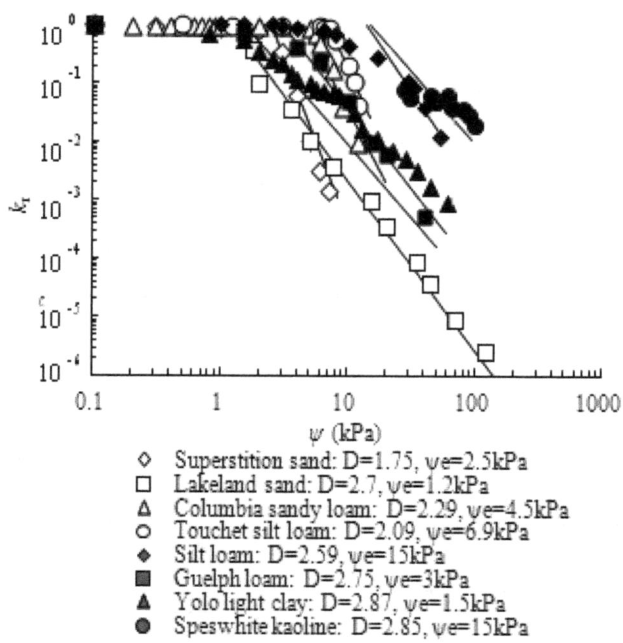

Figure 24. Comparison between predictions of Eq. (44) experiments of RHC

The relative hydraulic conductivity of the above mentioned eight soils were predicted using the fractal model with the same fitting parameters listed in Table 4. The fitting parameters were obtained by matching the experimental SWCC. Fig. 24 illustrated the prediction of Eq. (44) and experimental data of relative hydraulic conductivity. It should be noted that the fractal model parameters were not calibrated or adjusted to match the measured hydraulic conductivity values. The predictions of Eq. (44) were comparable for sand and loam, as shown in Fig. 24. The fractal model showed slight deviations from the measured data of clay. In the case of Yolo light clay and Speswhite kaolin, the difference between the predictions and the experimental data increased as the suction increases (Fig. 24). The fractal model showed better predictions at lower suction range. However, the accuracy of fractal model in the higher suction range could not be verified because the experimental results were available only for the lower suction ranges.

It is seen that it is feasible to estimate the relative hydraulic conductivity of unsaturated soils using the fractal dimension of the soil pore surface from Figs. 19-24. A practical method to express the relative hydraulic conductivity was proposed to use the fractal dimension and the air-entry value, which can be obtained from the fitting of the soil-water characteristic curve (SWCC).

3. Prediction of RHC and SWD from GSD

The feasibility to determine the soil hydraulic properties using the fractal dimension of the grain-size distribution (GSD) is examined by the experimental data of McGee Ranch soil. The grain-size distribution (GSD) was measured by Hunt and Gee (2002) and was shown in Fig. 25. The fractal dimension of the grain-size distribution (GSD) is 2.51, obtained from Fig. 25.

Hunt and Gee (2002) gave the values of θ_s and θ_r were 0.4 and 0.01, respectively, the air-entry value was 4.6kPa. Using the fractal dimension of the grain-size distribution (GSD), the soil-water characteristic curve (SWCC) and the relative hydraulic conductivity (RHC) can be determined. The parameters of the grain-size distribution (GSD) used to determine the hydraulic properties are listed in Table 3. Comparisons between the predictions of Eq. (29) using the fractal dimension of the grain-size distribution (GSD) and the experimental data of soil-water characteristic curve (SWCC) are shown in Fig. 26. It was seen that a good result is obtained to predict the soil-water characteristic curve (SWCC) using the fractal dimension of the grain-size distribution (GSD).

Hunt and Gee (2002) gave the values of saturated hydraulic conductivity was 0.001cm/s. Comparisons between the predictions of Eqs. (44) and (45) and experimental data were shown in Fig. 27. The prediction of Eqs. (44) and (45) nearly satisfied the experimental data of relative hydraulic conductivity in Fig. 27. In Fig.27, the predictions of Eq. (45) deviated from the experimental data near saturation, and extend to equal to the experimental results at low water content. The soil hydraulic conductivity can be approximately determined using the fractal dimension of the grain-size distribution (GSD).

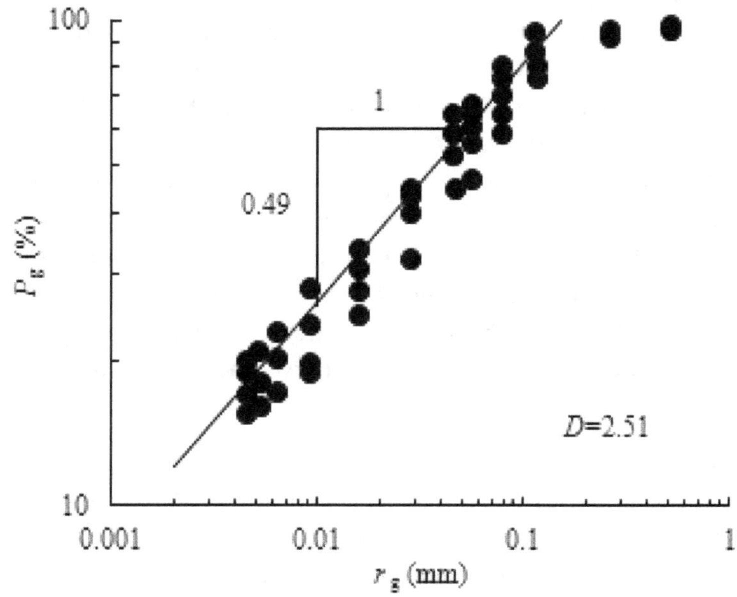

Figure 25. Fractal dimension of GSD for McGee Ranch soil (Data from Hunt and Gee, 2002)

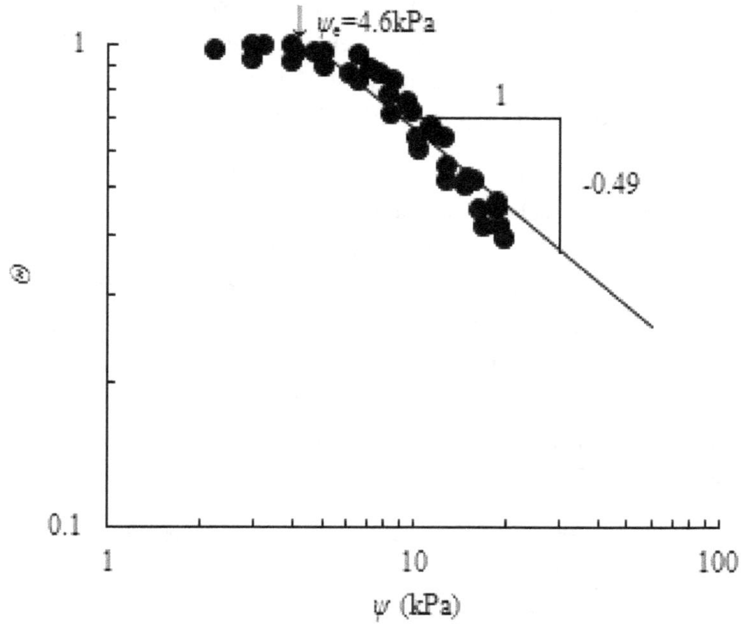

Figure 26. Comparisons between predictions and experiments of SWCC for McGee Ranch soil (Data from Hunt and Gee, 2002)

6. Comparison of the fractal model with the van Genuchten–Mualem (G-M) model

6.1. van Genuchten–Mualem (G-M) model for relative hydraulic conductivity (RHC)

van Genuchten (1980) derived an empirical relationship to describe the soil-water characteristic curve (SWCC):

$$\Theta = \frac{1}{(1+|\alpha\psi|^n)^m} \tag{49}$$

where α, n and m are the van Genuchten curve-fitting parameters, and m=1-1/n.

Burdine (1953) and Mualem (1976) presented a similar model to estimate the unsaturated hydraulic conductivity from pore size distributions inferred from soil water retention characteristics. The models share a number of similarities, allowing them to be written in a general form as (Hoffmann-Riem et al., 1999)

$$k_r = \frac{k_w}{k_s} = \Theta^l \left(\frac{\int_0^{\Theta} \psi^{-\chi} d\Theta}{\int_0^1 \psi^{-\chi} d\Theta} \right)^{\gamma} \tag{50}$$

where k_w is the unsaturated hydraulic conductivity at any water content, k_s is the saturated hydraulic conductivity, Θ is the normalized water content, ψ is matric suction. Generally accepted parameter values for the parameters (l, χ, γ) are (0.5, 1, 2) and (2, 2, 1) for the Mualem (1976) and the Burdine (1953) models, respectively.

Using the Mualem model, van Genuchten (1980) derived a closed-form to determine the relative hydraulic conductivity (RHC) at a degree of saturation,

$$k_r = \frac{\left[1-|\alpha\psi|^{n-1}\left(1+|\alpha\psi|^n\right)^{-m}\right]^2}{\left(1+|\alpha\psi|^n\right)^m} \tag{51}$$

$$k_r = \Theta^{\frac{1}{2}} \left[1-\left(1-\Theta^{\frac{1}{m}}\right)^m\right]^2 \tag{52}$$

Equations (51) and (52) are the representations of the G-M model, which is the most widely and popularly used to predict the hydraulic conductivity.

The other models for relative hydraulic conductivity (RHC) were also introduced as follows.

6.2. van Genuchten–Burdine (G-B) model for relative hydraulic conductivity (RHC)

Hydraulic conductivity resulting from the Burdine capillary models (Burdine, 1953) was expressed as

$$k_r = \Theta^2 \left[1 - \left(1 - \Theta^{\frac{1}{m}} \right)^m \right] \tag{53}$$

Replacing Eq. (49) in Eq. (53) yielded

$$k_r = \frac{\left[1 - |\alpha\psi|^{n-2} \left(1 + |\alpha\psi|^n \right)^{-m} \right]}{\left(1 + |\alpha\psi|^n \right)^{2m}} \tag{54}$$

where m=1-2/n.

Note that for both Eqs. (52) and (54), because m is less than 1, the derivative dk/dθ is infinite for θ_s, whatever the value of n.

6.3. van Genuchten–Fatt & Dykstra (G-FD) model for relative hydraulic conductivity (RHC)

The capillary model proposed by Fatt and Dykstra (1951) was expressed as

$$k_r = \frac{\int_0^\Theta (\Theta - x)\psi(x)^{-2-c} \mathrm{d}x}{\int_0^1 (1 - x)\psi(x)^{-2-c} \mathrm{d}x} \tag{55}$$

Evaluating $\psi(x)$ from Eq. (49), the integrals of Eq. (55) with the condition of m>-(2+c)/2n and n>2+c resulted in:

$$k_r = \frac{\Theta B(a_1, b) I_{\Theta^{1/m}}(a_1, b) - B(a_2, b) I_{\Theta^{1/m}}(a_2, b)}{B(a_1, b) - B(a_2, b)} \tag{56}$$

where a_1=m+(2+c)/n, a_2=2m+(2+c)/n, b=1-(2+c)/n, $I_x(a, b)$ and $B(a, b)$ are the Incomplete Beta and Beta functions of positive arguments a and b, respectively, and given by:

$$B(a,b)=\int_0^1 x^{a-1}(1-x)^{b-1}\mathrm{d}x,\quad I_y(a,b)=\frac{\int_0^y x^{a-1}(1-x)^{b-1}\mathrm{d}x}{B(a,b)} \tag{57}$$

The results presented by Touma (2009) were obtained with c=0.5 and a_1=1.

Even though the conductivity predicted by the capillary model of Fatt and Dykstra (1951) gave the best results compared with both the quasi-analytical solution and observations, the resulting expression was not easy to use because it was necessary to evaluate Incomplete Beta and Beta functions. In order to simplify G-FD model, Θ was expressed by Eq. (49) and k_r was according to Eq. (31) with ω=2+2.5/mn. The fitted result in a value of b was close to -mn, and a conductivity curve close to that predicted by Eq. (56).

6.4. Brooks & Corey-Brutsaert model for relative hydraulic conductivity (RHC)

Brutsaert (2000) presented a capillary model expressed as Eq. (37). Combining this capillary model with the soil-water characteristic curve (SWCC) of BC model (Brooks and Corey, 1966), relative hydraulic conductivity was written as Eq. (31) in form, and the parameter ω was given by

$$\omega = 2\text{-}\frac{2}{b} \tag{58}$$

Combining of BC model for the soil-water characteristics with the Mualem and the Burdine capillary models, the relative hydraulic conductivity curves were also expressed as Eq. (31) in form, and the parameter ω was given by

$$\omega = 3+\frac{2}{mm} \tag{59}$$

when BC is applied with the Burdine condition, and

$$\omega = 2.5+\frac{2}{mm} \tag{60}$$

when applied with the Mualem condition.

Here we focused on the G-M model, as it was the most widely used in soil science and hydrology.

van Genuchten (1980) found that the predictions of the G–M model were less accurate for Beit Netofa clay. The fractal dimension and the air-entry value evaluated from the fitted soil-water characteristic curve (SWCC) are 2.89 and 60kPa, respectively (Fig. 28) The fitted parameters of the van Genuchten model, α and n are 0.0015 and 1.17, respectively from Fig.28. The parameters for the determination of the relative hydraulic conductivity (RHC) are listed in Table 5.

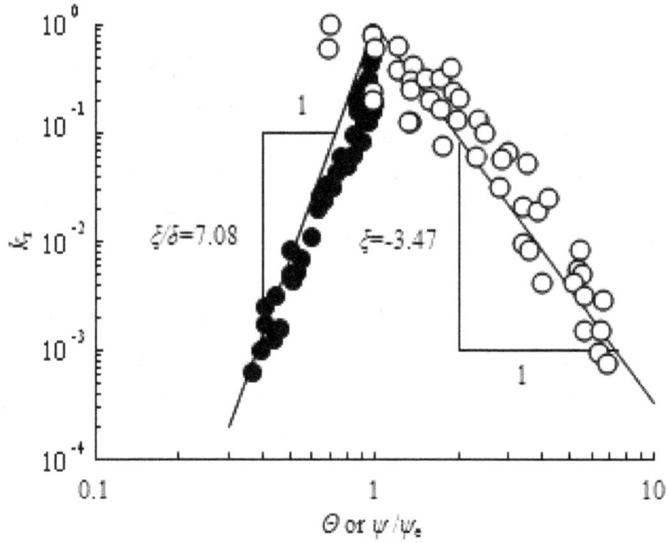

Figure 27. Comparisons between predictions and experimental data of RHC for McGee Ranch soil (Data from Hunt and Gee, 2002)

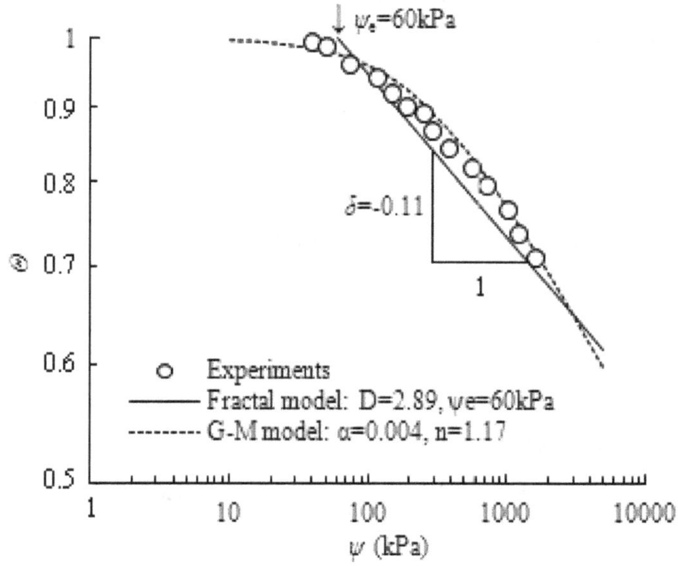

Figure 28. Fractal dimension and the parameters of G-M model obtained from SWCC for Beit Netofa clay (Data from van Genuchten, 1980)

Soil type	Fractal model				G-M model	
	δ	D	ξ	ξ/δ	$a\ (kPa^{-1})$	n
Beit Netofa clay	-0.11	2.89	-2.33	21.2	0.004	1.17
Hanford glacial soil 0-099	-0.62	2.38	-3.86	6.23	0.015	1.71
Hanford glacial soil 2-1637	-0.82	2.18	-4.46	5.44	0.076	1.89
Grenoble sand	-0.64	2.36	-3.92	6.13	0.4	2.17

Table 5. Parameters obatined from SWCC for prediction of RHC

The saturated and residual water contents of Beit Netofa clay are 0.446 and 0, respectively. Comparisons between the predictions of both the fractal model (Eq. (44)) and the G–M model (Eq. (51)) and the experimental data of the relative hydraulic conductivity (RHC) for Beit Netofa clay are shown in Fig. 29. It is seen that predictions of the fractal model (Eq. (44)) satisfactorily agree with the experimental data, especially at high matric suction, while the predictions of the G–M model (Eq. (51)) are found to deviate from the experimental data. However, the fractal model shows a sharp corner at the air-entry value point, which is disagree with the measured data that usually show smooth transition after the air-entry value point. This default does not reduce the advantages to predict the relative hydraulic conductivity (RHC) using the fractal dimension obtained from the pore-size distribution (PSD), because the hydraulic conductivity near the point of the air-entry value nearly equal to the saturated conductivity.

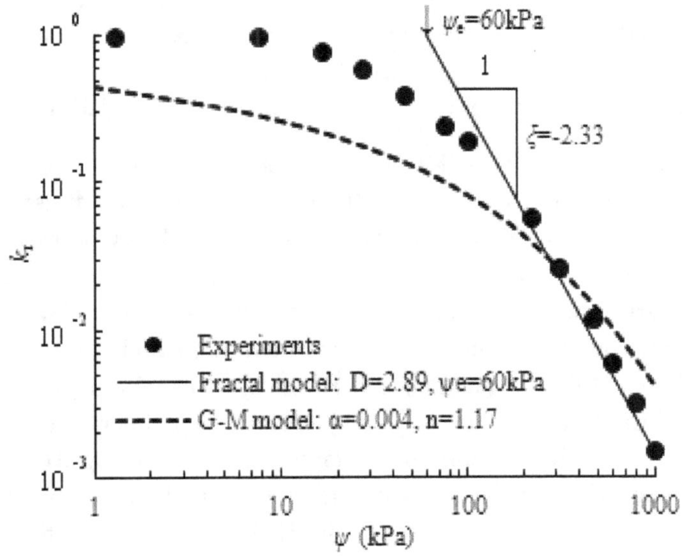

Figure 29. Comparison between the predictions of the RHC using fractal model and G-M model (van Genuchten, 1980)

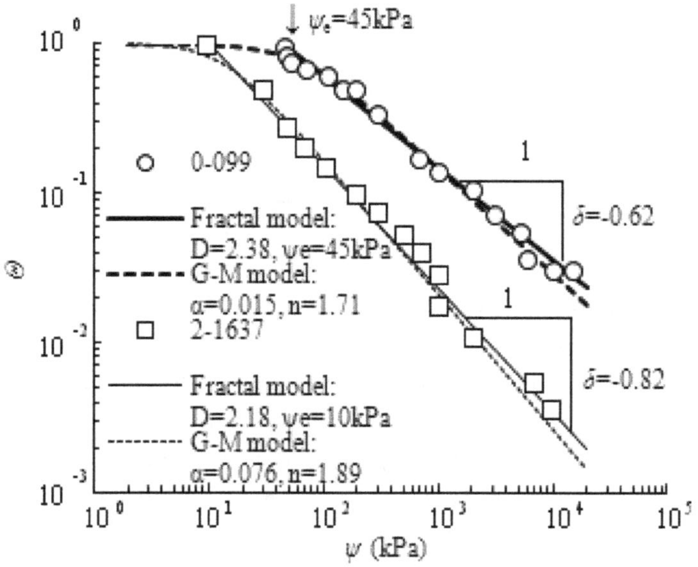

Figure 30. Fractal dimension and G-M parameters obtained from SWCC for Hanford glacial (Khaleel and Relyea, 1995)

The measured soil–water characteristic curves of Hanford glacial soil by Khaleel and Relyea (1995) were shown in Fig. 30. The saturated water contents (θ_s) are 0.338 and 0.303 for the samples 0–099 and 2–1637, respectively. The residual water contents (θ_r) are 0.039 and 0.025 for the samples 0–099 and 2–1637, respectively. From Fig. 30, it is seen that the soil-water characteristic curves (SWCC) of Hanford glacial soil are in good accord with the fractal model. The fractal dimensions (D), according to Eq. (29), are 2.35 and 2.18 obtained from the soil-water characteristic curves for the samples 0–099 and 2–1637 of Hanford glacial soil, respectively.

From the experimental data of the soil-water characteristic curve (SWCC), the parameters for determining relative hydraulic conductivity (RHC) using the fractal dimension are listed in Table 5. The larger the fractal dimension, the larger the complexity of the pore connectivity will be. The larger the fractal dimension, the larger the change in the water content with the same change in matric suction will occur. From the fittings of the soil–water characteristic curve in Fig. 30, the parameters (a and n) in the van Genuchten model (1980) for the soil–water characteristic curve (SWCC) are obtained, and are also listed in Table 5. Comparisons between the predictions of the fractal model (Eq. (45)) and the experimental data of the relative hydraulic conductivity (RHC) for Hanford glacial soil are shown in Fig. 31. The predictions of the fractal model (Eq. (45)) nearly agree with the experimental data of the relative hydraulic conductivity (RHC) for Hanford glacial soil in Fig. 31. The predictions of the G–M model (Eq. (52)) deviate from the experimental data of the relative hydraulic conductivity (RHC) in Fig. 31. It is seen that the predictions of the fractal model are better than those of the G–M model for Hanford glacial soil in Fig. 31.

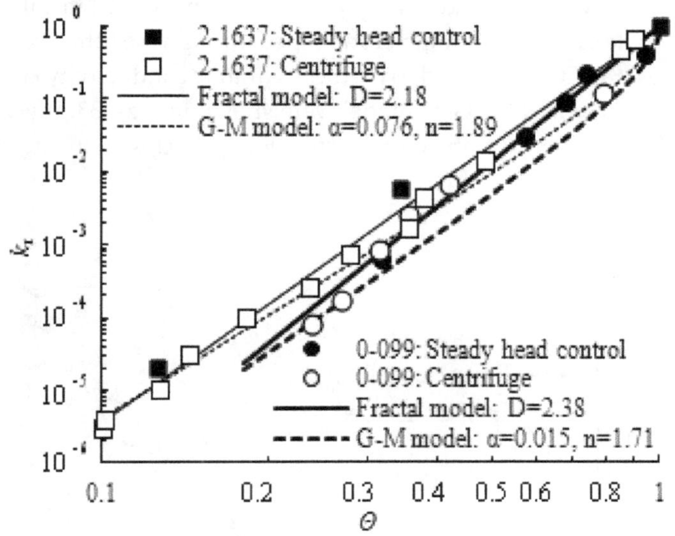

Figure 31. Comparison between the predictions using fractal model and G-M model (Data from Khaleel and Relyea, 1995)

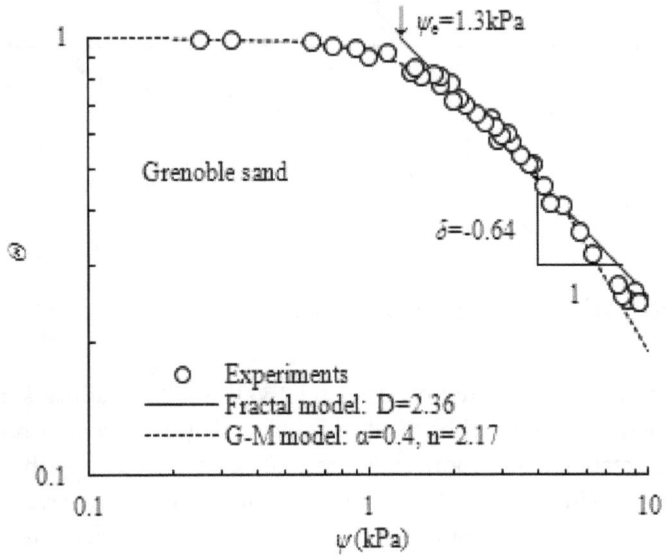

Figure 32. Fitted SWCC curves

The combinations of Eq. (29) with fractal capillary model were tested on Grenoble sand. Data points for the soil-water characteristic curve (SWCC) and hydraulic conductivity were taken from Touma (2009). The infiltration experiment was conducted under a constant head of 2.3cm (Touma & Vauclin, 1986). The solid lines in Figure 37 were the fitted curves for Eq. (29), and the resulting predicted hydraulic conductivity were shown in Fig. 32. Fractal model gave good results for the Grenoble sand. The parameters obtained through the fitting results of soil-water characteristic curves (SWCC) were used to predict hydraulic conductivity curve. Comparison between the fractal model and the G-M model were shown in Figs. 33. It was found that the prediction of hydraulic conductivity curve using Eq. (45) closed to the experiments for Grenoble sand.

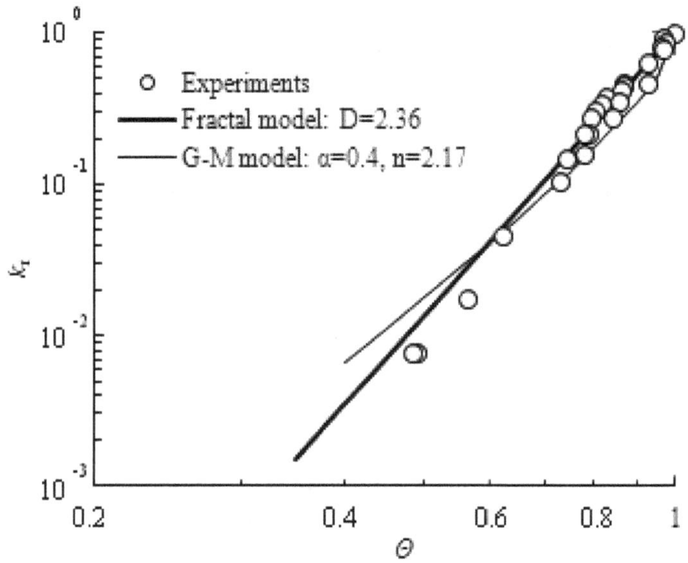

Figure 33. Comparison between fractal model and G-M model

7. Rainfall infiltration of unsaturated soils

Rainfall induced slope failures often occur as relatively shallow failure surfaces orientated parallel to the slope surface and are observed and analyzed by different mechanisms. The effect of seepage on slope stability is introduced in the analyses by calculating the critical depth for an infinite slope with rainfall infiltration, while the effect of both negative and positive pore water pressures on the stabilities of initially unsaturated slopes are explained and coupled with infinite slope analysis and pore-air flow analysis methods in order to present a predictive formulation of slope failures that occurs in rainfall evens and is derived from a fractal model on unsaturated soil. The formulation serves as a baseline analysis method for evaluating potentially unstable slopes.

The pore water pressure pattern that develops in the initially unsaturated soil will occur as a transient process as the infiltration moves downward into the soil profile. Several factors should been taken into account in unsaturated analyses. The shear strength of the soil mass and the development of seepage forces which both depend on the evolution of the pore water pressure profile must been addressed in detail.

Here, an individual soil slice can be treated as a one-dimensional column with vertical infiltration. Suppose that any lateral flow between the adjacent slices will be equal on the up-slope and down-slope boundaries. Then considering vertical infiltration only could meet the flow continuity requirement. These one-dimensional infiltration analyses are addressed by the saturated/unsaturated seepage finite element software SEEP/W (GEO-SLOPE International Ltd.2007) coupled with the air flow software AIR/W (GEO-SLOPE International Ltd.2007).

For a homogenous, isotropic soil, one-dimensional water flow equation (GEO-SLOPE International Ltd.2007) is given by:

$$m_{\mathrm{w}}\gamma_{\mathrm{w}}\frac{\partial H_{\mathrm{w}}}{\partial t}=\frac{\partial}{\partial y}\left(k_{\mathrm{w}}\frac{\partial H_{\mathrm{w}}}{\partial t}\right)+m_{\mathrm{w}}\frac{\partial P_{\mathrm{a}}}{\partial t}+Q_{\mathrm{w}} \tag{61}$$

where H_w is the total water energy potential comprised of both pressure and elevation potentials; P_a is the pore air pressure; k_w is hydraulic conductivity; y is y coordinate; m_w is the slope of SWCC; t is time; and Q_w has units of length per time.

For the air conversation of mass, we can arrive at the pore air general mass balance equation (GEO-SLOPE International Ltd.2007) as like:

$$\left(\rho_{\mathrm{w}}m_{\mathrm{w}}+\frac{\theta_{\mathrm{a}}}{\bar{R}T}\right)\frac{\partial P_{\mathrm{a}}}{\partial t}=\frac{\partial}{\partial y}\left[\frac{\rho_{\mathrm{a}}k_{\mathrm{a}}}{\gamma_{0\mathrm{a}}}\frac{\partial P_{\mathrm{a}}}{\partial y}+\frac{\rho_{\mathrm{a}}{}^{2}k_{\mathrm{a}}}{\rho_{0\mathrm{a}}}\right]+\rho_{\mathrm{a}}\lambda_{\mathrm{w}}m_{\mathrm{w}}\frac{\partial H_{\mathrm{w}}}{\partial t} \tag{62}$$

where k_a is air permeability; θ_a is volumetric air content; ρ_a is air density; γ_{oa} is initial air unit weight; ρ_{oa} is initial air density; $\bar{R}$is ideal air constant and $\bar{R}$=287J/(kg K) for dry air; T is temperature; H_w is total head and P_a is pore air pressure. These two governing equations can be used to determine total pressure head profile for the water and the pore air pressure, and then suction is attained.

The relationships relating volumetric water content and hydraulic conductivity to suction must be known to solve the Eq. (61). And similarly to calculate the pore air pressure the pore air permeability function is also wanted. To find the effect of different fractal dimensions and air-entry values of different soil types, series of parameter studies have conducted. Different types of Soil-water characteristic curves and hydraulic conductivity curves were shown in Figs. 34 and 35. In the legends of these pictures, Soil 10, 2.1,-5, for example, means a type of soil with air-entry value, ψ_e=10kPa, the fractal dimension, D=2.1 and saturated hydraulic conductivity k_s=1×10^{-5}m/s. It is assumed that θ_r=0. The pore air permeability function curve used in these studies was shown in Fig.36, and it remained a constant (GEO-SLOPE International Ltd.2007).

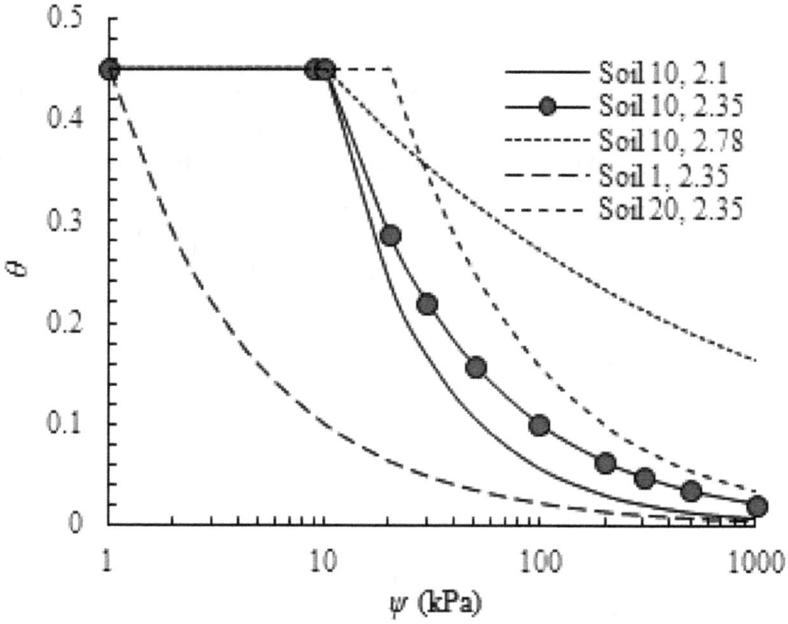

Figure 34. Soil-water characteristic curves for soils

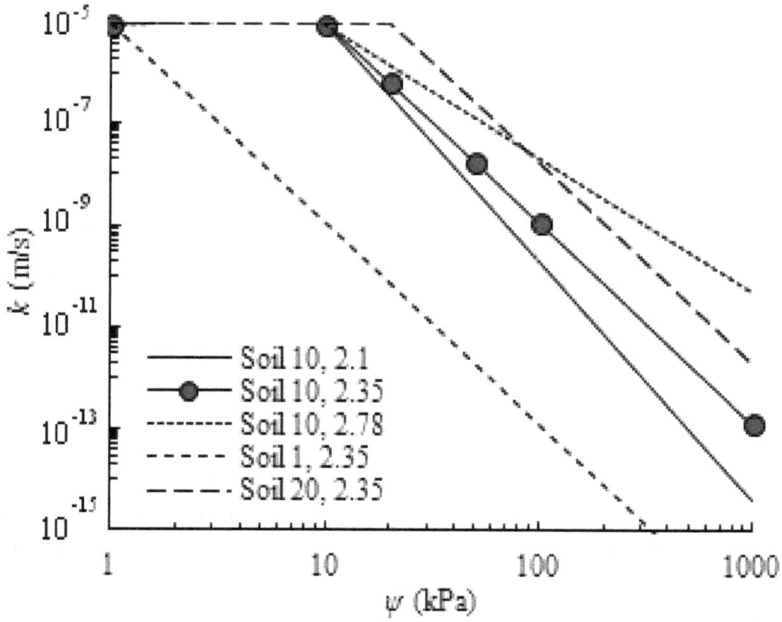

Figure 35. Hydraulic conductivity curves f

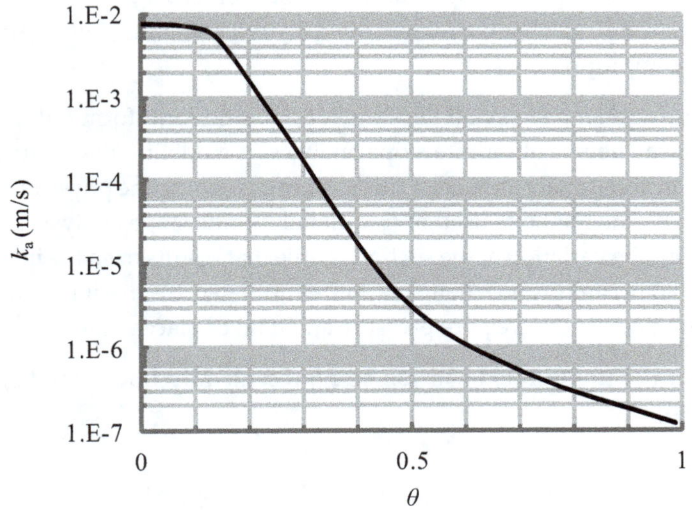

Figure 36. Pore air permeability function used in study

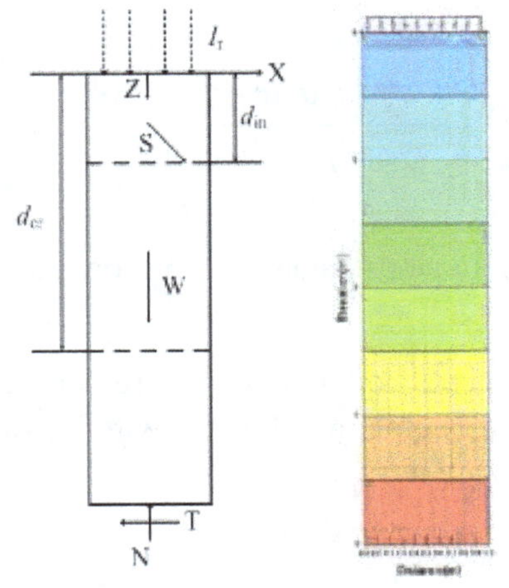

Figure 37. Sketch map of the one-dimensional infiltration model

A 4m-deep soil column was modeled for the one-dimensional analyses, as shown in Fig. 37, and infiltration occurs from the top of the column, while the bottom was set to be a pervious boundary. A linear hydrostatic suction distribution presented for the initial conditions. Analyses for the rainfall infiltration into different types of soils were performed using the unsaturated characteristic curves shown in Figs. 34-36. In applying a top boundary condition to simulate the infiltration of rainfall, it is important to realize that an influx boundary depends on the relationship between the saturated hydraulic conductivity and the rainfall intensity. In the following analyses, if a rainfall intensity greater than or equal to k_s with the non-infiltrating rainfall running off the slope, the top boundary condition is set to a total head equal to 4m; while if field results indicate that the rainfall intensity is less than k_s, then a flux type boundary condition is more appropriate.

Since initial failures often have small depth-to-length ratios and form failure planes parallel to the slope surface, the use of infinite slope analysis in modeling the infiltration process by vertical one-dimensional analysis makes it justified in describing the physical process of failure initiation. However, the methods used in traditional infinite slope analysis must be modified to take into account the variation of the suction profile that results from the infiltration process. There are two distinct failure mechanisms can be initiated by the infiltration process. The failure takes place due to positive pore pressure and it takes place while the suction still exists.

The shear strength of unsaturated soil can be represented by fractal model, and written as follows (Xu, 2004):

$$\tau = c' + (\sigma - u_{\mathrm{a}})\tan\phi' + \psi_{\mathrm{e}}^{3-D}\psi^{D-2}\tan\phi' \qquad (63)$$

where τ is the shear strength; c' is the effective cohesion; $(\sigma\text{-}u_a)$ is the total normal stress; φ' is effective friction angle; ψ is the suction; ψ_e is the air-entry value. The stability envelope can be represented as (Collins and Znidarcic, 2004):

$$d_{\mathrm{cr}} = \frac{c' + \gamma_{\mathrm{w}} h_{\mathrm{c}} \tan\phi' - \gamma_{\mathrm{w}} h_{\mathrm{p}} \tan\phi'}{\gamma\cos^2\beta(\tan\beta - \tan\phi')} \qquad (64)$$

where d_{cr} is the critical depth; β is slope angle; h_p is the pore passive pressure and h_c is the pore negative pressure.

We can get suction with analyses of both pore water pressure and pore air pressure in the soil slope under rainfall infiltration. Combining Eq. (63) with Eq. (64), the new stability envelope can be derived and written as follows:

$$d_{\mathrm{cr}} = \frac{c' + \psi_{\mathrm{e}}^{3-D}\psi^{D-2}\tan\phi' - \gamma_{\mathrm{w}} h_{\mathrm{p}} \tan\phi'}{\gamma\cos^2\beta(\tan\beta - \tan\phi')} \qquad (65)$$

The critical depth for infinite slope failure is now a function of the suction, ψ, pressure head h_p, the given material and slope characteristics, c', φ', γ, ψ_e, ξ, γ_w, and β. In this way, the slope stability issues related to the decrease in shear strength from a loss of soil suction and the development of seepage forces from positive pressure head generation can be clearly understood.

Because the infiltration results and the slope stability results were both presented in terms of the pressure head and the suction profile, the two analyses can be coupled to yield a comprehensive method for determining the location and time of failure for a slope if the soil, slope, and rainfall parameters are given. The methodology of the coupled analysis involves plotting a stability envelope as defined by Eq. (65) for specified soil and slope parameters, over a given infiltration profile generated from the particular rainfall and unsaturated characteristic curves. Coulomb failure and the initiation of slope mobilization are defined at the points where the infiltration trace intersects the stability envelope. Intersections of the infiltration trace and the stability envelope indicate points at which the slope is unstable and can be thought of as "critical depths" of failure. It must therefore be assumed that if the slope is initially stable, the point at which the stability envelope inter sects the initial suction distribution line will not contribute any further information about the failure mechanisms from infiltration.

Here, factors controlling the unsaturated soil slope under infiltration are studied and the results which combine infiltration analyses and stability analyses are shown in Figs.43-47. Infiltration analyses were performed by software SEEP/W coupled with AIR/W in order to calculate the suction file and the stability envelope as defined by Eq.(65) for slope parameters, β=40°, γ=19.6kN/m^3, γ_w=10kN/m^3, and shear strength parameters c'=5kPa, φ'=15°. Other parameters are studied in a systematic way.

7.1. Effect of fractal dimension

Soil column was saturated due to rainfall infiltration from the initial hydrostatic suction profile. It was shown that infiltration took place in unsaturated soil much faster for the soil with higher fractal dimension. Three fractal dimensions (D=2.1, 2.35 and 2.78) were calculated respectively and the results were shown in Fig.38. It was seen from Fig. 38 that the higher initial hydraulic conductivity existed on the top of soil column which had a higher fractal dimension. On the other hand, the stability envelope for the soil with higher fractal dimension was steeper. There was an obvious inflection point in the stability envelope of the soil with the low fractal dimension. The results also indicated that it was the relative shape of the unsaturated characteristic curves (Fig.34 and Fig. 35) that has a controlling effect on the suction distribution.

7.2. Effect of air-entry value

Three air-entry values (ψ_e=1kPa, 10kPa and 20kPa) are calculated respectively and the results as shown in Figs.39. Compared with the results of three different air-entry values, both the infiltration trace and the stability envelope have changed with the different air-entry values. When the air-entry value is relatively small (ψ_e=1kPa), as shown in Fig.39a, passive pore water pressure was generated on the upper soil column at the time t=32h. Due to the low initial permeability on the top of soil column and a state of lower saturation, fewer pores initially filled with water and there are fewer channels available for fluid transport and consequently the flow of water is hindered. With the commencement of infiltration at the top boundary, water is forced into a soil which is not capable of transporting it efficiently, which results in the development of positive pressure heads as the infiltration front progresses downward. While the air-entry value is relatively larger, as shown in Fig.39c, infiltration moves downward much faster and suction has reached to the air-entry value. Passive pore water pressure occurs at the bottom of the soil column which means that there is little unsaturated zone in the middle of soil column during the rainfall infiltration. Besides, the stability envelope is much smooth when the soil has a relatively low air-entry value, and consequently the critical depth is smaller.

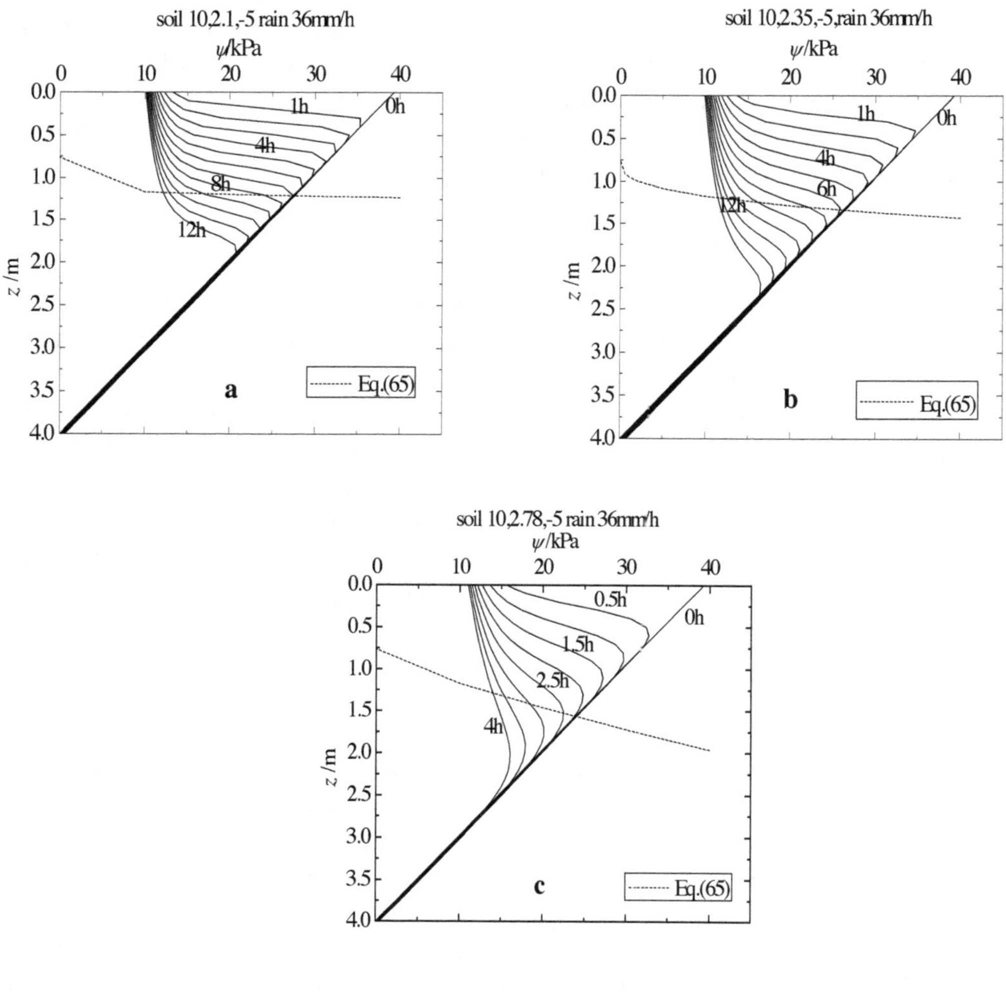

Figure 38. Infiltration results for soils with different fractal dimension and the new stability envelope

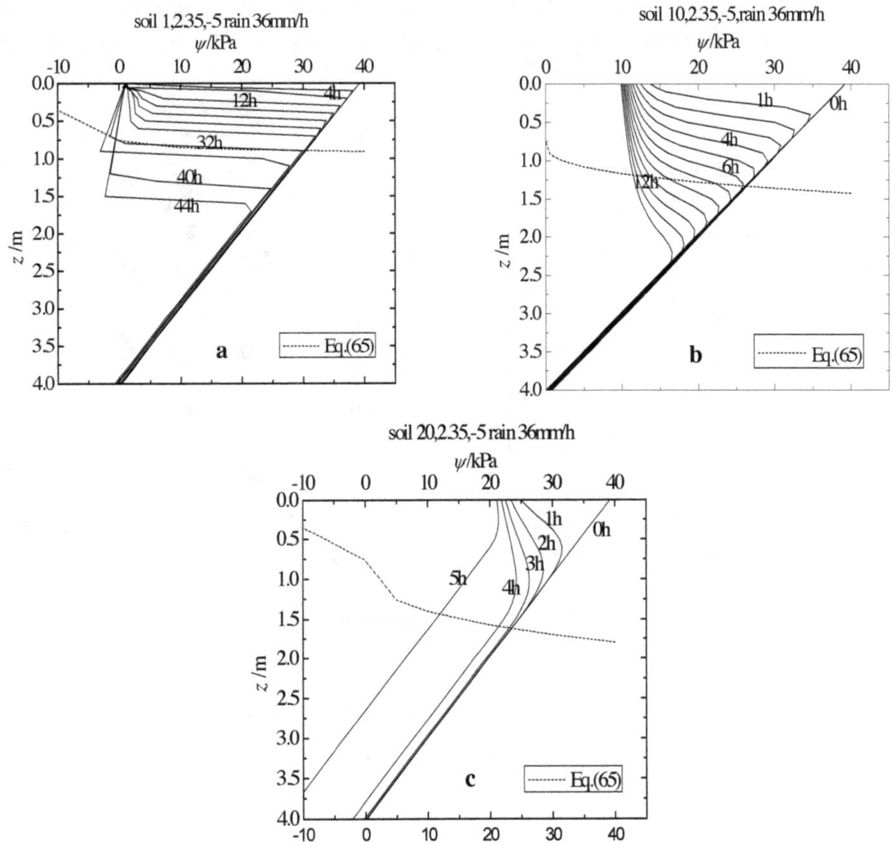

Figure 39. Infiltration results for soils with different air-entry value and the new stability envelope

7.3. Effect of saturated hydraulic conductivity

The dominate factors which control the stability of unsaturated soil slope under rainfall infiltration are the soil saturated hydraulic conductivity and rainfall intensity. While the soil saturated hydraulic conductivity determines the water transportation ability of soil and the infiltration quantity most depends on the rainfall intensity. In this parameter study series, three different saturated hydraulic conductivity values (k_s=1×10^{-5}m/s, 1×10^{-6}m/s, 1×10^{-7}m/s) are taken into account and the results are shown in Fig.40.

When the rainfall intensity is much larger than the saturated hydraulic conductivity, soil in the top region of the column has been saturated soon after the infiltration begins and its suction almost reaches to zero at the time t=1.5h. It can be seen that the suction at the base is increasing over time as a positive number, which means that the soil is actually de-saturating at its base while wetting up above. This is caused by the air which becomes trapped when the top has saturated and starts to resist water infiltration. As the situation that the rainfall intensity is much smaller than the saturated hydraulic conductivity as shown in Fig.40a, the suction of the top region decreases with the infiltration developing, however, the top region has remained unsaturated and the infiltration depth is also small due to the small rainfall quantity into the soil.

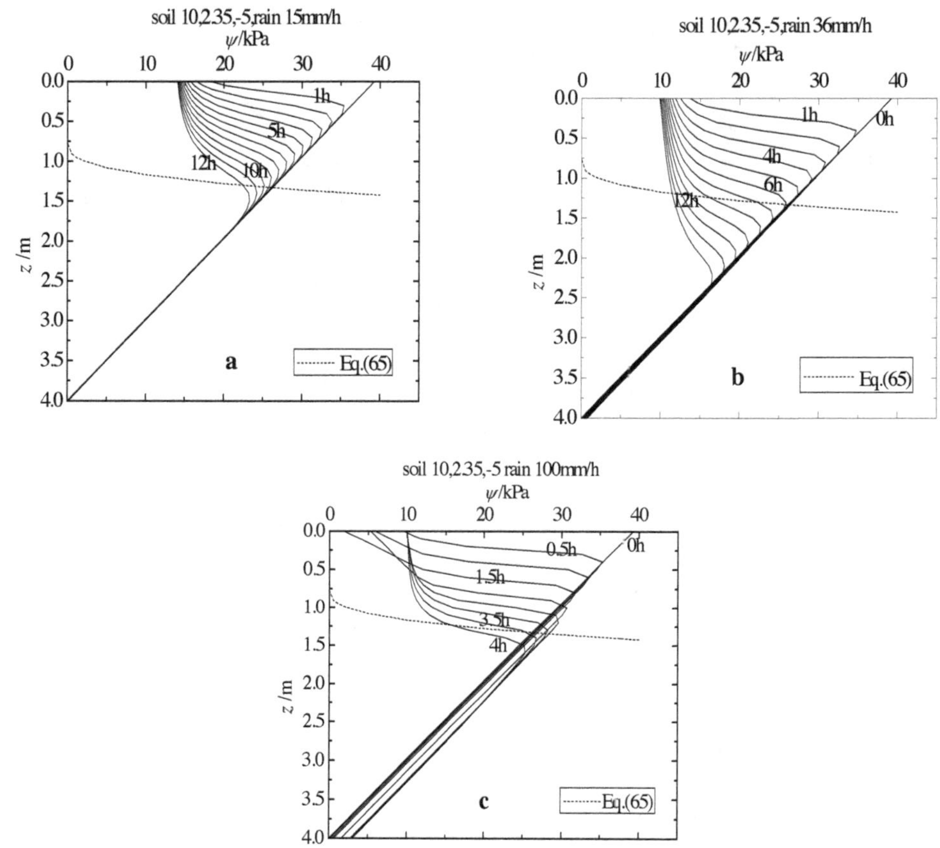

Figure 40. Infiltration results for soils due to different rainfall intensities

7.4. Effect of rainfall intensity

The rainfall intensity for a certain saturated hydraulic conductivity has taken into considered to simulate the infiltration process. The range is changed from 0.3mm/h to 100mm/h and the relationship between the time of failure and rainfall intensity for the three soil types of saturated hydraulic conductivity has plotted in Fig.41. These curves decrease as exponent forms similarly while the magnitude of different curves has changed dramatically. The time at which the soil slope failure occurs can be about ten hours for the soil saturated hydraulic conductivity k_s equals to 1×10^{-5}m/s, and for the situation that k_s equals to 1×10^{-6}m/s, the time can range from 10 to 100 hours. While much more time is needed for the lower saturated hydraulic conductivity soil type.

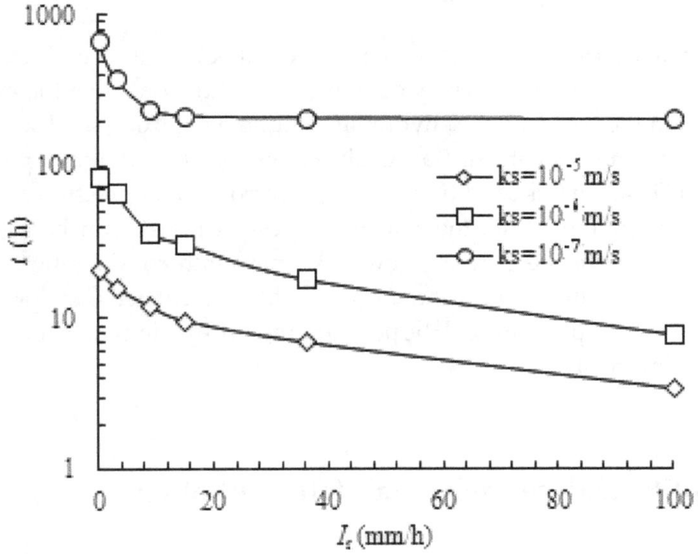

Figure 41. The time of failure vs. rainfall intensity

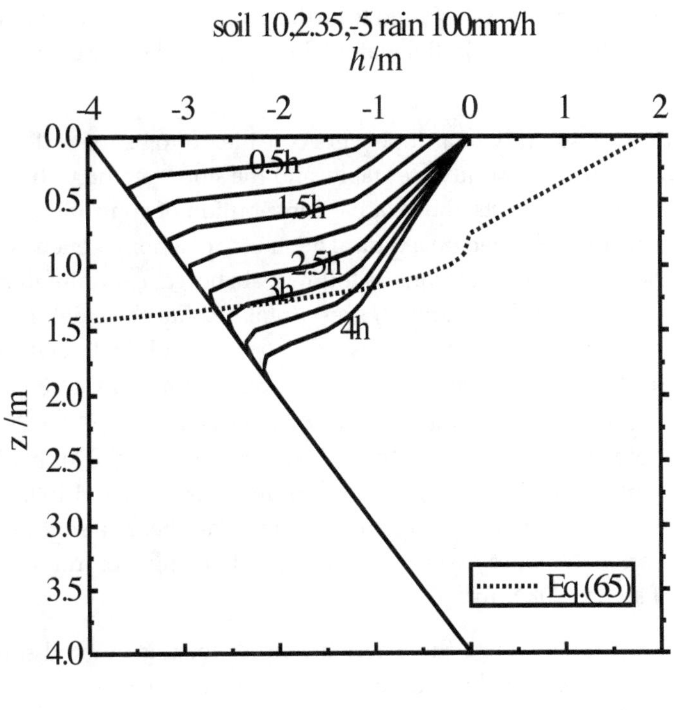

Figure 42. Infiltration results without air phrase consideration

7.5. Infiltration analyses coupled with the air flow analyses

In the infiltration analyses, air flow movement in the soil pore is also conducted. The effect of the pore air flow movement is especially obvious in the situation when the rainfall intensity is much larger than the soil saturated hydraulic conductivity. Compared with the results of infiltration analyses coupled with air flow analyses, the results without air phrase consideration are shown in Fig.42. In this condition, the negative pore water pressure gradually reaches zero at the top of soil column. While the pore water pressure remains unchanged at the bottom of the column. If we take the pore air flow movement into consideration, the results are different as shown in Fig.40c. Besides, the suction could be calculated more precisely when we attain both the pore water pressure and the pore air pressure by simulating infiltration analyses coupled with air flow analyses.

8. Slope stability analyses due to rainfall infiltration

It was widely recognized that the rainfall infiltration took a great role in causing landslides, while the relative importance of soil properties, rainfall intensity, initial water table depth and slope geometry in inducing instability of a homogenous unsaturated soil slope under different rainfall was investigated through a series of parametric studies (Rahardjo, et al, 2007). Soil properties and rainfall intensity were found to be the primary factors controlling the instability of slopes due to rainfall, while the initial water table depth and slope geometry only played a secondary role.

The factors affecting the stability of a slope were considered to be the soil properties, rainfall intensity, initial depth of the groundwater table, and the slope geometry (i.e., slope angle and slope height). To assess the effects and relative contribution of controlling factors, a series of parametric studies were performed on a typical geometry of a homogeneous soil slope shown in Fig. 43. It had the boundary conditions as follows: ab, bc, cd=q=I_r(rainfall intensity); ah, de, fg=q=0m^3/s (i.e. no flow boundary); and ef, gh=h_t(total head at the side). Four slope heights, H_s,(3m, 4m, 5m and 6m), three slope angles, α,(26.6, 33.7 and 45.0), eight initial depths of groundwater table (GWT), H_w,(0.5m, 1m, 2m, 3m, 5m, 7.5m, 10m and 15m), three fractal dimensions D (2.1, 2.35 and 2.78), four air-entry values ψ_e (1kPa, 10kPa, 20kPa and 50kPa), five values of saturated hydraulic conductivity, k_s (10^{-4}m/s, 10^{-5}m/s, 10^{-6}m/s, 10^{-7}m/s and 10^{-8}m/s) and five rainfall intensities I_r(3mm/h, 9mm/h, 15mm/h, 36mm/h and 100mm/h each for 24h duration) were used in six cases of parametric studies. The shear strength parameters of the soils used in the parametric study were c'=10kPa, effective angle of internal friction, φ'=26°, and unit weight of soil, γ=20kN/m^3.

The fractal model of the soil-water characteristic curves for unsaturated soil was written as Eq. (29). The function of relative hydraulic conductivity was used as Eqs. (44) and (45). The derived SWCC and RHC for the all six soils were shown in Figs. 34 and 35. In the legends of these pictures, the symbol S in the soil names represent "soil", the first number means the air-entry value, and the second one means the fractal dimension.

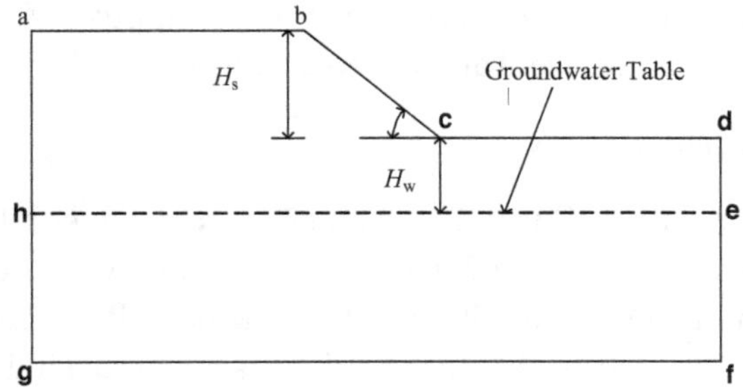

Figure 43. Sketch map of the slope in study

In the seepage analysis, the governing partial differential equation (GEO-SLOPE International Ltd.2007) for a two-dimensional transient water flow used in the finite element seepage model was written as follows:

$$m_w \gamma_w \frac{\partial (H-y)}{\partial t} = \frac{\partial}{\partial y}\left(k_w \frac{\partial H_w}{\partial t} \right) + m_w \frac{\partial P_a}{\partial t} + Q \tag{66}$$

where H is the total head; k_x are k_y ars hydraulic conductivity in x, y direction respectively ; Q is the applied boundary flux; m_w is the slope of the SWCC and γ_w is the unit weight of water. Eq. (66) was solved by SEEP/W software (Geo-Slope 2007). The boundary conditions used in the transient seepage analysis were shown in Fig. 43. And the initial condition for the analyses was a hydrostatic condition with a limiting pore water pressure of -75kPa. Then the SEEP/W software generated the negative pore water pressure as the initial condition. The pore-water pressures obtained from the seepage analysis were then used in the slope stability analyses to calculate the factor of the slope safety, F_s.

The equation for unsaturated shear strength was used as Eq. (63). In the study, we chose c'=10kPa, φ'=26°. The Morgenstern-Price Method was adopted in the slope stability analysis. This method considered both shear and normal inter-slice forces, satisfied both moment and force equilibrium, and allowed for a variety of user-selected inter-slice force function. The slope stability analysis using Morgenstern-Price Method was performed using SLOPE/W software (Geo-Slope 2007). The pore-water pressures, u_w, obtained from the transient seepage analyses using SEEP/W were added to SLOPE/W to be incorporated in the slope stability analyses.

The results were presented with attention on the effects of the factors which impacting on the soil slope stability under rainfall for 24h with a combination of various controlling factors.

8.1. Effect of soil properties

The variation in factor of safety with time for a homogeneous soil slope of constant slope height H_s=6m, initial groundwater table depth H_w=2m, subjected to rainfall intensities of 3, 9 and

15mm/h of respective soil for 24h with a combination of various soil types (different fractal dimensions and air-entry values) were analysized as follows.

1. Effect of fractal dimension

In Figs.44, the fractal dimension, D was 2, 2.35 and 2.78, respectively. The air-entry value, ψ_e remained 10kPa. It is shown that the reduction of slope stability is larger in the condition of larger fractal dimension under the condition of the same rainfall intensities and the saturated hydraulic conductivity. This means that the soil slope is much safe if the fractal dimension is low while the saturated hydraulic conductivity remains constant. The factor of safety decreases in a short time under rainfall infiltration when the fractal dimension is relatively large, as shown in Fig.44c, while the decrease occurs at a longer time for the low fractal dimension soil type, as shown in Fig.44a.

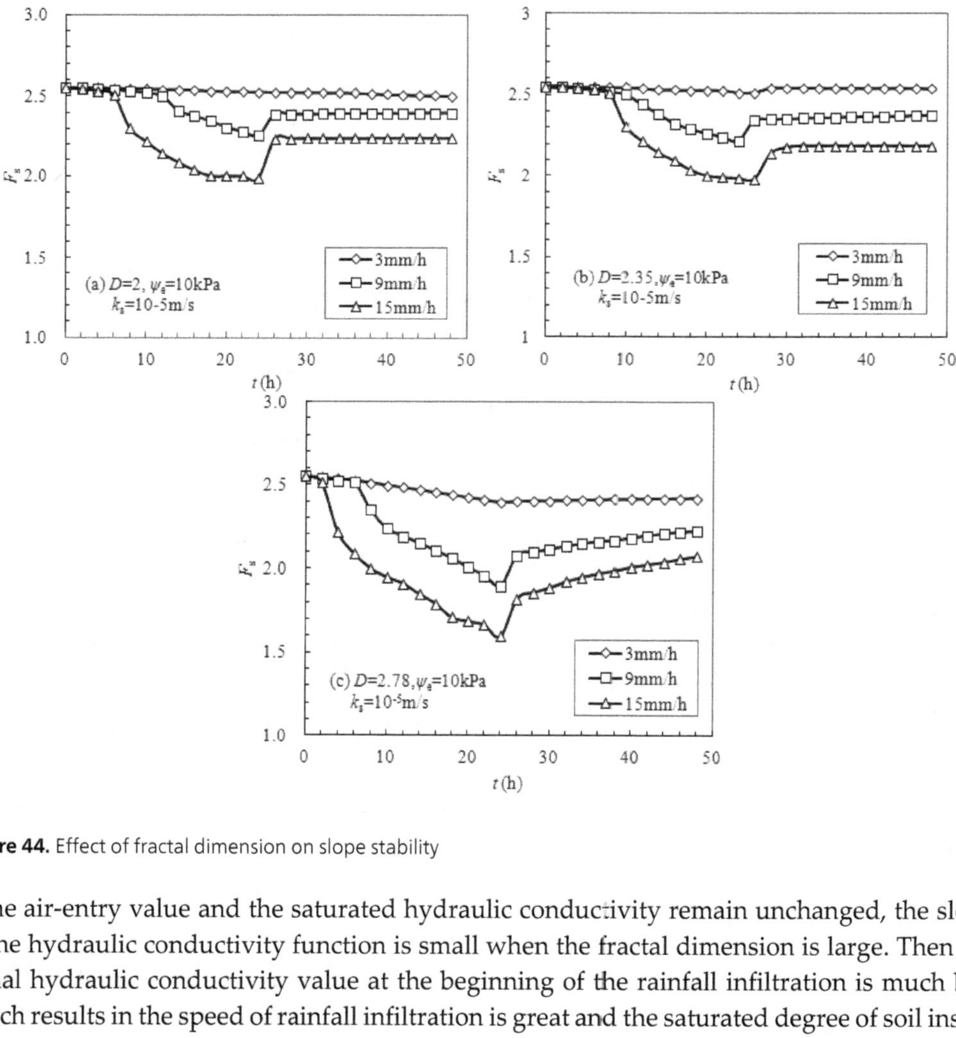

Figure 44. Effect of fractal dimension on slope stability

If the air-entry value and the saturated hydraulic conductivity remain unchanged, the slope of the hydraulic conductivity function is small when the fractal dimension is large. Then the initial hydraulic conductivity value at the beginning of the rainfall infiltration is much big, which results in the speed of rainfall infiltration is great and the saturated degree of soil inside of slope changes quickly leading unsaturated soil to become saturated. The passive pore water pressure generated by the increase of degree of soil saturation makes a negative impact on the instability of soil slope under rainfall.

2. Effect of air-entry value

The effects of air-entry value on the stability of soil slope due to rainfall infiltration are shown in Figs. 45. The air-entry value, ψ_e is 1kPa, 10kPa, 20kPa and 50kPa respectively and the fractal dimension, D remains 2.35. The style of the slope safety curve changes a lot with different air-entry values in Figs. 45. In Fig. 45a, the air-entry value, ψ_e is 1kPa and the decrease of factor of safety occurs during 16~20h after rainfall began. As the air-entry value, ψ_e is 10kPa, shown in Fig. 45b, the decrease of slope safety factor occurs at 8h after rainfall began. At the beginning of rainfall infiltration, the decrease happens in Figs. 45c-d, in which the air-entry value is 20kPa and 50kPa respectively. This means that a delay time appears which is depended with the air-entry value. And it is obvious that small air-entry value decides a relative long delay time. Besides, the factor of slope safety remains minimum during another 24 hours after rainfall stops in Fig. 45a, which has a low air-entry value (ψ_e=1kPa). While ψ_e=50kPa, the recovery of safety factor curve in Fig. 45c occurs and dramatically. This is also due to the style of hydraulic conductivity function, as shown in Fig. 35. The initial hydraulic conductivity value becomes large when the air-entry value increases with fractal dimension, saturated hydraulic conductivity and saturation degree remain the same.

3. Effect of saturated hydraulic conductivity

The effects of saturated hydraulic conductivity on stability of a homogenous soil slope are reflected through the relationship between F_s and k_s, shown in Fig. 46. The saturated hydraulic conductivity, k_s, was varied with five different values of 10^{-8}m/s, 10^{-7}m/s, 10^{-6}m/s, 10^{-5}m/s and 10^{-4}m/s for a homogeneous soil slope of constant soil type S10, 2.35 (ψ_e=10kPa, D=2.35), H_w=2m, H_s=6m, α=33.7°, subjected to rainfall for 24h with six rainfall intensities of 3, 9, 15, 36, 100 and 360mm/h. All the plots in Fig. 47 shows that soil with k_s values of 10^{-8}m/s and 10^{-7}m/s, respectively, are less affected by rainfall. Contrarily, soil with k_s values of 10^{-5}m/s and 10^{-4}m/s are greatly affected by rainfall. It is suggested that soil slopes with a low saturated hydraulic conductivity are relatively safe under short-duration rainfall infiltration and for the soil slopes with a high saturated hydraulic conductivity the stability is affected by the short-duration rainfall greatly. This means that slopes with low saturated hydraulic conductivity need a long-duration rainfall to intrigue the instability.

The F_s(min)-k_s critical curve was presented in the broken line in Fig. 47 as follows:

$$F_s(\min) = \frac{a}{1 + be^{-ck_s}} \tag{67}$$

where F_s(min) is minimum factor of safety; k_s is saturated hydraulic conductivity; a, b and c are the fitting parameters and e is natural number (i.e., 2.71…). Here, a=1.293, b=-0.492, c=1.2×10^5, and the corresponding coefficient of correlation, r^2 is 0.9998.

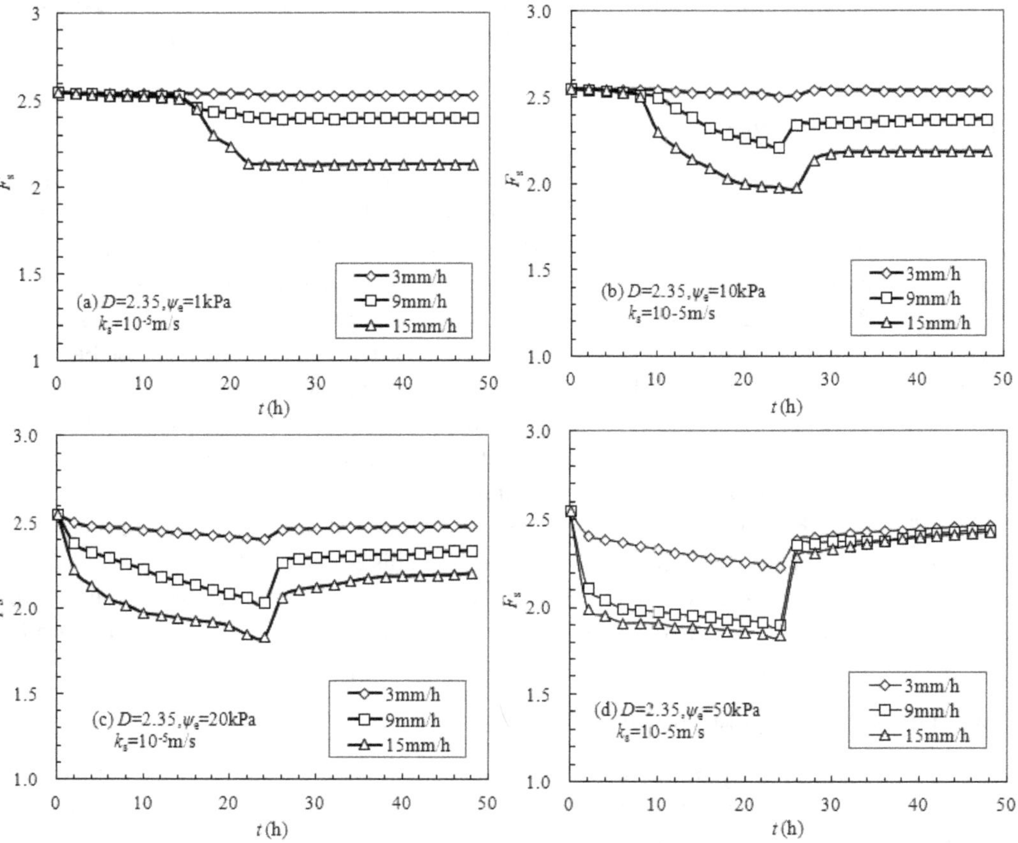

Figure 45. Effect of air-entry value on the slope stability

8.2. Effect of rainfall intensity

The relationship between minimum factor of safety, F_s(min), versus logarithmic of rainfall intensity, I_r are plotted in Fig. 47. The semi log plot shows that generally the F_s(min) and I_r relationships follow a sigmoid shape in Fig. 47. There are two inflect points for the sigmoid shape line. The F_s(min) is almost constant at very low rainfall intensities for all soil types. And it starts to decrease rapidly when the first inflection point is reached. The trend in F_s(min) versus I_r relationship observed in Fig. 47 can be described by a sigmoid equation (MMF Line) as the form of

$$F_s(\min) = \frac{ab + cI_r^{\,d}}{b + I_r^{\,d}} \tag{68}$$

where F_s(min) is minimum factor of safety; I_r is rainfall intensity; and *a, b, c, d* are fitting parameters. The values for the fitting parameters and the corresponding coefficient of correlation, r^2, for the sigmoid line in Fig.47.

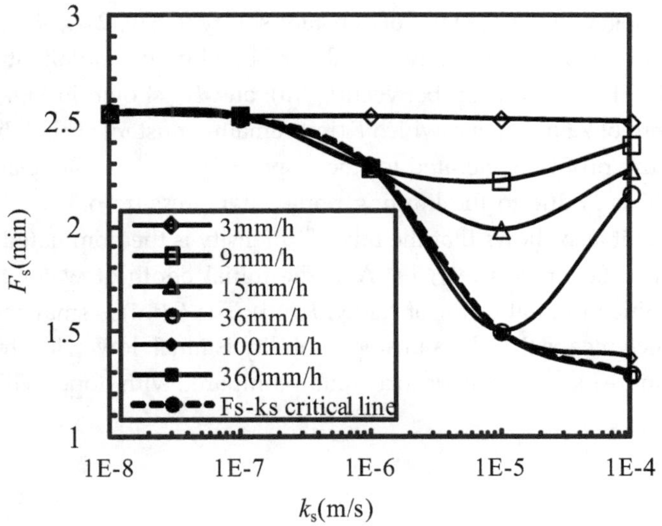

Figure 46. Relationship between K_s and minimum factor of safety

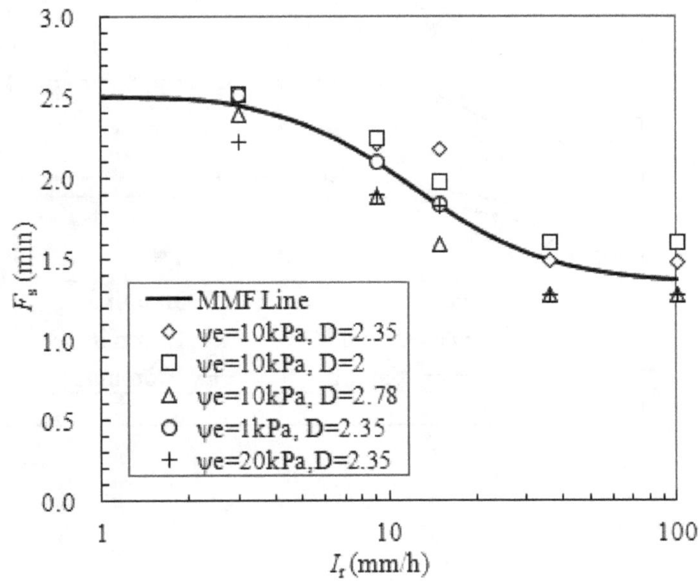

Figure 47. Effect of rainfall intensity on variation of minimum factor of safety

8.3. Effect of initial water table location

The effects of initial groundwater table location on stability of a homogenous soil slope are reflected through the relationship between F_s(ini) and F_s(min) with H_w, shown in Fig. 48. The initial depth of water table, H_w, was varied with five different values of 0.5, 1, 2, 2.5, 5, 7.5, 10 and15m for a homogeneous soil slope of constant soil type S10, 2.35, -5 (ψ_e=10kPa, D=2.35, k_s=10^{-5}m/s), H_s=6m, α=33.7°, subjected to rainfall for 24h with five rainfall intensities of 3, 9, 15, 36, and 100mm/h. The relationship between F_s(ini), and H_w shown in Fig. 48 appears to be linear up to a depth of 7.5m beyond which F_s(ini) remains constant. This is because the initial pore-water pressure profiles generated for the slope at H_w=7.5m as same as those generated when H_w=7.5m. This is due to the limiting pore-water pressure of −75kPa adopted in the analyses. In Fig. 54 it also shows that the rainfall intensity is the dominated factor impacting on the reduction in factor of safety, Fs. And the initial depth of water table, H_w, mainly determines the value of initial factor of safety, F_s(ini). The F_s(ini) is smaller for slopes with a shallower H_w which means that the soil slope stability is much lower. Therefore, slopes with a shallow H_w are more likely to fail due to a rainfall compared with slopes which has a deep H_w.

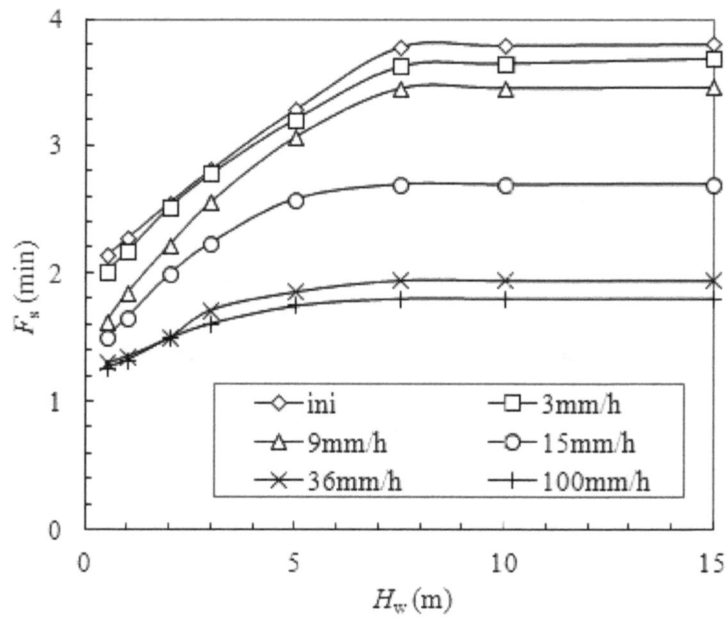

Figure 48. Effect of initial groundwater table depth on factor of safety

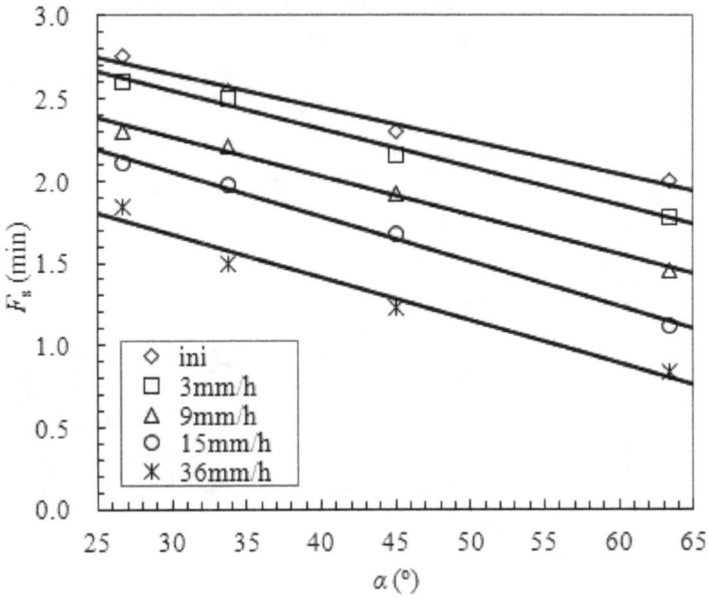

Figure 49. Effect of slope angle on factor of safety

8.4. Effect of slope angle

The effect of slope geometry on the stability of a homogenous soil slope is evaluated in terms of slope angle (α) and slope height (H_s). The effects of slope angel on the stability of a homogenous soil slope are reflected through the relationship between F_s(ini) and F_s(min) with α, shown in Fig. 49. The slope angel, α was varied with three different values of 26°, 33.7°, 45°and 63° for a homogeneous soil slope of constant soil type S10, 2.35, -5 (ψ_e=10kPa, D=2.35, k_s=10^{-5}m/s), H_s=6m, H_w=2m, subjected to rainfall for 24h with five rainfall intensities of 3, 9, 15 and 36mm/h. The relationship between F_s(min) and α shown in Fig. 49 appears to be negative linear. In general, the higher the slope angle, the lower the initial factor of safety and the minimum factor of safety. Because a steep slope will yield a lower factor of safety compared with a flat slope.

8.5. Effect of slope height

The effects of slope angel on the stability of a homogenous soil slope are reflected through the relationship between F_s(ini) and F_s(min) with H_s, shown in Fig. 50. The slope height, H_s, was varied with four different values of 3m, 4m, 5m and 6m for a homogeneous soil slope of constant soil type S10, 2.35, -5(ψ_e=10kPa, D=2.35, k_s=10^{-5}m/s), H_w=2m, α=33.7°, subjected to rainfall for 24h with four rainfall intensities of 3mm/h, 9mm/h, 15mm/h and 36mm/h. Fig. 50 shows that initial factor of safety decreases exponentially as the slope height increases. It also suggests that high slopes are generally easier to fail under rainfall due to the low initial factor of safety. The reduction in factor of safety for a high slope is smaller and occurs at a slower rate compared with a low slope.

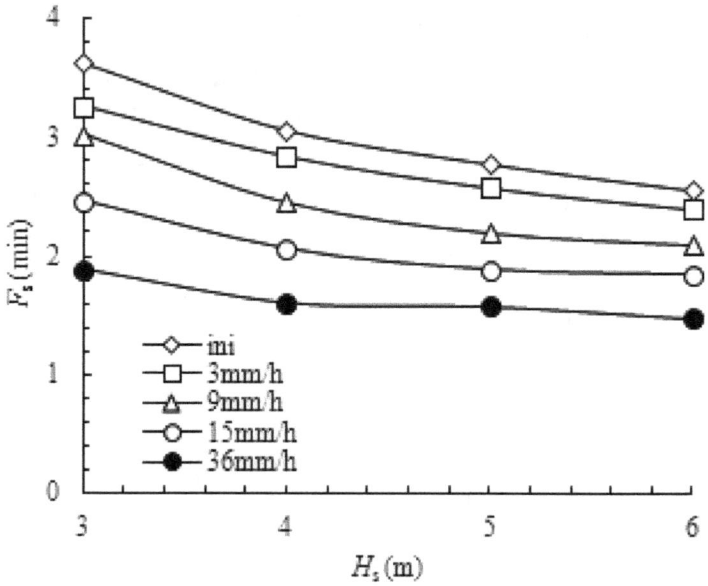

Figure 50. Effect of slope height on factor of safety

Acknowledgements

The National Natural Science Foundation of China (Grant No. 41272318) and State Key Laboratory of Ocean Engineering are sincerely acknowledged for their financial support.

Author details

Yongfu Xu*

Address all correspondence to: yongfuxu@sjtu.edu.cn

Department of Civil Engineering, Shanghai Jiao Tong University, Shanghai, China

References

[1] Abramowitz M, Stegun I. Handbook of Mathematical Functions. Dover Publications, New York, NY. 1970

[2] Avnir D, Jaroniec M, An isotherm equation for adsorption on fractal surfaces of heterogeneous porous materials. Langmuir, 1989, 5, 1431–1433.

[3] Brooks RH, Corey AT. Properties of porous media affecting fluid flow. ASCE J Irrig Drain Div 1966, 92:61–68.

[4] Brutsaert W. Some methods of calculating unsaturated permeability. Trans Am Soc Agr Engrs 1967;10:400-404.

[5] Brutsaert W. The permeability of a porous medium determined from certain probability laws for pore size distribution. Water Resour Res 1968; 4: 425-34.

[6] Brutsaert W. A concise parameterization of the hydraulic conductivity of unsaturated soils. Advances in Water Resources, 2000, 23, 811-815.

[7] Burdine NT. Relative permeability calculations from pore-size distribution data. Trans Am Inst Min Engrs 1953;198:71-7

[8] Burdine NT. Relative permeability calculation from pore size distribution data. Transactions of the American Institute of Mining Engineers, 1953, 198, 71–78.

[9] Carman PC. Flow of gases through porous media. New York: Academic Press, 1956.

[10] Childs EC, Collis-George N. The permeability of porous materials. Proc Roy Soc A 1950; 201: 392-405.

[11] Collins,B.D., Znidarcic. Stability analyses of rainfall induced landslides [J].Journal of geotechnical and geoenvironmental engineering, 2004,4,362-372.

[12] Corey AT. Pore-size distribution. In: van Genuchten MTh, Leij FJ, Lund LJ, editors. Proceedings of the International Workshop, Indirect Methods for Estimating the Hydraulic Properties of Unsaturated Soils, University of California, Riverside, 1992. p. 37–44.

[13] Crawford JW. The relationship between structure and the hydraulic conductivity of soil. Eur J Soil Sci 1994, 45:493–501.

[14] Delage P, Audiguier M, Cui YJ, Howat MD. Microstructure of a compacted silt. Can Geotech J 1996;33:150-158.

[15] Dirksen C. Unsaturated hydraulic conductivity. In: Smith KA, Mullins CE, editors. Soil analysis physical methods. New York: Dekker; 1991. p. 209–69.

[16] Fatt I, Dykstra H. Relative permeability studies. Transactions of the American Institute of Mining Engineers, 1951, 192: 249-255.

[17] Fuentes C, Vauclin M, Parlange J-I. A note on the soil–water conductivity of a fractal soil. Transp Porous Media 1996, 23:31–36.

[18] Fuentes C, Haverkamp R, Parlange J-Y. Parameter constraints on closed-form soil water relationships. Journal of Hydrology, 1992, 134, 117-142.

[19] Gates JI, Tempelaar-Lietz W. Relative permeabilities of California cores by the capillary pressure method. Drilling and Production Practice, Am Petrol Inst 1950:285-298.

[20] GEO-SLOPE International Ltd. Seepage Modeling with SEEP/W2007 [M].Third Edition, Canada, 2008,3.

[21] GEO-SLOPE International Ltd. Air Flow Modeling with AIR/W2007 [M].Third Edition, Canada, 2008,3.

[22] GEO-SLOPE International Ltd. Stability Modeling with SLOPE/W2007 [M].Third Edition, Canada, 2008,3.

[23] Gimenez D, Perfect E, Rawls WJ, Pacheoaky Ya. Fractal models for predicting soil hydraulic properties: a review. Eng Geo 1997;48:161-183.

[24] Hoffmann-Riem H, van Genuchten M.Th, Flühler H. A general model of the hydraulic conductivity of unsaturated soils, in: van Genuchten M.Th, Leij FJ, Wu L. (Eds.), Proceedings of International Workshop, Characterization and Measurements of Hydraulic Properties of Unsaturated Porous Media. Riverside, CA. 22–24th Oct. 1997. University of California, Riverside, 1999, pp. 31–42.

[25] Hunt AG, Gee GW. Application of critical path analysis to fractal porous media:comparison with examples from the Hanford site. Adv in Water Resources 2002;25:129–46.

[26] Irmay S. On the hydraulic conductivity of unsaturated soils. Trans Am Geophys Un 1954;35:463-467.

[27] Jarvis NJ, Messing I. Near-saturated hydraulic conductivity in soils of contrasting texture as measured by tension infiltrometers. Soil Sci Soc Am J 1995;59:27–34.

[28] Kahr G., Kraehenbuehl F, Stoeckli HF, Muller-Vonmoos M. Study of the water –bentonite system by vapour adsorption, immersion colorimetry and X-ray technique: Part II. Heats of immersion, swelling pressures and thermodynamic properties. Clay Miner. 1990, 25, 499–506.

[29] Khaleel R, Relyea JF. Evaluation of van Genuchten–Mualem relationships to estimate unsaturated hydraulic conductivity at low water contents. Water Resour Res 1995; 31: 2659–1668.

[30] Kozeny J. Ueber kapillare leitung des wassers im boden, sitzungsberichte, akad der wissensch. Wien, Math-Naturw Klass Abt IIa 1927;136:271-306.

[31] Mandelbrot BB, The Fractal Geometry of Nature, W.H. Freeman, New York, 1982.

[32] Matsuoka H, Soil Mechanics. Morikita Shuppan Co., Ltd., 1999, p. 61.

[33] Mualem Y. A new model for predicting the hydraulic conductivity of unsaturated porous media. Water Resources Res 1976, 12:513–22.

[34] Polubarinova-Kochina PYa. Theory of ground water movement. Princeton, NJ: Princeton University Press, 1952. p. 613 (translated from the Russian by DeWiest, JMR 1962).

[35] Purcell WR. Capillary pressures-their measurement using mercury and the calculation of permeability therefrom. Trans Am Inst Min Met Engrs Petrol Devel Technol 1949;186:39-46.

[36] Rahardjo H, Ong TH, Rezaur RB. Factors controlling instability of homogeneous soil slopes under rainfall [J]. Journal of geotechnical and geoenvironmental engineering, 2007,12,1532-1543.

[37] Ravichandran N, Krishnapillai S. A Statistical Model for the Relative Hydraulic Conductivity of Water Phase in Unsaturated Soils. International Journal of Geosciences, 2011, 2, 484-492.

[38] Rieu M, Sposito G. Fractal fragmentation, soil porosity and soil water properties. Soil Sci Soc Am J 1991; 55: 1483–1489.

[39] Schaap MG, Leij FJ, van Genuchten MT. Rosetta: A computer program for estimating soil hydraulic parameters with hierarchical pedotransfer functions, J. Hydrol., 2001, 251, 163-176.

[40] Simunek T, Wendroth O, van Genuchten MT, et al., Water Resources Res. 35 (1999) 2965.

[41] Smettem KRJ, Kirkby C. Measuring the hydraulic properties of a stable aggregate soil. J Hydro 1990, 117: 1-13.

[42] Stingaciu LR, Weihermüller L, Haber-Pohlmeier S et al. Determination of pore size distribution and hydraulic properties using nuclear magnetic resonance relaxometry: A comparative study of laboratory methods. Water Resources Research, 2009, 46: W11510.

[43] Sugii T, Uno T, Hayashi T, Proceedings of the 31st Japan National Conference on Geotechnical Engineering, Kitami, Japan, 1996, pp. 2075

[44] Toledo PG, Novy RA, Davis HT, Scriven LE. Hydraulic conductivity of porous media at low water content. Soil Sci Soc Am J 1990;54:673–9.

[45] Touma J, Vauclin M. Experimental and numerical analysis of two-phase infiltration in a partially saturated soil. Transport in Porous Media, 1986, 1, 27-55.

[46] Touma J. Comparison of the soil hydraulic conductivity predicted from its water retention expressed by the equation of van Genuchten and different capillary models. European Journal of Soil Science, 2009, 60: 671-680

[47] Tyler SW, Wheatcraft SW. Fractal process in soil water retention. Water Res Res 1990, 26:1047–54.

[48] Uno T, Kamiya K, Tanaka K. The distribution of sand void diameter by air intrusion method and moisture characteristic curve method. Proc Japan Soc Civil Engrg, 1998, 603(III-44): 35-44.

[49] van Genuchten MTh, Leij FJ. On estimating the hydraulic properties of unsaturated soils. In: van Genuchten M.Th, Leij FJ, Lund LJ, editors. Proceedings of the International Workshop, Indirect Methods for Estimating the Hydraulic Properties of Unsaturated Soils, University of California, Riverside, 1992. p.1-14.

[50] van Damme H. Scale invariance and hydric behaviour of soils and clays. CR Acad Sci Paris 1995, 320: 665–81.

[51] van Genuchten MTh. A close form equation for predicting the hydraulic conductivity of unsaturated soils. Soil Sci Soc Am J 1980;44:892–7.

[52] Vogel HJ, Roth K. A new approach for determination effective soil hydraulic function. Eur J Soil Sci 1998;49:547–56.

[53] Watabe Y, Leroueil S, Le Bihan J-P, Influence of compacted conditions on pore-size distribution and saturated hydraulic conductivity of a glacial till, Can Geotech J, 2000, 37:1184-1192.

[54] Wheeler SJ, Karbue D, Constitutive modeling, In: Alonso EE, Delage P, Proc. 1st Int Conf Unsat Soils Rotterdam: AA Balkema, 1996.

[55] Wyllie MRJ, Spangler MB. Application of electrical resistivity measurements to problem of fluid flow in porous media. Bull Am Ass Petrol Geologists 1952, 36:359-403.

[56] Xu YF, Sun DA. A fractal model for soil pores and its application to determination of water permeability. Physica A 2002;316(1-4): 56–64.

[57] Xu YF, Dong P. Fractal approach to hydraulic properties in unsaturated porous media. Chaos, Solitons & Fractals, 2004, 19(2): 327-337

[58] Xu YF, Sun DA, Yao YP. Surface fractal dimension of bentonite and its application to determination of swelling properties. Chaos, Solitons & Fractals, 2004, 19(2): 347-356.

[59] Xu YF, Calculation of unsaturated hydraulic conductivity using a fractal model for the pore-size distribution, Computer and Geotechnics, 2004, 31(7):549-557.

[60] Xu YF. Fractal approach to unsaturated shear strength. Journal of Geotechnical and Geoenvironmental Engineering, 2004, 130(3): 264-273

2

Data Assimilation Application to the Subsurface Flow and Solute Transport

Bill X. Hu and Juxiu Tong

1. Introduction

Natural media are generally heterogeneous. The medium properties (hydraulic parameters such as hydraulic conductivity and porosity) are often treated as random space functions. Without an adequate description of the spatial distributions of the hydraulic parameters, such as hydraulic conductivity and porosity, the ability of the numerical model to predict groundwater flow and solute transport is limited and tends to deteriorate over time. However, characterization of field heterogeneity is always difficult due to economical and technical limitations, which make the predictions, deteriorate over time.

Many inverse algorithms have been developed in the field of hydrogeology (Constantinescu et al, 2007; Hoeksema and Kitanidis, 1984; Poeter and Hill, 1997; Sun, 1994; Sun and Yeh, 1992; Vrugt et al., 2005a; Vermeulen et al., 2006; Yeh and Liu, 2000; Yeh and Zhang 1996, Zhu and Yeh, 2005). With newly developed techniques, more observations become available. But the observation data from these techniques are usually indirectly related to the model parameters, which are of the most interest in practice. Moreover, there may be different kinds of measurements. Therefore, it is necessary to develop approaches to dynamically reconcile the observation information of different kinds. The data assimilation method is established in this situation (Thomsen and Zlatev, 2008).

Data assimilation is different from the traditional parameter estimation methods. Originating from meteorology and oceanography (Daley, 1991; LeDimet and Talagrand, 1986), the data assimilation method has been developed for improvement of operational weather forecasts and ocean dynamics prediction (Evensen, 2003, 2004; Fang et al., 2006; Houtekamer and Mitchell, 1998, 2001; Thomsen and Zlatev, 2008). Data assimilation methods have been applied to many study areas (Andreadis and Lettenmaier, 2006; Aubert et al., 2003; Clark et al., 2006; McLaughlin and Townley, 1996; Mclaughlin, 2002; Natvik and Evensen, 2003; Reichle, 2008).

In geophysics, data assimilation methods have been used to assimilate geophysical data to characterize medium heterogeneity (Christakos, 2005). Data assimilation methods have been used in hydrology for uncertainty and sensitivity analysis (Liu and Gupta, 2007; Van Geer et al., 1991; Vrugt et al., 2005b) and in the petroleum industry (Oliver et al., 2008).

The traditional Kalman filter (KF) is an efficient sequential data assimilation method for linear dynamics and measurement processes with Gaussian error statistics (Kalman, 1960; Kalman and Bucy, 1961; Drecourt, 2003; Drecourt et al, 2006; Gelb, 1974; Maybeck, 1979; Tipireddy et al., 2008; Zou et al., 1991; Zhang et al., 2007). To assimilate data for nonlinear dynamics and measurement processes, the extended Kalman filter (EKF) was developed (Jazwinski, 1970). However, the EKF is very unstable if the nonlinearities are strong. Furthermore, this method is not computationally feasible for large-scale environmental systems.

To overcome these limitations, the ensemble Kalman filter (EnKF) is proposed (Evensen, 1994). The Ensemble Kalman filter (EnKF) method is a Monte-Carlo variant of the KF, consisting of a set (or ensemble) of parallel short-term forecasts and data assimilation cycles. Statistics derived from the short-term forecasts are used to estimate background error covariances during the subsequent data assimilation step. Further, ensemble-based techniques have two potential advantages over the traditional EKF (Cohn, 1997): (1) the computational cost of ensemble techniques are significantly less than the EKF because covariances are estimated using a limited-size random sample, while the error covariances of EKF for each model component are propagated using linear tangent and adjoints of the fully nonlinear model, and exorbitant computational expense for a high-dimensional model [however, it may be possible to reduce the computations by computing in a reduced-dimensional subspace, e.g., Fisher (1998)]; (2) in terms of accuracy, ensemble filters may be more accurate than the EKF since covariances are estimated by propagating model states with a fully nonlinear model rather than under assumptions of linearity. Moreover, it is also able to account for the possible model noise/error.

Recent development of the EnKF is not limited to updating system state variables (such as hydraulic head in transient groundwater flow modeling and solute concentration in solute transport modeling) as in conventional data assimilation, but also allow updating state variables and model parameters (such as hydraulic conductivity in transient groundwater flow modeling or in transient solute transport on the basis of the steady state groundwater flow) to yield more accurate model predictions. The application of the EnKF in hydraulic modeling (including groundwater flow and solute transport) has also received more and more attention. Chen and Zhang (2006) have successfully applied data assimilation for transient groundwater flow in geologic formations via EnKF, but in their study, they did not consider the influence of different boundaries and the locations of observation wells on the data assimilation results. Huang et al. (2009) used the EnKF data assimilation method to calibrate a heterogeneous conductivity field using steady state groundwater flow data and calibrate solute plume initial condition using tracer test data.

All of the above works are done by applying the EnKF to calibrate the hydraulic conductivity by assimilating the hydraulic head. For the solute transport modeling, though Liu et al. (2008) applied the EnKF to calibrate the conductivity fields by assimilating tritium concentration at the MADE site, they first assimilated hydraulic head observations to update the hydraulic conductivity and obtained the flow field for the solute transport process. They also incorporated the updated variable longitudinal dispersivity during the solute transport modeling, which will influence the solute prediction and the corresponding inversed results including the updated hydraulic conductivity field.

All uncertainty sources, including parameters and output observations, need to be considered for improvement of forecast capability (Wang et al., 2009). In most physical systems, states and/or parameters are bounded, which introduces constraints on their estimates. While in sequential data assimilation techniques like the EnKF, the state variables are updated by minimizing the mean square error rather than following physical principles, which may result in the violation of physical constraints and many methods have been proposed to solve these problems (Prakash et al., 2008; Wang et al., 2009). In the conservative solute transport modeling, solute concentrations have their own physical meaning. So it is better to pose some constraints to them in case they violate the physical meaning after the application of the EnKF method (Hemant and Olive, 2009, Sridhar et al., 2007; Thacker, 2007). We call this method as constrained EnKF method.

According to the above introduction about the EnKF, the EnKF application in groundwater is classified into two types: using groundwater flow data or conservative solute transport data to calibrate the hydraulic parameters, e.g. hydraulic conductivity and corresponding state variables, e.g., hydraulic head and conservative solute concentrations. The rest is arranged as follows:

In section 2, the EnKF data assimilation method is introduced based on the numerical forward groundwater flow and conservative solute transport models.

In section 3, the research design for the groundwater flow and conservative solute transport is described. The synthetic or illustrative examples or cases for the groundwater flow and conservative solute transport will be depicted in detail. In our research, we will apply EnKF method to update a hydraulic conductivity field through assimilating transient head data in the groundwater flow modeling part. The study focuses on investigating the capability of EnKF to update a hydraulic conductivity field through assimilating transient head data, and explores the influences of different factors on assimilation results. In the conservative solute part, we will assimilate solute transport data to calibrate conductivity fields via the constrained EnKF method, where the conservative solute concentration is constrained to be non-negative. The constrained EnKF method is used to identify a two-dimensional (2D) heterogeneous hydraulic conductivity field with mixed Neumann/ Dirichlet boundary conditions through assimilating conservative solute concentration observations. The study focuses on investigating the capability of constrained EnKF to update a hydraulic conductivity field, and explores the influences of different factors on assimilation results.

In section 4, the updated results through assimilating groundwater flow and conservative solute transport data will be provided, and then discussion and analysis of the influences of various factors on the results will be provided in detail. Conclusions for the inverse modeling of the groundwater flow and conservative solute transport will be made.

2. Data assimilation method

Since data assimilation is a process in which estimates of model parameters and variables are updated by requiring consistency with observations and governing equations, a data assimilation system is composed of a model operator, an observation operator and a data assimilation algorithm.

2.1. Transient stochastic flow model

The governing equation for transient groundwater flow in unsaturated-saturated medium is (Bear, 1972; Arlen, 2005):

$$\nabla \cdot \mathrm{q}(\mathrm{x},t) + g(\mathrm{x},t) = S_s \frac{\partial h(\mathrm{x},t)}{\partial t} \tag{1}$$

$$\mathrm{q}(\mathrm{x},t) = -K(\mathrm{x}) \nabla h(\mathrm{x},t) \tag{2}$$

where q $[LT^{-1}]$ is the Darcy's flux; $h(\mathrm{x}, t)$ $[L]$ is the hydraulic head; x $[L]$ is the spatial parameters, *i.e.*, in the three-dimensional domain, x means x, y, z; $g(\mathrm{x}, \mathrm{t})$ $[T^{-1}]$ is the external sink or source term; $K(\mathrm{x})$ $[LT^{-1}]$ is the hydraulic conductivity; ∇ stands for $(\partial/\partial x, \partial/\partial y, \partial/\partial z)$; S_s is the specific storage. In this study, the hydraulic conductivity $K(\mathrm{x})$ is considered as a log-normally distributed random space function while specific storage S_s is treated as a deterministic constant. Since $K(\mathrm{x})$ is a random function, the flow equations become stochastic nonlinear partial differential equations.

2.2. Transient conservative solute transport model

When the right side in Equation (1) is 0, then it becomes a governing equation for the steady state flow. On the basis of the mass balance and Fick's law, the governing equation for the transport of conservative solute in groundwater flow can be written as (Zheng, 1990):

$$\frac{\partial \theta c(\mathrm{x},t)}{\partial t} = \nabla \cdot \left(\theta D \cdot \nabla c(\mathrm{x},t)\right) - \nabla \cdot \left(\mathrm{q}(\mathrm{x},t) \cdot c(\mathrm{x},t)\right) - g(\mathrm{x},t) c_s \tag{3}$$

Where θ $[L^3L^{-3}]$ is the porosity of the medium; $c(\mathrm{x}, \mathrm{t})$ $[ML^{-3}]$ is the solute concentration; D $[L^2T^{-1}]$ is the hydrodynamic dispersion tensor; $c_s[ML^{-3}]$ is the solute concentration of water entering from sources or sinks.

2.3. Ensemble Kalman filter

EnKF is a Monte Carlo method. The EnKF method is briefly introduced with perturbed observation proposed by Burgers et al. (1998). In this algorithm, predictions of model variables

S, including both the variables (hydraulic head in the groundwater flow modeling or solute concentration in the solute transport modeling) and parameters (hydraulic conductivity), are given by their ensembles. Assuming normal distribution of model predictions, the ensemble mean is supposed to be the best estimate of the true state, and prediction error around the mean is measured by covariance of the ensemble (Evensen, 2003).

The covariance, *P*, of forecast and analysis error of a random variable *S* are defined as

$$P^{\mathrm{f}} \cong P_e^{\mathrm{f}} = \overline{\left(S^{\mathrm{f}} - \overline{S}^{\mathrm{f}}\right)\left(S^{\mathrm{f}} - \overline{S}^{\mathrm{f}}\right)^{\mathrm{T}}} = S^{\mathrm{f}\prime}\left(S^{\mathrm{f}\prime}\right)^{\mathrm{T}} / N - 1 \tag{4}$$

$$P^{\mathrm{a}} \cong P_e^{\mathrm{a}} = \overline{\left(S^{\mathrm{a}} - \overline{S}^{\mathrm{a}}\right)\left(S^{\mathrm{a}} - \overline{S}^{\mathrm{a}}\right)^{\mathrm{T}}} = S^{\mathrm{a}\prime}\left(S^{\mathrm{a}\prime}\right)^{\mathrm{T}} / N - 1 \tag{5}$$

In the forecast step, forecasted model variables of each ensemble member are updated according to

$$S_{i,t+1}^{\mathrm{f}} = M(S_{i,t}^{\mathrm{a}}) + e_{1i} \quad e_{1i} \sim N(0, Q_s) \tag{6}$$

Where $S_{i,t+1}^{f}$ is the forecasted model variable of the *i*th ensemble member at time *t*+1;$S_{i,t}^{a}$ the analyzed model variable of the *i*th ensemble member at time *t*; *M* (.) is model operator, which is the flow model in groundwater modeling and transport model in solute modeling; $e_{1i}$is the model error vector, which is assumed to satisfy a Gaussian distribution with zero mean and covariance matrix Q_s.

In the analysis step, the observation data are perturbed by adding random observation errors. The observation vector at the time step t+1 for each ensemble member is given by:

$$d_{i,t+1} = HS_{i,t+1}^{\mathrm{true}} + e_{2i} \tag{7}$$

Where e_{2i} is random error vector of observation with zero mean and covariance matrix R_e.

The forecast of each ensemble member is updated as follows (Burgers et al., 1998):

$$S_{i,t+1}^{\mathrm{a}} = S_{i,t+1}^{\mathrm{f}} + P_e^{\mathrm{f}} H^{\mathrm{T}}\left(HP_e^{\mathrm{f}} H^{\mathrm{T}} + \mathrm{R_e}\right)^{-1}\left(d_{i,t+1} - S_{i,t+1}^{\mathrm{f}}\right) \tag{8}$$

Where *H* is the observation operator used to convert the model state variables to observations; $\overline{S}_{i,t+1}^{f}$is the mean of forecast state vector of ensemble members at time *t*+1.

The analysis state estimate at time $t+1$ is given by the mean of the ensemble members. The analyzed ensemble is then integrated forward until the next observation (hydraulic head or solute concentration) is available and the process is repeated. In comparison with commonly used inverse methods (e.g., general least squares and maximum likelihood methods), the EnKF can dynamically adjust the system estimate, without reprocessing existing data when new observations become available. The data assimilation procedure using the EnKF in this paper will be introduced later combined with the description of the study cases below.

3. Illustrative examples for the groundwater flow and conservative solute

3.1. Transient state groundwater flow

In this section a two-dimensional, transient flow in a heterogeneous saturated medium is used to demonstrate the capability of EnKF to estimate the hydraulic conductivity field by assimilating hydraulic head measurements and to explore the sensitivity of EnKF to various factors. The flow domain is a square with $L_x = L_y = 50.0$ [L] (L is any consistent length unit) and uniformly discretized into 50×50 square grid cells. To investigate the effect of boundary condition on parameter calibration, two different boundary conditions for the study domain are proposed in this study. In the first case (referred to as Case 1 hereinafter), all four sides of the flow domain are prescribed as no-flow boundaries. There is a pumping well with continuous and constant pumping rate of 10 [L^3/T] (T is any consistent time unit) in the middle point of the flow domain (25, 25). In the second case (referred to as Case 2 hereinafter), the no-flow condition is still assumed for the top and bottom boundaries, but constant head boundary conditions are used on the left and right sides. Similarly, there is a pumping well in the middle point of the flow field (25, 25), but the initial pumping rate is 500 [L^3/T] and the pumping rate increases 5 [L^3/T] every 5 [T] to speed up the head decrease in the domain.

The conductivity field generator scheme developed by Hu et al. (2009) was used in this study to generate the reference (or real) conductivity field and the initial guess conductivity field in this study:

$$\log K = -0.457 - 0.0300M + 7.76n + 2.03\log d_{10} \tag{9}$$

Where M is the matrix volume fraction of the grain size distribution in the medium; n is the porosity of the medium; d_{10} is the 10th percentile of grain size distribution. The distributions of M, n and d_{10} were obtained from the field experiment. In this study, a special case is considered in which d_{10} is a spatial random variable, its distribution is represented by a fractal function according to the field measurement, while the other two parameters, M and n are assumed to be constant and their mean values are used in Eq. (9). Using Hu et al.'s (2009) method, multiple, equally-probable realizations of the hydraulic conductivity field is generated based on Eq. (9). One of the generated realizations is shown in Figure 1(a), where it is apparent that the field is in the log form. One can see that the realized field is heterogeneous, but statistically stationary. An ensemble of a sufficient number of realizations should reach a constant field. To generate a "real" or reference conductivity field, one realization is randomly picked and modified the field by inserting a high conductivity layer in the middle of the domain by setting the conductivity value at every point in the layer 100 times the original value. The constructed figure with log form is shown in Figure 1(b) and the red band part is the modified part. This field will be used to generate hydraulic head data at selected measurement points, while the other realizations, without modification, will be used as the initial "guess" conductivity field by averaging them. For example, In Figure 1(c), the 500 realizations mean log$K$ field is taken as the representative initial "guess" conductivity field.

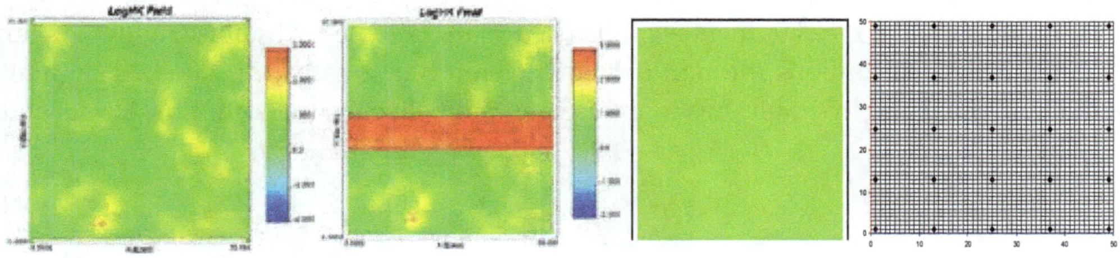

(a) (b) (c) (d)

(Note: all the Figures of log hydraulic conductivity fields for the transient groundwater flow here and next have the same color scale)

Figure 1. Distribution of log conductivity field: (a) One realization of the initial guess field; (b) Real conductivity field; (c) Ensemble mean of 500 realizations; (d) Uniform distribution of 25 observation wells in the study domain

Based on the hydraulic boundary conditions and the conductivity fields described above, the flow equations are solved with the Galerkin finite-element method (FEM).

The simulated results based on the reference conductivity field at the chosen "measurement" locations will be used as the real values of the hydraulic heads at the locations. Considering possible measurement error in reality, the real head values are perturbed by white noises and the perturbed head results serve as the observed head values. In this study, the noise has a mean equal to 0 (indicating unbiased observations) and the standard deviation is 1% of the hydraulic head measurements. It should be pointed out that the EnKF method is not limited to this simplification of measurement error. In fact, the errors, assumed or inferred from the measurements, may vary with measurement types, times, and locations. In this study, the simulation time is evenly subdivided into many time steps of size 1 [T].

In case 1, in the vertical (z) direction, the bottom level of the study domain is -100 [L], and the upper level is 2.5 [L]. The initial hydraulic head throughout the domain is assumed to be 3.0 [L], so the saturated groundwater is confined at first, and the specific storage for this confined study medium is assumed to be 0.0001. After some time of the pumping the groundwater table goes below the upper level of the medium in some areas, as the groundwater becomes unconfined. The specific storage for the unconfined study medium is set to 0.1.

In case 2, bottom level of the study domain is -100 [L], and the upper level is 2.0 [L]. The constant hydraulic head on the left border is 3.0 [L], and the constant hydraulic head on the right border is 2 [L]. All the initial hydraulic head except the constant head boundaries are 0 [L]. The specific storage and specific yield for the confined or unconfined study medium are the same as in case 1.

In the case study, the observation wells are distributed evenly in the study domain (e.g. 25 wells shown in Figure 1(d)), and continuous data can be obtained from the observation wells.

The procedure of the data assimilation using the EnKF is shown in a flowchart in Figure 2. First, ensembles of initial hydraulic conductivity fields are generated, and the generated hydraulic conductivity fields are used in transient flow model to calculate hydraulic head in the study area. These hydraulic conductivity fields and heads serve as the first-step forecast results. Second, if there are the hydraulic head measurements at the observation wells at the current time step, the measurements will be used to update the hydraulic conductivity and hydraulic head distributions through the EnKF algorithm. Finally, the updated hydraulic conductivity field and hydraulic heads are used to reinitialize the transient flow model at the next time step.

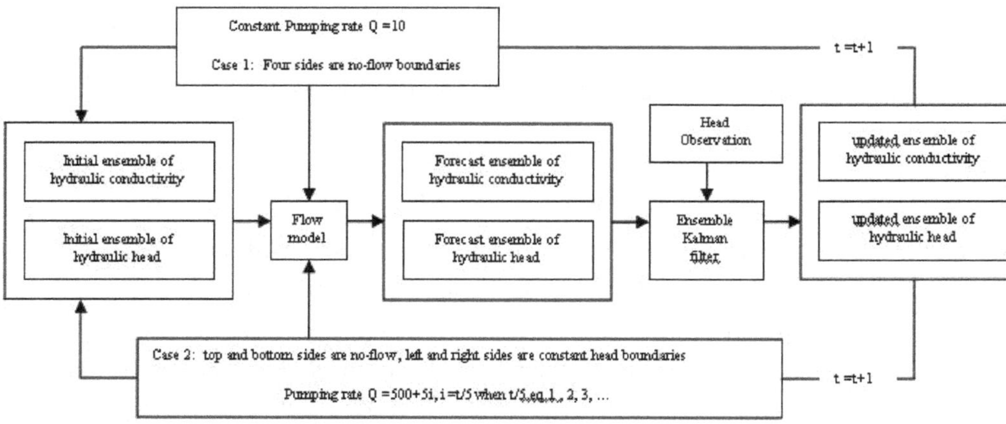

Figure 2. Flowchart of hydraulic head assimilation scheme for transient groundwater flow

3.2. Transient state conservative solute transport

The conservative solute transport modeling is based on the steady state groundwater flow. In our study, the same study case is chosen as Case 2 ($L_x = L_y = 50.0$ [L], 50×50 square grid cells, the top and bottom boundaries are the second Neumann condition with no-flow, but constant head boundary conditions on the left and right sides.) in part 3.1 except that there is no pumping well at all. In this way, the steady state flow can be obtained. With the same idea as in part 3.1, the ensemble realizations of the initial guessed hydraulic conductivity fields are produced by Hu et al.'s (2009) method. One of the generated realizations of log form is directly picked out and shown in Figure 3(a), which is also used as the true or reference hydraulic conductivity field in the solute transport modeling. The ensemble number is selected as 300, and these realizations mean log$K$ field is taken as the representative initial "guess" conductivity field, which is shown in Figure 3(b). Uniform distributions of 25 and 81 observation wells are studied here, and the former distribution have been presented in Figure 1(c), while the later distribution is presented in Figure 3(c). Continuous solute concentration data can be obtained from the observation wells where the solute reaches.

Based on the hydraulic boundary conditions and the conductivity fields described above, the forward groundwater flow equation is solved with the Galerkin finite-element method (FEM). Then the Darcy's velocity can be calculated and the transport equation is solved with the Galerkin FEM based on the initial instant injection of solute concentration somewhere.

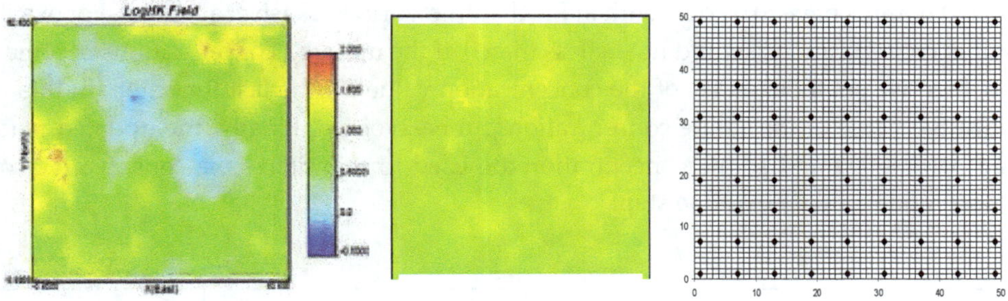

(a) (b) (c)

(Note: all the Figures of log hydraulic conductivity fields for the conservative solute transport here and next have the same color scale)

Figure 3. Distribution of log conductivity field: (a) One realization or real conductivity field; (b) Ensemble mean of 300 realizations of the initial guess field; (c) Uniform distribution of 81 observation wells in the study domain

The observed concentration values are obtained by adding white noises to the real solute concentration values, which is the simulated results based on the reference conductivity field. The noise is unbiased and the standard deviation is 1% of the solute concentration measurements. The simulation time for the solute transport modeling is evenly subdivided into many time steps of size 0.1 [*T*] after the steady water velocity field is obtained by solving the governing flow equation.

To purely verify the ability of the EnKF method to update the hydraulic conductivity field through assimilating the solute concentration, all the other related parameters are assumed to be known except the hydraulic conductivity in the study domain. The solute longitudinal and transverse dispersivity is 0.1 [*L*] and 0.01 [*L*], respectively. The corresponding random porosity filed is also one realization produced in Hu et al. (2009). In order to explore the EnKF ability to update the hydraulic conductivity field through assimilating the solute concentration, the random porosity filed is the same for every realization.

Models usually have relevant physical laws or settings, but the EnKF formulation appears to be a promising approach for dealing with a wide class of nonlinear state estimation problems, it cannot handle bounds on the state variables (the solute concentrations) and/or the parameters (hydraulic conductivity here) that are being estimated (Hemant and Olive, 2009; Prakash et al., 2008; Thacker, 2007; Wang et al., 2009), which will results in the violation of physical meaning. So it is necessary to account for bounds on state and parameters being estimated. The procedure of the data assimilation method using the constrained EnKF is shown in a flowchart as Figure 4. First, ensembles of initial parameters fields are generated, which are used in steady-state flow model to calculate solute concentration. If there are the solute concentration measurements at the observation wells at the current time step, the measurements will be used to update the hydraulic conductivity field and solute concentration distributions through the EnKF algorithm. If it is desired to apply the ensemble Kalman filter for state estimation when states are bounded, then it become necessary to modify the EnKF. Therefore, constraints are treated in a post processing after applying the EnKF (Wang et al., 2009), and bounds have to be deal with in EnKF framework (Prakash et al., 2008). For example, the solute concentrations should not be less than 0, if the updated concentrations become less than 0, then we set them as 1% of the concentration value for the last time step. In this way, the negative value of the solute concentration can be avoided. Finally, the updated reactive related parameters and solute concentration are used to reinitialize the conservative solute transport model at the next time step.

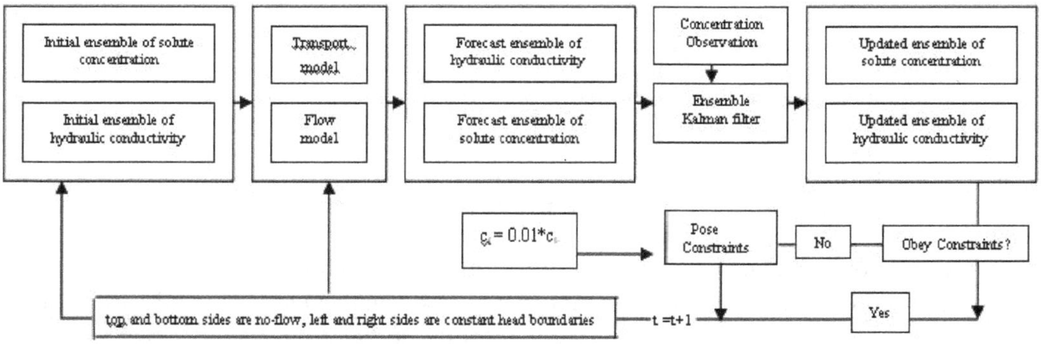

Figure 4. Flowchart of the constrained EnKF for solute concentration assimilation scheme with steady state groundwater flow

4. Application of data assimilation method to the groundwater flow and conservative solute transport

To determine the accuracy of the updated state vectors, the reference (also called true) simulation is compared to the EnKF simulation using the root mean squared error (*RMSE*) as the following:

$$RMSE = \sqrt{\frac{1}{M}\sum_{i=1}^{M}\left(\bar{S}_i^f - S_i^t\right)^2} \tag{10}$$

where *RMSE* is the criteria for the degree of the data assimilation capability, *M* is the total number of numerical grid elements in the study domain, which is 2601 in this hypothetic study case of section 3 or 4; $\overline{S}_i^f$ and S_i^t stands for the estimated and reference hydraulic conductivities at grid *i*, respectively.

4.1. Transient state groundwater flow

In this part, the observations are assumed to be available for us for every simulation time, so the observed data can be assimilated for every current simulated time via the EnKF method.

4.1.1. *Influence of ensemble size*

To determine the ensemble size that should be used in our study, it is needed to study the influence of ensemble size on the data assimilation results. Here, it is assumed that there are 25 (5×5) observation wells, and they are distributed uniformly within four no-flow boundaries (Case 1).

The mean and variance values of the hydraulic conductivity field with and without data assimilation for different ensemble size is shown in Figures 5 and 6. The simulated hydraulic conductivity is compared with the real hydraulic conductivity fields for different ensemble size. The mean and variance values of the updated hydraulic conductivity fields for the ensemble size of 70, 90, 100, 200 are almost the same with the *RMSE K* more than 10 [*L*/*T*] at some assimilation steps, I only plot the ensemble size of 90 as a representative for simplicity in Figures 5 and 6. Similarly, I only plotted the ensemble size of 300 and 500 as representatives. The initial ensemble mean hydraulic conductivity for different ensemble size (the value is an approximate constant as 13.10 [*L*/*T*]) is apparent at the first time step in Figure 5. It is shown that the first 2 or 3 time steps of data assimilation are the most efficient for every ensemble size simulation, with decrease of *RMSE K* value during these time steps. Also, I found that after the first 5 time step, the effect of assimilating observation data became worse (Figure 6) for different ensemble sizes 300 and 500. So, it is better for us to assimilate only a few time step observation data. However, in Huang et al. (2009), the more data assimilation time steps, the more improved the updating hydraulic conductivity fields become. The criterion to end the process of data assimilation is introduced as:

$$DifK = \sqrt{\frac{1}{NM}\sum_{i=1}^{M}\sum_{j=1}^{N}\left(K_i^j - \overline{K}_i\right)^2} \leq 10^{-2} \tag{11}$$

Where *DifK* is the mean difference of the updated value and the corresponding ensemble mean value over the domain; K_i^j is the model result of the *j*th ensemble for the *i*th grid; $\overline{K}_i$ is the ensemble mean value at the *i*th grid. After the end of the data assimilation process, I consider the latest update as providing the most acceptable results.

It is very clear that the mean and variance values of hydraulic conductivity fields with data assimilation for the ensemble size of 300 and 500 are much closer to the real hydraulic conductivity field. This demonstrates that the data assimilation method for the transient flow via EnKF in this study can update the model parameters (hydraulic conductivities here) well. Moreover, the larger the ensemble size, the better the updated hydraulic conductivity field, which agrees with the results from Chen and Zhang (2006), Evensen (2003, 2004) and Huang et al. (2009). It is well known that with the increase of ensemble size, more effort will be needed for computation. Since the ensembles consisting of 300 and 500 members (realizations) yield similar results (the mean and variance values are very close to the real hydraulic conductivity field at a sufficient number of time steps), an ensemble consisting of 300 members is considered to be an appropriate number of realizations for both the results and computational efficiency. To compare the spatial distribution of the updated hydraulic conductivity fields under different assimilation steps with the real field, the spatial distributions of the log hydraulic conductivity field under the first and fourth time steps are plotted in Figures 7(a) and 7(b), respectively, with ensemble members of 300 and 25 observation wells. According to the plot, it can be seen that after four time step data assimilation, the updated mean hydraulic conductivity field is much better than the initial one shown in the Figure 1(c). Furthermore, it is very like the real field (Figure 1(b)), which suggest that the 2D saturated transient flow simulation by assimilating easily observed hydraulic head data can not only update the hydraulic head in time, but is also an efficient method to update other hydraulic parameters such as the hydraulic conductivity.

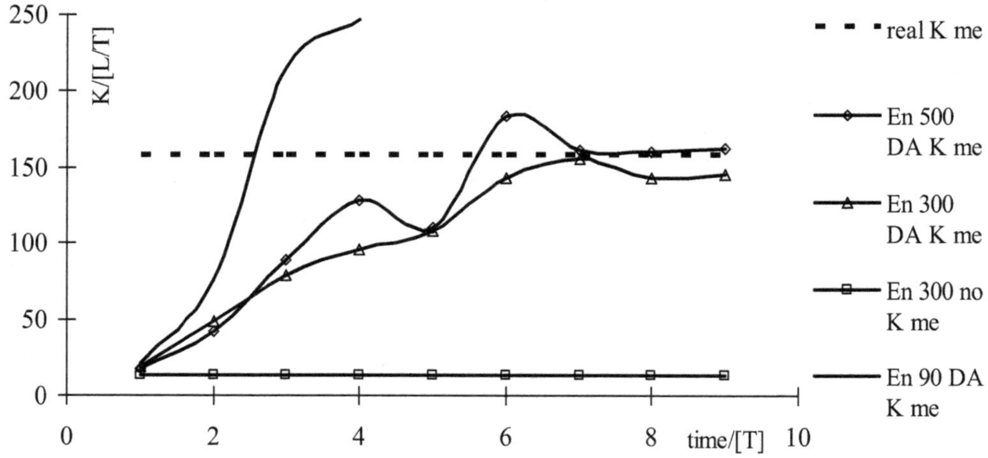

Figure 5. Hydraulic conductivity mean values of the real field, initial guessed field and calibrated fields through data assimilations with various ensemble sizes and under various time steps

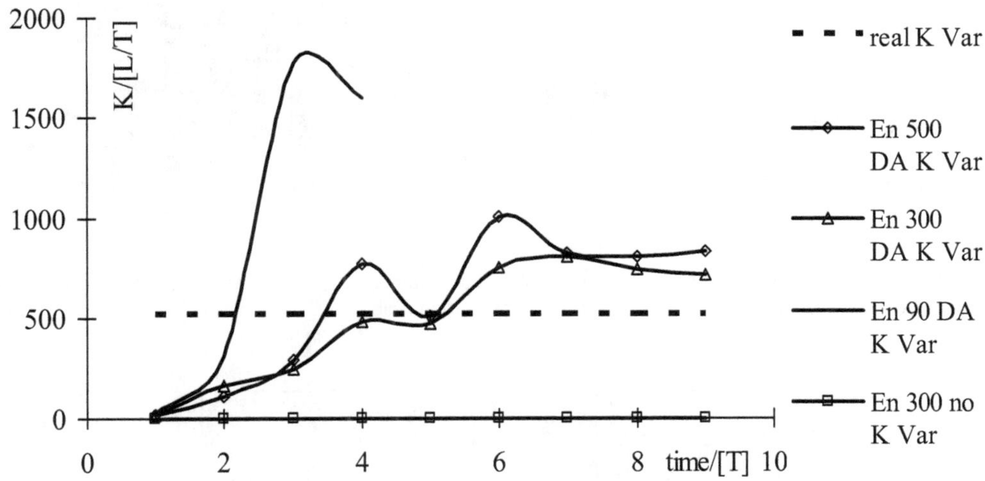

Figure 6. Hydraulic conductivity variance values of the real field, initial guessed field and calibrated fields through data assimilations with various ensemble sizes and under various time steps

4.1.2. Influence of boundary condition

In this section, the ensemble size of 300 is chosen as discussed above, 25 observation wells is selected to update the hydraulic conductivity by assimilating the observed hydraulic head data for different boundary conditions. The first condition is the one used in Case 1, four no-flow boundaries, the assimilation results have obtained and shown in Figure 7(a), (b).

Since the no-flow boundaries are not common in reality, the effects of non-zero flow boundary condition on data assimilation results are explored. In the second condition, we assume that there is a constant flow flux at each Neumann boundary: 2 [*L/T*] at the top boundary, 1 [*L/T*] at the bottom boundary, 1.5 [*L/T*] at the right boundary, -1.5 [*L/T*] at the left boundary, where the positive flux value means flow out of the study domain, while the negative value means flow into the domain. Under this boundary condition and with the same pumping rate as the first condition, the updated conductivity fields after the first and the fourth time step assimilations are shown in Figures 7(c) and 7(d). In comparison of Figure 7(b) and 7(d) with the real conductivity field in Figure 1(b), one can conclude that the updated conductivity field is similar to the real field only after four data assimilation steps when the boundaries are Neumann boundary condition, whether it is no flow or non-zero flow condition.

The third boundary condition is the one used in the Case 2, two no-flow boundaries and two constant hydraulic head boundaries. The updated hydraulic conductivity fields after 3 different time steps of data assimilation are shown in the Figure 8. It is shown from the figure that the updated hydraulic conductivity field is not similar to the real field even after the 50th assimilation step. The results shown in Figures 7 and 8 indicate that this data assimilation method for transient flow works better for the study domain with Neumann boundaries than that with mixed no-flow and constant head boundaries. In other words, data assimilation

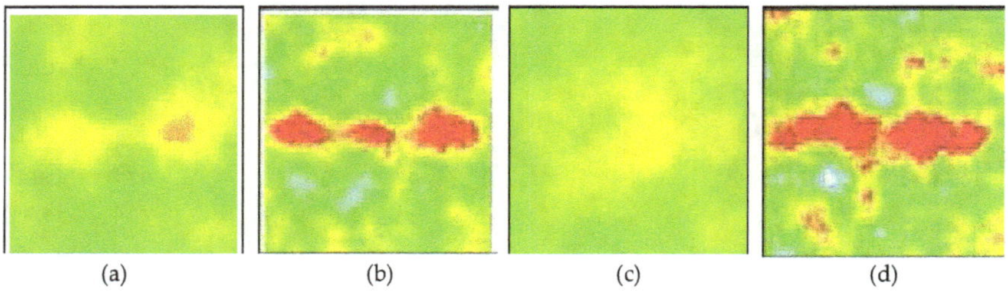

Figure 7. Assimilated log$K$ field at 1th (a) and 4th step (b) in Case 1, and at 1th (c), 2th (d) and 4th (e) step with non-zero flow at the boundaries for ensemble size of 300 with 25 observation wells

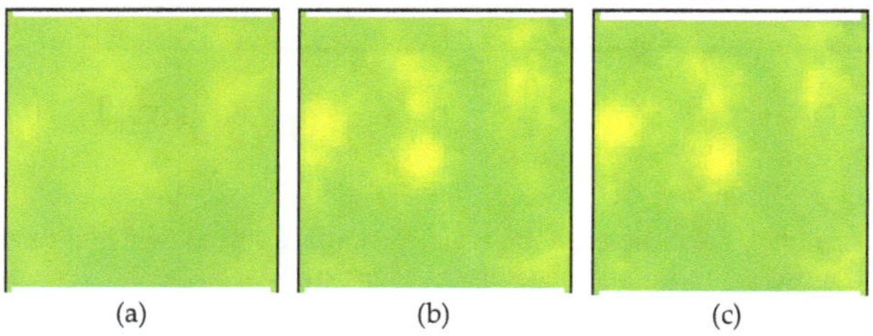

Figure 8. Identified log$K$ in Case 2 for ensemble size of 300 with 25 observation wells at different assimilation steps: (a) the first step; (b) the 5th step; (c) the 50th step

method for transient flow does not work well for the study domain with mixed no-flow and constant head boundaries. These results can be explained by the boundary hydraulic characteristics. For the Neumann boundary condition, hydraulic head variation is directly related to the hydraulic conductivity distribution, without being constrained by constant head boundaries. While, for transient flow with a constant-head boundary, the hydraulic head variation will not be fully determined by the hydraulic conductivity field due to the head constraints imposed at the boundaries. Using extreme cases as examples, when the hydraulic conductivity is extremely large or small in the study domain with a constant-head boundary, the hydraulic head in the boundary is still constant. Therefore, the hydraulic conductivity field is not the only factor that determines the hydraulic head distribution in the transient flow domain, and the constant-head boundary is also a strong influence on the hydraulic head change in the study domain. Therefore, the constant head boundary condition will decrease the sensitivity of the transient hydraulic head change with hydraulic conductivity.

4.1.3. Impact of number of observation wells

To investigate the effect of the number of observation wells on the data assimilation, in addition to the conductivity field in Figure 7 calibrated through 25 observation wells, the data assimilation method is applied to Case 1 with 9, 16, 36, 49, 64 and 81 observation wells, where the observation wells are uniformly distributed in the study domain. The study results are shown in Figure 9.

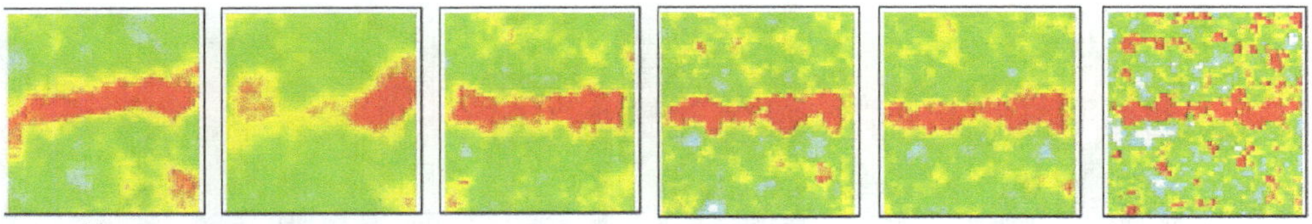

(a) (b) (c) (d) (e) (f)

Figure 9. Calibrated hydraulic conductivity field with various observation wells: (a) 9 wells; (b) 16 wells; (c) 36 wells; (d) 49 wells; (e) 64 wells; and (f) 81 wells.

Comparing the real hydraulic conductivity field, the initial mean hydraulic conductivity field with the calibrated hydraulic conductivity fields with various number of observation wells shown in Figure 7 and 9, one can tell that with the increase of the observation wells, the updated hydraulic conductivity distribution will become closer to the real hydraulic conductivity field until the number of observation wells reaches 64. As shown in Figure 9 that when the observation wells are limited, such as 9 and 16 wells, the updated hydraulic conductivity is not close to the real hydraulic conductivity field, and the high conductivity channel is not well captured. However, when the observation wells become too many, such as 81 wells, the true information is too much, which will disturb the assimilation results and lead to "bad" calibration results. This phenomenon can be explained as one observed data from one observation well will influence a certain space around the well via EnKF by data assimilation. If the observation wells are too close to each other, some area will be calibrated by data from two wells at the same time. However, the observation wells and their influencing regions are considered to be independent in the data assimilation method, so, too many observation wells lead to over calibration of the hydraulic conductivity field. Moreover, the suitable uniform distance of two observation wells is still a problem and how to define the appropriate data assimilation frequency and space is an issue that requires further study.

From the assimilation results it is told that 25, 36, 49 and 64 observation wells are suitable for the calibration of the conductivity field with an ensemble size of 300, which is composed of 2601 numerical grids cells in the study area. In general, these results demonstrate that the data assimilation method using an appropriate number of observation wells for transient flow to calibrate a heterogeneous conductivity field for an ensemble size of 300 via EnKF is effective and convenient. Furthermore, it displays similarities with the real field.

4.1.4. Influence of boundary condition

To further investigate the influence of the observation wells on identification of a conductivity field from transient flow data via EnKF, three different observation well locations are set in case 1 with the ensemble size of 300. As shown in Figure 10(a)-(c), there are the same number of 16 hydraulic head observation wells in the study field, but their locations are distributed with three different patterns, denoted 16, 016, 0016, and their corresponding updated hydraulic conductivity field distributions through assimilating the observed hydraulic head data are displayed in the Figures 9(b), 11(a) and 11(b).

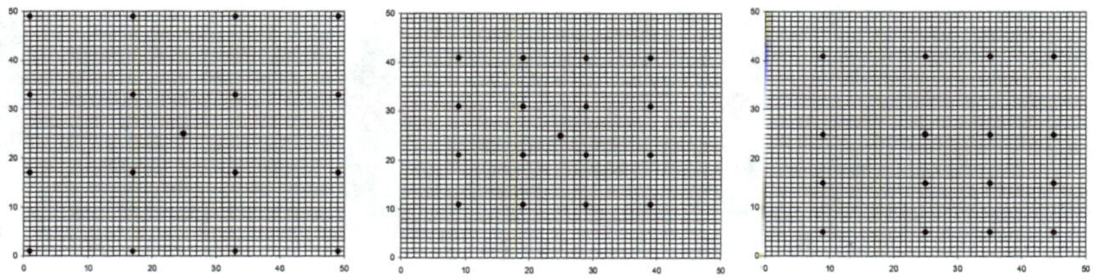

(a) (b) (c)

Figure 10. Different distribution patterns for 16 observation wells: (a) 16; (b) 016; and (c) 0016 (Diamonds are observation wells while the circle is the pumping well.)

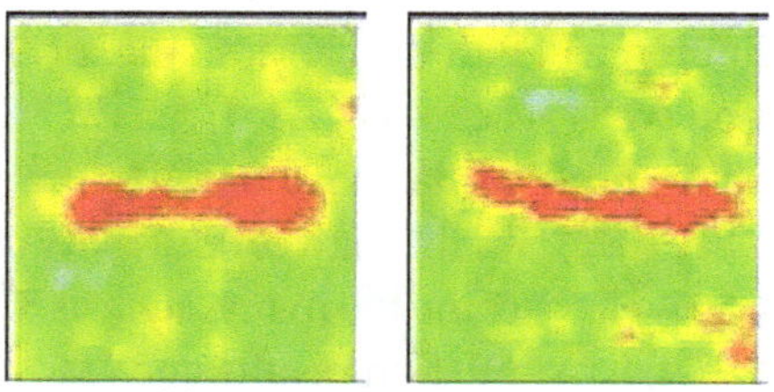

(a) (b)

Figure 11. Calibrated hydraulic conductivity field with different distributions of 16 observation wells: (a) 016; and (b) 0016

In the first pattern shown in Figure 10(a), the 16 hydraulic head observation are uniformly distributed in the whole study domain, but there is not a single observation well located within the high hydraulic conductivity zone. As discussed above, the updated hydraulic conductivity field shown in Figure 9(b) does not represent the existence of the high conductivity zone well. The observed data will influence the updated data around the observation locations (Ott et al., 2004), so the updated hydraulic conductivity distribution cannot capture the zone of high hydraulic conductivity.

For the observation wells shown in Figure 10(b), they are still uniformly distributed. Although the observation well locations are still not in the high hydraulic conductivity zone, they are around that zone. Therefore, the updated hydraulic conductivity field shown Figure 11(a) is similar to the real field. The third distribution pattern of the observation wells is shown in Figure 10(c), and the wells are not distributed uniformly, and are located closer to each other on the lower right side while some are in the high hydraulic conductivity zone. Similar to the Figure 11(a), the updated hydraulic conductivity field shown in Figure 11(b), is very similar to the real conductivity field.

The study results indicate that the observation well locations will influence the assimilation results of the hydraulic conductivity field. Better results are obtained if the observation wells sample zones of distinct conductivity. Optimal design of observation well locations to maximize data efficiency is a research issue requiring further research.

4.1.5. Conclusions

In this study, the ensemble Kalman Filter (EnKF) is used to estimate heterogeneous hydraulic conductivity distributions by assimilating hydraulic head measurements in a transient flow pumping field. The measurement error is considered, but the model error is not. The forecast model is assumed to be known. Two synthetic cases with different boundaries are designed to investigate the capacity and effectiveness of the proposed data assimilation method to identify a conductivity distribution. The developed method is suitable for 2 or 3-D flow field, but only demonstrated on 2-D transient flow. In case 1, the four sides are Neumann boundaries with a constant pumping rate. While in case 2, the left and right sides are constant head boundaries and the top and bottom sides are still no-flow boundaries and the pumping rate changes every 5 [T]. The head measurements are used to identify a heterogeneous hydraulic conductivity field. Based on the study results, the following conclusions are obtained.

1. The EnKF can be used to effectively calibrate a heterogeneous conductivity distribution by assimilation of hydraulic head measurements in a transient flow field with no-flow boundaries. Only after a few (no more than 10) assimilation steps, the spatial distributions of hydraulic head and hydraulic conductivity become significantly closer to the real field distributions than those without data assimilation.

2. The model predictability depends on ensemble size. For our study cases, 300 realizations were found to be sufficient for transient flow calculation.

3. Using the EnKF to calibrate a heterogeneous conductivity field through assimilating transient head data, the calibration is very efficient for the study domain with all Neumann boundaries, while it is not efficient with any constant boundaries in study domain. The boundary condition will significantly affect the assimilation results.

4. The appropriate number of observation wells for our case study is 25, 36, 49 and 64 respectively. Compared with the total 2601 grids in the study area, the observation well number is still small. When well number is too small, the data cannot capture the variation, but when the number is too big, the data will be redundant and cause over calibration.

5. The location of the observation wells will also significantly influence the calibration results. Better results will be obtained if the observation wells adequately sample the heterogeneities. For the study case, if the observation wells are located in both high and low conductivity regions, the calibration results will be better than if they are only located in one region.

6. A criterion is introduced to determine the end of the data assimilation process, but how to develop a more suitable criterion needs further study. That criterion will allow the identification of the point where continued data assimilation will deteriorate the quality of the results.

7. The simulation time of the assimilation steps is very short, and this is a problem requires our deeper thinking. The model error is not considered here, which will pose too much confidence to the model estimate. Maybe the latter one is the cause of the former problem.

However, the model error is unknown here, so how to choose a model and determine the model error is another issue needs further thinking.

4.2. Transient state conservative solute transport

Similar to part 4.1, it is assumed that the solute concentration samples can be obtained in the observation wells with every simulation time step, so in this part the observed data will be assimilated through the constrained EnKF method (described in section 3.2) at every simulated time. Since the ensemble size of 300 is enough in part 4.1, 300 realizations will be used in this part too.

4.2.1. Small initial distribution of solute concentration

The solute concentration can be predicted after the water velocity field is obtained, and the initial instant injection is given. The solute concentration distribution of the initial instant injection is shown in Figure 12. The initial injection area is very small in Figure 12(a), and the larger on is in Figure 12(b). There are 81 observation wells as shown in Figure 3(c). I plot the mean and variance values of the hydraulic conductivity field with and without data assimilation under 300 ensemble sizes with small area of initial solute concentration in Figures 13 and 14, respectively. I compare the real conductivity field with simulated fields with and without the constrained EnKF method. The initial ensemble mean hydraulic conductivity under 300 ensemble size is 13.35 [*L*/*T*], as shown in Figure 13, while the real field mean is 7.92 [*L*/*T*]. It is shown in Figure 13 that in comparison with the initial conductivity field, the updated mean conductivity through the data assimilation method is much closer to the real mean value. However, some updated results could change from "good" at the first several steps to "bad" after many assimilation steps. The results at the first five or seven assimilated time steps are the most efficient. In Figure 14, the variance value of the updated result approaches the real value after several assimilation steps, however, the variance will continue increasing and deviates from that in the real field as time goes. The results are similar to the results in Figure 13. The *RMSE K* value also decreases in the first several time steps and then increases. The results suggest that it is only meaningful to use the data assimilation method to update parameters for the first several assimilation time steps. If we just take the results from the first several assimilation steps, the updated hydraulic conductivity field is significantly improved in comparison with the initial guessed field.

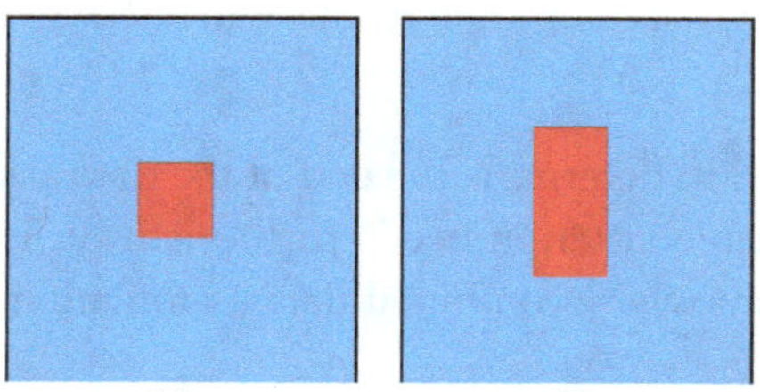

(a) (b)

Figure 12. Solute concentration distribution of the initial instant injection (red zone): (a) small distribution; and (b) large distribution

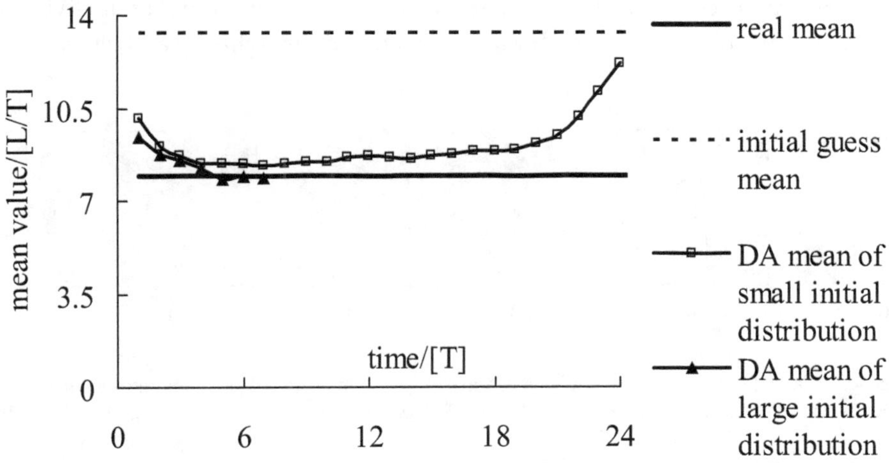

Figure 13. Mean values of hydraulic conductivity field: the real field, without and with data assimilation with small and large area of initial solute concentration under 81 observation wells

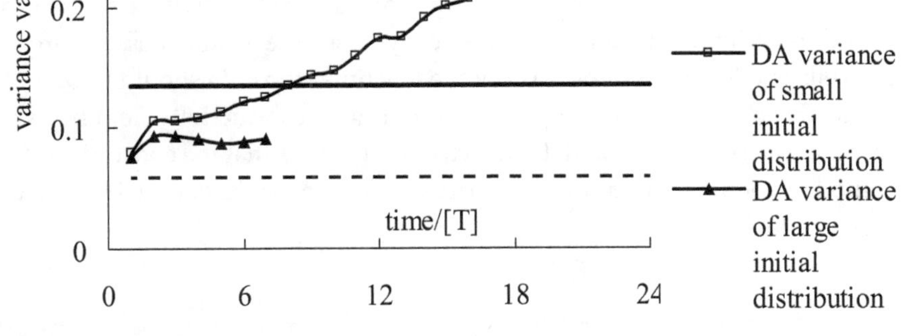

Figure 14. Variance values of hydraulic conductivity field: the real field, without and with data assimilation with small and large area of initial solute concentration under 81 observation wells

We plot the spatial distributions of the updated hydraulic conductivity field at various time steps in Figure 15, and the corresponding absolute hydraulic conductivity difference between the updated result and the real one is presented in Figure 16. The initial ensemble mean guess of hydraulic conductivity field (Figure 3(b)) is uniform, while the real field is heterogeneous and shown in Figure 3(a). The spatial distribution of updated hydraulic conductivity field in Figure 15 have captured the feature of the real field, two low hydraulic conductivity zones in the middle and two high hydraulic conductivity zones at the right-top and left-middle areas.

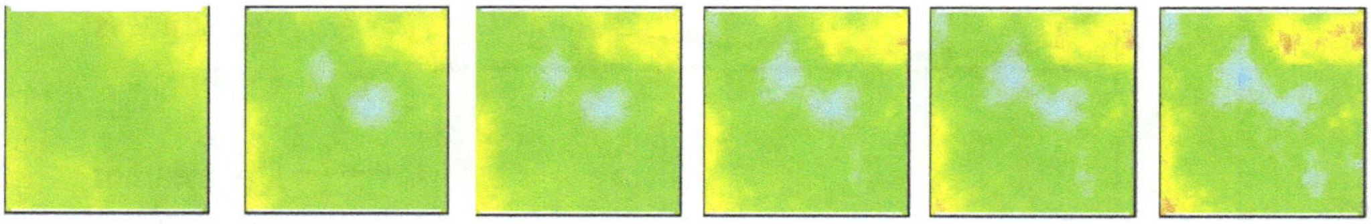

(a) (b) (c) (d) (e) (f)

Figure 15. Assimilated log*K* field at different assimilation steps with small area of initial solute concentration under 81 observation wells: (a) the 1th step; (b) the 4th step; (c) the 5th step; (d) the 7th step; (e) the 8th step; (f) the 11th step

From Figure 16, it is can be seen that the updated hydraulic conductivity fields are better than the initial guess, and the hydraulic conductivity has been significantly improved in the middle area. This is explained as that the observed data are mainly obtained from the middle area during the simulation time. However, the constrained EnKF method overestimates the high hydraulic conductivity zone in the left-bottom and the right-top areas (Figures 15 and 16). The reason for the over estimation will be explained later, which leads to the deterioration of the results of the mean and variance values shown in Figure 13 and 14. It is also noticed that observed data of only 7 to 9 can be obtained from the observation wells because the solute does not spread to all domains at any time step, and solute duration time at any location is limited. Therefore, for any given well, if too many time step observations are used to update the conductivity field, the concentration values in some steps will be zero, such as in the time period before the solute reaches the well point or the period after the solute passes that point. If these void concentration data are used to update the conductivity field, the data will deteriorate the assimilation results. For the study case, the solute passes through some observation wells in a few time steps, so only the few time step data should be assimilated. In general, the EnKF can significantly improve the hydraulic conductivity field by assimilating 7 to 9 observed solute concentration data, which demonstrates that the constrained EnKF is an efficient method to update the hydraulic conductivity field via assimilating the observed solute concentration available.

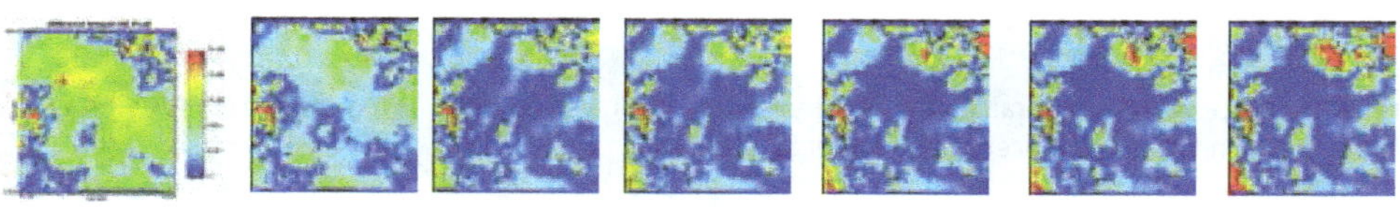

(a) (b) (c) (d) (e) (f) (g)

Figure 16. Absolute difference between real and updated or initial guess hydraulic conductivity field with small area of initial solute concentration under 81 observation wells: (a) initial guess; (b) 1th assimilation step; (c) 4th assimilation step; (d) 5th assimilation step; (e) 7th assimilation step; (f) 8th assimilation step; (g) 11th assimilation step

(Note: all the Figures of absolute difference of hydraulic conductivity fields for the conservative solute transport here and next have the same color scale)

Moreover, compared with 2601 grid elements in the study domain, the number of the observation data (7 to 9) is very small, which is very convenient in the reality. In part 4.1, From the study on updating conductivity field through assimilating hydraulic head data, it is found that 25 observed data (observation wells) are needed to capture major feature of the same conductivity field as the one used in this study. Therefore, the concentration data are more efficient than hydraulic head data. This is because the solute concentration in the middle part will be influenced by the groundwater velocity in the upstream and downstream areas. The hydraulic conductivity field will affect the hydraulic head field through the flow model, and the velocity is controlled by both the hydraulic head and conductivity fields through the Darcy's law (see Equation (2)). Thus, the groundwater velocity is directly influenced by the hydraulic conductivity field. In this study, the solute transport is dominated by the advection, so the solute concentration will be very sensitive to the hydraulic conductivity in the upstream and downstream areas, and the observed solute concentration data will contain the information of the hydraulic conductivity distribution. Furthermore, the dispersivity parameters are also related to the groundwater velocity closely in the solute transport modeling, which suggests that the dispersion part of the solute transport is also sensitive to the hydraulic conductivity indirectly. So only 7 to 9 observed solute concentration data in the middle part can be used to improve the hydraulic conductivity distribution not only in this middle part but can also in the left and right areas.

In part 4.1, it was already shown that the EnKF method by assimilating the hydraulic head data is not very efficient for the study domain with the mixed boundaries, no-flow boundary at the top and bottom borders and constant head boundary at the left and right borders. However, the updating results are very significant with the same hydraulic boundary conditions in this study. Therefore, it is can be concluded that updating a conductivity field through the constrained EnKF method via assimilating the solute concentration data is much more efficient than via assimilating the hydraulic head data.

4.2.2. Large initial distribution of solute concentration

To further explore the capability of the constrained EnKF to update the hydraulic conductivity field via assimilating solute concentrations, it is necessary to add more observation data. So the initial distribution of the solute area is extended from the one in Figure 12(a) to the one in Figure 12(b). In this way, the solute plume will cover a larger area and more observation wells can sample the solute concentration. The mean and variance values of the hydraulic conductivity field with and without data assimilation are plotted in Figures 13 and 14 at the first few and different assimilation steps, and compare them with the values in the real hydraulic conductivity field.

In Figure 13 and 14, it is shown that both the mean and variance values of the updated hydraulic conductivity field by the constrained EnKF become closer to the real ones than those of the initial guess field. Moreover, the *RMSE K* value at different assimilation time is less than that of the initial guess, which indicates that the updated hydraulic conductivity field is better than the initial guess hydraulic field. The number of the wells from which the plume can be observed during the application of the constrained EnKF by assimilating the solute concentration varies from 13 to 16.

The spatial distributions of the updated hydraulic conductivity field and the corresponding absolute hydraulic conductivity difference between the updated result and the real field at some time steps are shown in Figures 17 and 18. From the figures, it can be seen that the updated hydraulic conductivity field captures well the main feature of the real field, and the updated results are much better than the initial guess. The low hydraulic conductivity zones of the real field are together in the middle area, while it is also indicated in the updated fields as shown in Figure 17. Like the real field, the high hydraulic conductivity zones can also be seen in the right-upper and the left-middle areas of the updated ones. In comparison with the 7th step updated results with a smaller initial plume size, shown in Figures 15(d) and 16(e), and with a larger initial plume size, Figures 17(c) and 18(c), the larger the initial plume size, the slightly better the updated results.

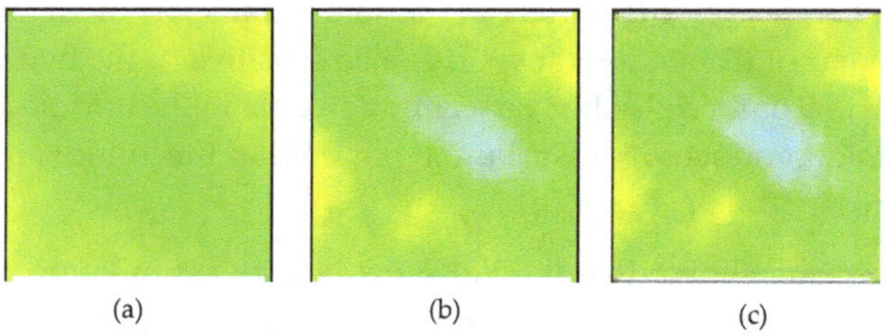

Figure 17. Assimilated log*K* field at different assimilation steps with large area of initial solute concentration under 81 observation wells: (a) the 1th step; (b) the 4th step; (c) the 7th step

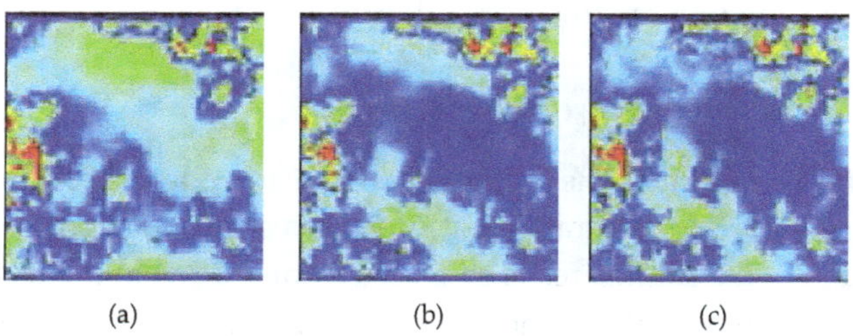

Figure 18. Absolute difference between real and updated hydraulic conductivity field with large area of initial solute concentration under 81 observation wells: (a) 1th assimilation step; (b) 4th assimilation step; (c) 7th assimilation step

What we want to mention is that the observation wells are uniformly distributed in space, which is very easy to conduct in practice, and the hydraulic conductivity field can be improved by obtaining solute concentration from these wells. If the concentration data are available in all the observation wells, it is not necessary to get the data at every observation well. They can be gotten selectively, that is to say that some observed data at these wells can be gotten at this time, and some observed data at some different wells (not the same observation wells) cab also be gotten at the next time. But how to determine the criteria which time is better to obtain observed data from which observation wells is an issue requiring further research.

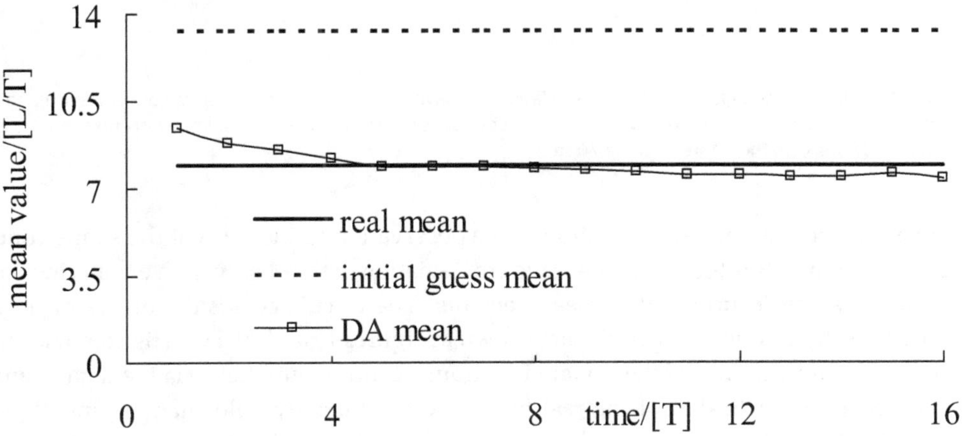

Figure 19. Mean values of hydraulic conductivity field: the real field, without and with data assimilation and large area of initial solute concentration under 81 observation is 5% of the solute concentration measurements

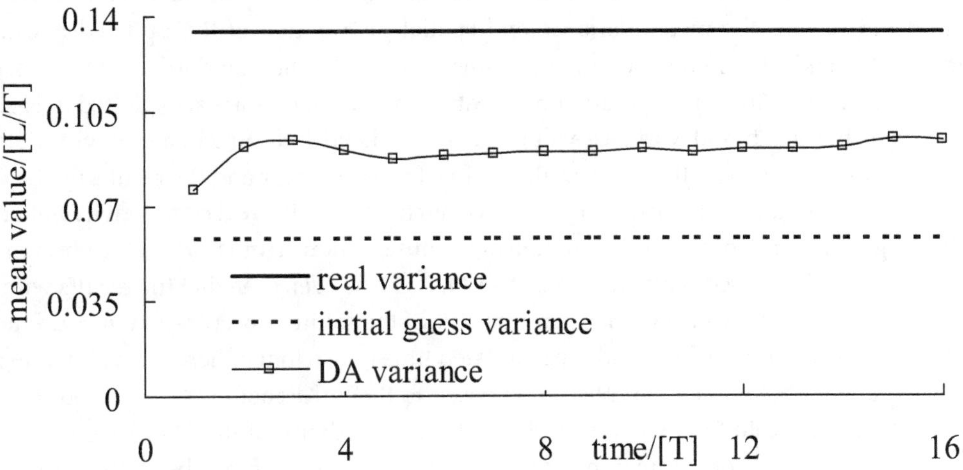

Figure 20. Variance values of hydraulic conductivity field: the real field, without and with data assimilation and large area of initial solute concentration under 81 observation wells while the standard deviation of the observation error is 5% of the solute concentration measurements

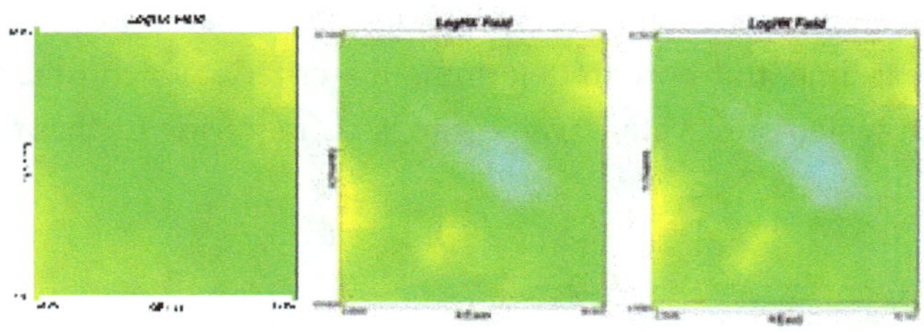

(a) (b) (c)

Figure 21. Assimilated log*K* field at different assimilation steps with large area of initial solute concentration under 81 observation wells while the standard deviation of the observation error is 5% of the solute concentration measurements: (a) the 1th step; (b) the 4th step; (c) the 7th step

However, maybe the standard deviation of the observation error as 1% of the solute concentration measurements is too small, the standard deviation of the observation error is increased to 5% of the solute concentration measurements. The calculated results of the mean and variance of the hydraulic conductivity are shown in Figures 19 and 20. From figures, it is found that the mean and variance of the updated hydraulic conductivity field via the data assimilation method is closer to that of the real hydraulic conductivity field than the initial guess conductivity field. the spatial distribution of the updated hydraulic conductivity field at different assimilated time steps are plotted in Figure 21. Comparing Figure 21 with Figure 17, it is found that these results are very similar. So the standard deviation of the observation error does not affect the simulation results very much here from the first view. Therefore, the absolute hydraulic conductivity difference between the updated result and the real field are not plotted for simplicity.

To have a further look at the influence of the standard deviation of the observation error on the updated results, the mean, variance and spatial distribution of the updated hydraulic conductivity field at different assimilated time steps with the standard deviation of the observation error be 10% of the solute concentration measurements are also plotted in Figures 22-24. From the mean and variance values of the updated hydraulic conductivity field in Figures 22 and 23, it is seen that the simulated time become longer and the results by the data assimilation method via the constrained EnKF is much closer to the real conductivity field than the initial guess even the standard deviation of the observation error is 10% of the observation measurements. Compared with Figures 13 and 14, it can be even seen that the results with the standard deviation of the observation error be 10% of the solute concentration measurements are much better than that of 1%, and the simulated time is also longer here. This phenomenon can be explained that maybe the constraints posed to the EnKF method have introduced too much potential error to the model, which lead to the avoidance of the filter divergence to a certain degree and longer simulation time, and also the increase of the observation error will give a tradeoff between the model error and the observation error. Furthermore, it is seen that the spatial distribution of the updated hydraulic conductivity fields still similar to that in Figure 17. Therefore, it is can proved that the standard deviation of the observation error does not affect the simulation results very much in a deeper step.

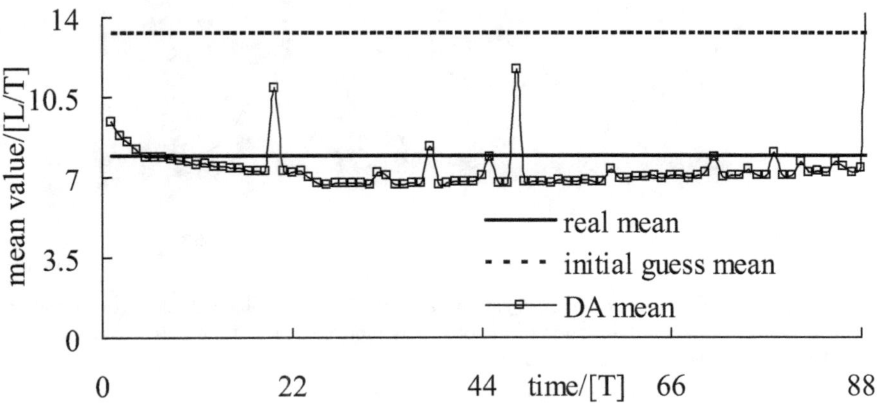

Figure 22. Mean values of hydraulic conductivity field: the real field, without and with data assimilation and large area of initial solute concentration under 81 observation wells while the standard deviation of the observation error is 10% of the solute concentration measurements

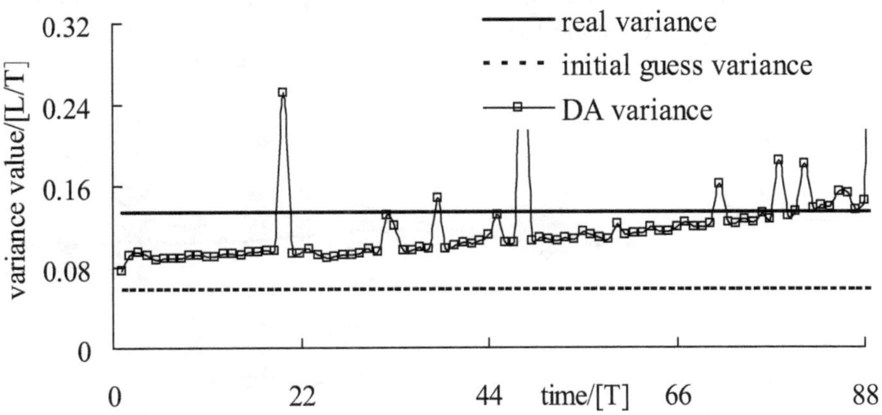

Figure 23. Variance values of hydraulic conductivity field: the real field, without and with data assimilation and large area of initial solute concentration under 81 observation wells while the standard deviation of the observation error is 10% of the solute concentration measurements

The *RMSE* values for the updated hydraulic conductivity field under different percentage (3%, 5%, 9%, 10%) standard deviation of the observation error of the solute concentration measurements are also plotted in Figure 25. From the figure, it can be seen that the *RMSE* values will increase as the assimilated time increase for different perturbations to the observation error. The larger perturbations added to the observation errors, the greater the *RMSE* values for the updated hydraulic conductivity fields. It is seen that if the standard deviation of the observation errors is greater than 5% of the measurements, the *RMSE* values for the updated hydraulic conductivity have the trend to increase. So it is better to use the standard deviation

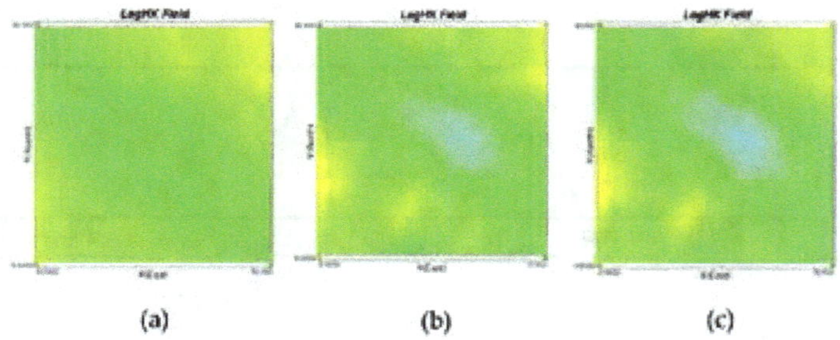

Figure 24. Assimilated log*K* field at different assimilation steps with large area of initial solute concentration under 81 observation wells while the standard deviation of the observation error is 10% of the solute concentration measurements: (a) the 1th step; (b) the 4th step; (c) the 7th step

of the observation errors no more than 5% of the measurements if I want to have less *RMSE*. Furthermore, it can also be found that the plots of the mean, variance and *RMSE* values for the updated hydraulic conductivity field with the standard deviation of the observation errors as 10% of the measurements in Figures 22, 23 and 25 are very identical, they almost fluctuate at the same time. This can also be seen from the former results in section 4.1, where the *RMSE* values become very high at the later assimilated time steps like the corresponding mean or variance values and they are not plotted there.

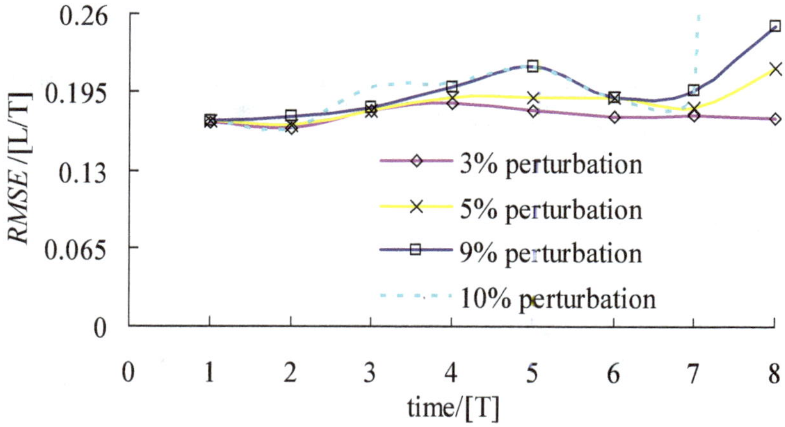

Figure 25. *RMSE* values of the assimilated hydraulic conductivity field with large area of initial solute concentration under 81 observation wells under different standard deviation of the observation errors

4.2.3. Conclusions

In this study, the constrained EnKF is used to estimate heterogeneous hydraulic conductivity distributions by assimilating solute concentration measurements for transient solute transport in a steady state flow field. The measurement error is considered, but the model error is not.

The forecast model is assumed to be known. A synthetic case of a rectangular field with the no-flow boundary at the upper and bottom borders and constant head boundary at the left and right borders is designed to investigate the capacity and effectiveness of the data assimilation method to identify a hydraulic conductivity distribution. The developed method is only demonstrated on 2-D transient solute transport with two different areas of initial instant solute source in the middle area of the field. The solute concentration measurements are used to identify the heterogeneous hydraulic conductivity field. Based on the study results, the following conclusions are obtained.

1. The constrained EnKF can be used to effectively calibrate a heterogeneous conductivity distribution by assimilating solute concentration measurements for a transient transport in a steady state flow field with the mixed no-flow boundary and constant head boundary. Only after a few (no more than 10) assimilation steps, the spatial distributions of hydraulic conductivity become significantly closer to the real field distribution than those without data assimilation.

2. The model predictability depends on the initial distribution of the solute concentration. The larger area of the initial distribution, the more observed data can be obtained, the better the calibration results.

3. Due to the strong sensitivity of solute concentration to the hydraulic conductivity field, less observation wells are needed to achieve the same calibration results for a hydraulic conductivity field through assimilating solute concentration data than assimilating hydraulic head data.

4. The data assimilation method can produce useful results in the first five or seven time step assimilation.

5. In study of part 4.1, the EnKF method was used to calibrate a conductivity field by assimilating hydraulic head data. Their study results indicate that the EnKF method cannot well capture a conductivity distribution in a field with constant head boundaries through assimilating hydraulic head data; while this study results indicate the EnKF method can through assimilating concentration data.

6. Maybe the constraints posed to the EnKF method have introduced too much potential error to the model, which lead to the avoidance of the filter divergence to a certain degree and longer simulation time, and also the increase of the observation error will give a tradeoff between the model error and the observation error. So the standard deviation of the observation error varies from 1% to 5% even to 10% of the solute concentration measurements, the simulated results by the data assimilation method are still very similar.

7. The assimilated simulation steps are not very long, this problem may be caused by ignoring the model error. Maybe in future, the model error (which is not known right now) can be added. Moreover, the data assimilation method just provides us the best estimate values for our study case, but it does not consider the mass balance during the simulation automatically.

Acknowledgements

This work is partly supported by the National Nature Science Foundation of China Major Research Project (Grant No. 91125024), the National Nature Science Foundation of China (Grant No. 51209187), and the Fundamental Research Funds for the Central Universities (Grant No. 2652011286).

Author details

Bill X. Hu[1,2] and Juxiu Tong[1]

1 Collage of Water Resources and Environmental Sciences, China University of Geosciences, Beijing, China

2 Department of Earth, Ocean and Atmospheric Sciences, Florida State University, Tallahassee, USA

References

[1] Andreadis KM, Lettenmaier DP (2006) Assimilating remotely sensed snow observations into a macroscale hydrology model. Adv. Water Resour 29(6): 872-886

[2] Arlen WH (2005) The U.S. Geological Survey Modular Ground-Water-the Ground-Water Flow Process. U.S. Geological Survey, Reston

[3] Aubert D, Loumagne C, Oudin L (2003) Sequential assimilation of soil moisture and streamflow data in a conceptual rainfall-runoff model. Journal of Hydrology 280: 145-161

[4] Bear J (1972) Dynamics of Fluids in Porous Media. New York, Elsevier Pub. Co., Inc.

[5] Burgers G, van Leeuwen PJ, Evensen G (1998) Analysis scheme in the ensemble Kalman filter. Monthly Weather Review 126: 1719-1724

[6] Chen Y, Zhang D (2006) Data assimilation for transient flow in geologic formations via Ensemble Kalman Filter. Adv. Water Resour. 29: 1107-1122

[7] Christakos G (2005) Methodological developments in geophysical assimilation modeling. Reviews of Geophysics 43: 2004RG000163, 1-10

[8] Clark MP, Slater AG, Barrett AP (2006) Assimilation of snow covered area information into hydrologic and land-surface models. Adv. Water Resour. 29(8): 1209-1221

[9] Cohn SE (1997) An introduction to estimation theory. J. Meteor. Soc. Japan 75(1B): 257-228

[10] Constantinescu EM, Sandu A, Chai TF, Carmichael GR (2007) Assessment of ensemble-based chemical data assimilation in an idealized setting. Atmospheric Environment 41: 18-36

[11] Daley R (1991) Atmospheric Data Analysis. New York: Cambridge University Press, pp. 457

[12] Drecourt JP (2003) Kalman filtering in hydrologic modeling. DAIHM Technical Report, May 20

[13] Drecourt JP, Madsen H, Rosbjerg D (2006) Calibration framework for a Kalman filter applied to a groundwater model. Adv. Water Resour. 29(5): 719-734

[14] Evensen G (1994) Sequential data assimilation with a nonlinear quasi-geostrophic model using Monte Carlo methods to forecast error statistics. Journal of Geophysical Research 99 (C5), 10.143-10.162

[15] Evensen G (2003) The ensemble Kalman filter: theoretical formulation and practical implementation. Ocean Dynamics 253: 343-367

[16] Evensen G (2004) Sampling strategies and square root analysis schemes for the EnKF. Ocean Dynamics 54: 539-560

[17] Fang F, Piggott MD, Pain CC, Gorman GJ, Goddard AJH (2006) An adaptive mesh adjoint data assimilation method. Ocean Modeling 15: 39-55

[18] Fisher M (1998) Development of a simplified Kalman filter. ECMWF Research Department Tech. Memo. 260, 16 pp [Available from European Center for Medium-range Weather Forecast, Shin-field Park, reading, Berkshire, RG2 9AX, United Kingdom.]

[19] Gelb A (1974) Applied optimal estimation. MIT Press, Cambridge, MA

[20] Hemant AP, Oliver DS (2009) Data assimilation using the constrained ensemble Kalman filter (SPE 125101). In SPE Annual Technical Conference, 4-7 October 2009, New Orleans, Louisiana, U.S.A.

[21] Hoeksema RJ, Kitanidis PK (1984) An application of the statistical approach to the inverse problem in two-dimensional groundwater modeling. Water Resour. Res. 20(7): 1003-1020

[22] Houtekamer PL, Mitchell HL (1998) Data assimilation using an ensemble Kalman Filter technique. Monthly Weather Review 126: 796-811

[23] Houtekamer PL, Mitchell HL (2001) A sequential ensemble Kalman filter for atmospheric data assimilation. Mon. Wea. Rev. 129: 123-137

[24] Hu BX, Meerschaert MM, Barrash W, Hyndman D, He C, Li X, Guo L (2009) Examining the influence of heterogeneous porosity fields on conservative solute transport. J. Contaminant Hydrology 108: 77-88.

[25] Huang C, Bill XH, Li X, Ye M (2009) Using data assimilation method to calibrate a heterogeneous conductivity field and improve solute transport prediction with an unknown contamination source. Stochastic Environment Research and Risk Assessment DOI 10. 1007/s00477-008-0289-4

[26] Jazwinski AH (1970) Stochastic Processes and Filtering Theory. Elsevier, New York

[27] Kalman RE (1960) A new approach to linear filtering and prediction problems, Transactions of the ASME—Journal of Basic Engineering 82 (Series D): 35-45

[28] Kalman RE, Bucy RS (1961) New results in linear filtering and prediction theory. Retrieved 2008-05-03

[29] LeDimet FX, Talagrand O (1986) Variational algorithms for analysis and assimilation of meteorological observations—theoretical aspects. Tellus Series A—Dynamic Meteorology and Oceanography 38(2): 97-110

[30] Liu GS, Chen Y, Zhang DX (2008) Investigation of flow and transport processes at the MADE site using ensemble Kalman filter. Advances in Water Resources 31: 975-986

[31] Liu Y, Gupta HV (2007) Uncertainty in hydrologic modeling: Toward an integrated data assimilation framework. Water Resour. Res. 43, W07401, doi: 10.1029/2006WR005756

[32] Maybeck PS (1979) Stochastic models, estimation and control. Academic Press, INC., London, LTD

[33] Mclaughlin D, Townley LR (1996) A reassessment of the groundwater inverse problem. Water Resour. Res. 32(5): 1131-1161

[34] Mclaughlin D (2002) An integrated approach to hydrologic data assimilation: interpolation, smoothing, and filtering. Adv Water Resour. 25: 1275-1286

[35] Natvik L-J, Evensen G (2003) Assimilation of ocean color data into a biochemical model of the North Atlantic Par 1. Data assimilation experiments. Journal of Marine System 40-41: 127-153

[36] Oliver DS, Reynolds AC, Liu N (2008) Inverse theory for petroleum reservoir characterization and history matching. July 28

[37] Poeter EP, Hill MC (1997) Inverse models: a necessary next step in ground-water modeling. Ground water 35(2): 250-260

[38] Prakash J, Patwardhan SC, Shah SL (2008) Constrained state estimation using the ensemble Kalman filter. 2008 American Control Conference, Westin Seattle Hotel, Seattle, Washington, USA, June 11-13

[39] Reichle RH (2008) Data assimilation methods in the Earth sciences. Advances in Water Resources doi:10.1016/j.advwatres.2008.01.001

[40] Sridhar U, Dolence E, Li KY (2007) Constrained extended Kalman filter for nonlinear state estimation. In Proceedings of the 8th International IFAC Symposium on Dynamics and Control of Process Systems

[41] Sun NZ (1994) Inverse problems in groundwater modeling. Kluwer Academic Publisher. Dordrecht, Netherlands

[42] Sun NZ, Yeh W-G (1992) A stochastic inverse solution for transient groundwater flow: Parameter identification and reliability analysis. Water Resour. Res. 28(12): 3269-3280

[43] Thacker WC (2007) Data assimilation with inequality constraints. Ocean Modelling 16(3-4): 264-276

[44] Thomsen PG, Zlatev Z (2008) Development of a data assimilation algorithm. Computers and Mathematics with Applications 55: 2381-2393

[45] Tipireddy R, Nasrellah HA, Manohar CS (2008) A Kalman filter based strategy for linear structural system identification based on multiple static and dynamic test data. Probabilistic Engineering Mechanics doi:10.1016/j.probengmech.2008.01.001

[46] Van Geer FC, Te Stroet CBM, Zhou X (1991) Using Kalman filtering to improve and quantifying the uncertainty of numerical groundwater simulations: 1. the role of system noise and its calibration. Water Resour. Res. 27(8): 1987-1994

[47] Vrugt JA, Robinson BA, Vesselinov VV (2005a) Improved inverse modeling for flow and transport in subsurface media: Combined parameter and state estimation. Geophys. Res. Lett. 32: L18408, doi:10.1029/2005GL023940

[48] Vrugt JA, Diks CGH, Gupta HV, Bouten W, Verstraten JM (2005b) Improved treatment of uncertainty in hydrologic modeling: Combining the strengths of global optimization and data assimilation. Water Resour. Res. 41: W01017, doi: 10.1029/2004WR003059

[49] Vermeulen PTM, Stroet CBM, Heemink AW (2006) Model inversion of transient nonlinear groundwater flow models using model reduction. Water Resour. Res. 42: W09417, doi: 10. 1029/2005WR004536

[50] Wang DB, Chen YG, Cai XM (2009) State and parameter estimation of hydrologic models using the constrained ensemble Kalman filter. Water Resources Research 45, W11416, doi:10.1029/2008WR007401

[51] Yeh T-C J, Liu S (2000) Hydraulic tomography: Development of a new aquifer test method. Water Resour. Res. 36(8): 2095-2105

[52] Yeh T-C J, Zhang J (1996) A geostatistical inverse method for variably saturated flow in the vadose zone. Water Resour. Res. 32(9): 2757-2766

[53] Zhang D, Lu Z, Chen Y (2007) Dynamic reservoir data assimilation with an efficient, dimension-reduced Kalman filter. SPE Journal 12(1): 108-117

[54] Zheng C (1990) A Modular Three-dimensional transport model for simulation of advection, dispersion and chemical reactions of contaminants in groundwater systems. S.S. Papadopulos & Associates, Inc. Rockville, Maryland 20852

[55] Zhu J, Yeh T-C J (2005) Characterization of aquifer heterogeneity using transient hydraulic tomography. Water Resour. Res. 41: W07028, doi:10.1029/2004WR003790

[56] Zou YX, Testroet CBM, van Geer FC (1991) Using Kalman filtering to improve and quantifying the uncertainty of numerical groundwater simulation: 2. Application to monitoring network design. Water Resour. Res. 27(8): 1995-2006

3

Hydraulic Efficiency and Safety of Xylem Sap Flow in Relation to Water Stress in Woody Plants

Katarzyna Marciszewska and Mirela Tulik

1. Introduction

Trees as long-living organisms can be looked as the archives of ecological events recorded in the annual rings of wood. These can play an important role in studying and reconstruction of climate variation. The ability of a tree genotype to adjust the phenotype over the life of a tree is a consequence of short-term to long-term physiological responses to environmental changes, the ability which can be used to link environment with wood structure. The anatomical features of tree rings such as conduits size and density, cell wall thickness and the share of different tissues are sources of information on global environmental changes throughout the last millennium and on the strategies of species' responses to the evaluation of the impact of predicted climate change on vegetation dynamics.

The aim of this chapter is to present the up to date knowledge about the changes in woody plant hydraulics in response to water stress. The understanding of drought impact on plant water conducting systems structure and functioning and their possible adjustments becomes especially essential in respect to climate change projections for Central Europe predicting lower summer precipitation combined with prolonged drought periods. The issue of structural basis for water flow in wood of woody plants and the methods of its measurement and assessment are presented and the adjustments to water stress in short- and long-time scale as well as at different levels of woody plant organizations discussed.

2. Structural basis for woody plants hydraulics

Wood anatomy can be studied in the view of different disciplines e.g. technical applications, tree pathology, ecology, dendrochronology but also in terms of hydraulics as a specific hydrosystem based on cells with lumina filled with water.

In perennial vascular plants the transport of water is to great extent determined by the hydraulic architecture. This term describes the xylem network within the plant and its variations with wood type, plant age and growth form (e.g. liana, shrub vs. tree) [1]. The xylem network integrates all main parts of the plant's body, i.e. roots, branches and leaves. It means that any root in this system is more or less directly connected with any branch and not with a single one. Moreover, the xylem network is redundant in two meanings: at a given level of the stem several xylem element are present in parallel and they develop lateral contacts with other tracks of vessels or tracheids.

The xylem network is mainly represented by tracheary elements (conduits) that arise from vascular cambium periodically or continuously. Mature tracheary elements are dead cells with lignified walls. The lignin makes the xylem cells strong and prevents them from collapsing. Moreover, due to lignification the cell walls are impermeable and in effect waterproof.

The tracheids, conducting elements in conifers and ferns wood, are elongated cells with tapering ends (Figure 1A). The water moves from one tracheid to another *via* bordered pits located mainly on their end walls. The pit is an "unlignified gap" within secondary cell wall while primary wall and middle lamellae are present but structurally and chemically modified to thin membrane, through which water flows from tracheid to tracheid. In conifers, bordered pits appear only on the radial walls of tracheids except in the late wood where can also occur in tangential walls.

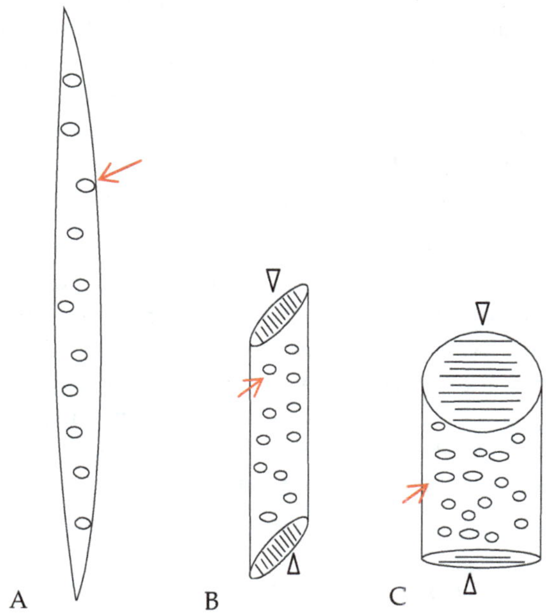

Figure 1. Schematic drawing of tracheary elements: A - tracheid, B -vessel member of late wood and C - vessel member of early wood. The bordered pits on the lateral walls of the tracheary elements are marked by arrows. Arrowheads depict perforation plates at the ends of vessel members.

Vessels, evolutionary younger in comparison to tracheids, are chief water-conducting elements in angiosperms xylem (Figure 1 B and C). They consist of the series of single cells, called vessel members. The vessel members are in general shorter than tracheids but with diameters usually larger. During the differentiation phase the end walls between two vessels members are hydrolised partly or completely and gap or gaps are formed in their end walls, thereafter called the perforation plates. The perforation plate facilitates water movement in the xylem. Vessels can be visualized as vertically aligned tubes or "pipes" made up of many cells with lignified walls that transport liquids in xylem. The vessels sidewalls have bordered pits to allow lateral movement of water.

In angiosperms two main types of wood patterns are recognized based on vessels diameter and distribution within annual ring (Figure 2). There are ring porous (Figure 2C) and diffuse – porous wood (Figure 2D). In the ring porous wood the vessels of earlywood are distinctly larger than those in the latewood of the previous and of the same growth ring, and they form a well-defined zone. In this type of wood the transition to the latewood within the same growth ring is abrupt. The diffuse-porous wood has vessels with more or less the same diameter throughout the growth ring.

In diffuse-porous species and in conifers all conduits within annual increments are responsible for water transport. In ring-porous trees water is transported mainly by vessels of earlywood whereas late wood vessels rather store water.

Generally, there is a continuous increase in the size of tracheary elements from leaves to roots. The increase concerns both tracheids length and diameter as one proceeds from branches to trunk and down into the roots. Similarly a gradual increase is found in vessel diameter and vessel length from twigs, down along the shoot and extending into the roots. The basipetal increase in vessel diameter is associated with decrease in vessel density, defined as a number of vessels per unit of cross-sectional area. The increase in conduit size can also be observed on transverse sections in a stem or a branch of a tree. The increase proceeds outward from the inner growth (Figure 3).

Tracheids and vessel members are dead cells, therefore the mechanism of water transport is based mainly on purely physical forces. Water moves spontaneously through the tracheary elements only from places of higher water potential [ψ] to places of lower water potential, i.e. along a decreasing ψ gradient (Figure 4).

In 1895, two Irish plant physiologists H. H. Dixon and J. Joly [2] proposed the cohesion-tension theory that water is pulled up the plant by tension gradient created by transpirational water lost from the leaf surface [3-5]. Although this theory is widely accepted, the alternative explanations of the mechanism of water transport are postulated. One of them is multiforce theory [6] according to which the tension gradient is supported by forces such as tissue pressure [7], or gradients in chemical activity of water [8] and absorption of water from the atmosphere adds to transpirational water uptake [9].

The basic pulling force resulting from transpiration process implies that the water in xylem is in a metastable state and thus vulnerable to cavitation that is the phase change from water to

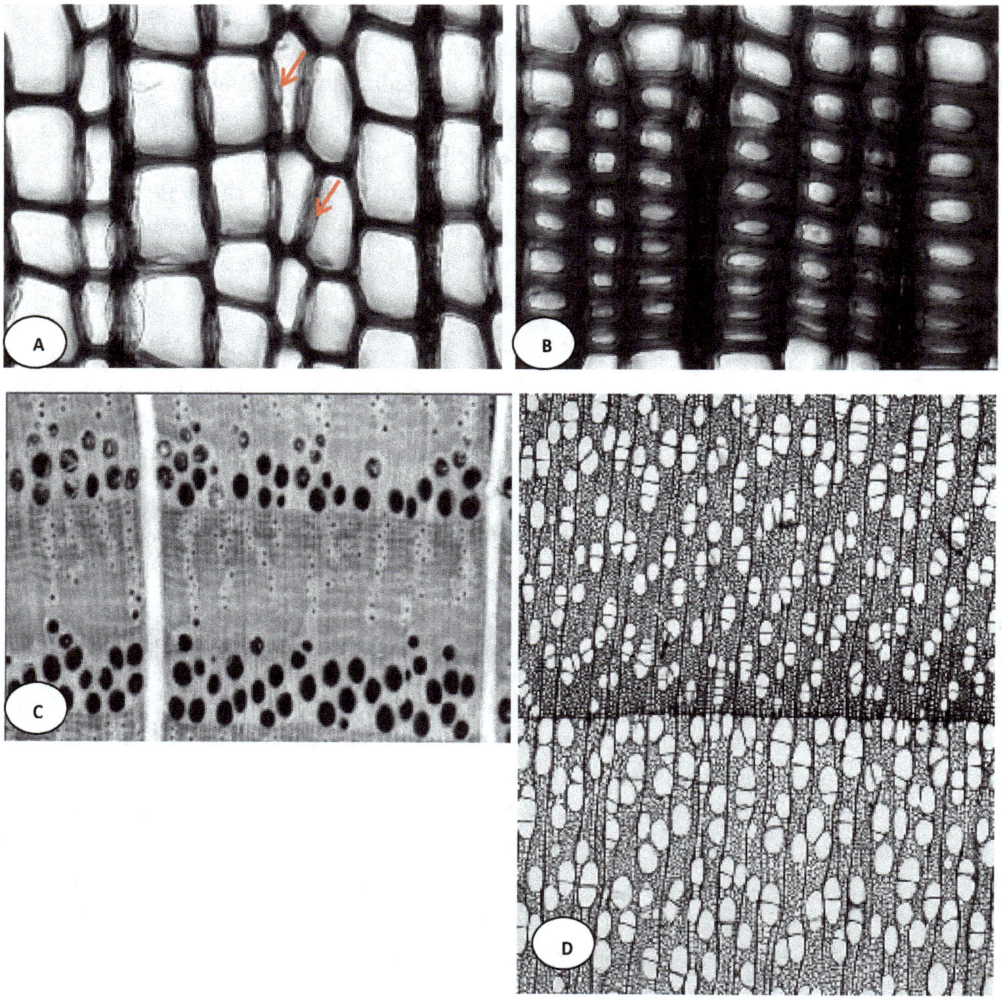

Figure 2. Microscopic cross sections of Scots pine (*Pinus sylvestris* L.) (A, B), pedunculate oak (*Quercus robur* L.) (C) and silver birch (*Betula pendula* Roth) (D) stem wood. Cross sections of: A - earlywood of pine tree (*Pinus sylvestris* L.) (x400). Tracheids with large diameter and thin cell walls are visible. The bordered pits on the radial cell walls are marked by arrows. Earlywood is mainly responsible for water transport in coniferous wood; B - latewood of pine tree (x400). Tracheids with smaller diameter and thicker cell walls are visible. The tracheids of latewood take on much of the plant support task; C - ring-porous wood of oak tree (*Quercus robur* L.) (x160). Large earlywood vessels are visible as a ring close to boundary of the annual increment; D - the diffuse-porous wood with vessels almost the same in diameter within entire growth ring (x200).

vapour. When the initially small bubble expands, blocking the ascent of water, the tracheary elements could be embolized [10]. The variety of tracheary elements in hydrosystem leads to differences in vulnerability to cavitation. More vulnerable to cavitation are vessel members than tracheids because of generally larger diameter. Therefore in vascular plants the conduction of water is a matter of compromises between safety and efficiency.

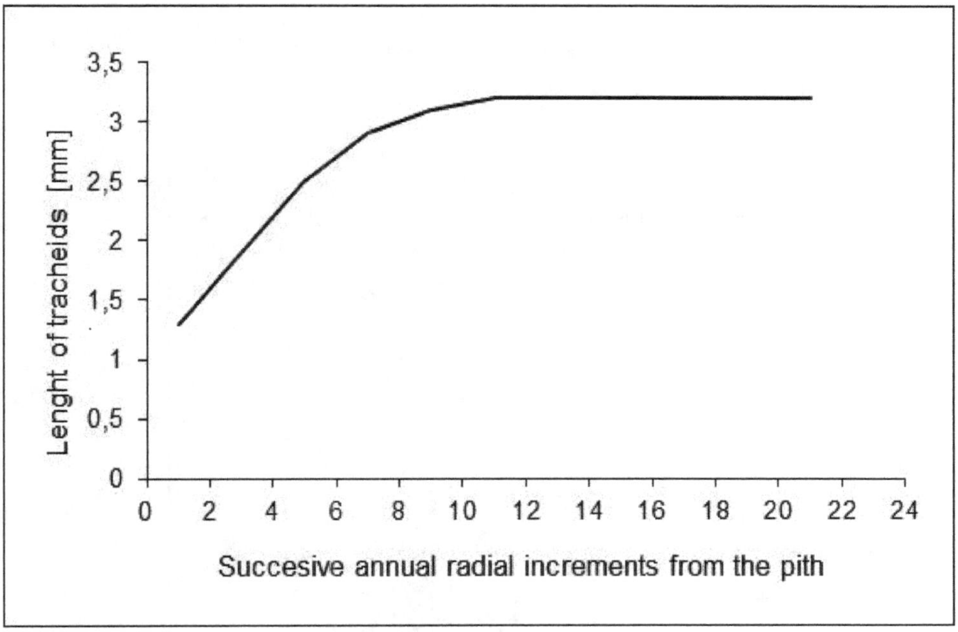

Figure 3. Variation of the tracheids length with age as the function of the distance from the pith in wood of pine tree [106, modified].

3. Measurement methods and estimators of hydraulic conductivity in plants

Our understanding of hydraulic conductivity in wood, similarly as hydraulic conductivity in soil, is based on Darcy's law that demonstrates the proportionality of water flux to hydraulic gradient and describes the effectiveness of the driving force behind the water movement. The law was formulated by Henry Darcy (1859) based on the results of experiments on the flow of water through beds of sand, thus describes the flow of a fluid through a porous medium. Darcy's law is analogous to Fourier's law in the field of heat conduction, Ohm's law in the field of electrical networks, or Fick's law in diffusion theory.

The soil and vascular plants are similar hydraulically [11], thus same physical law may be applied to describe soil and plants hydraulics. For the theoretical calculation of the volume of sap flow in the plant conducting elements the Hagen–Poiseuille equation (published in 1840) is applied, which is a special case of the Darcy-Weisbach formula. The assumptions of the equation are that the fluid is viscous and incompressible; the flow is laminar through a pipe of constant circular cross-section that is substantially longer than its diameter; and there is no acceleration of fluid in the pipe. The equation is also known as the Hagen–Poiseuille law, Poiseuille law and Poiseuille equation.

The flow of water through the tracheary element is usually compared with flow *via* smooth-walled capillary of circular transverse section. The rate of flow through such a capillary is

Figure 4. Scheme of water transport in Soil-Plant-Atmosphere Continuum (SPAC).

expressed in the light of Hagen-Poiseuille law to be proportional to the applied pressure gradient and the hydraulic conductivity and is written as follows [12]:

$$\frac{dv}{dt} = K_{capillary} \frac{\Delta P}{\Delta l} \tag{1}$$

where dV/dt is the rate of flow, $\Delta P/\Delta l$ is applied pressure gradient and $K_{capillary}$ is hydraulic conductivity.

Hydraulic conductivity (K) is the constant that defines the proportionate relationship of flux to hydraulic gradient and in accordance with Hagen- Poiseuille formula could be recorded as follows:

$$K_{capillary} = \frac{\rho r^4}{8\eta} \tag{2}$$

where ϱ is density of water, η the water viscosity coefficient, r the radius of the capillary.

In case of wood many conducts (capillaries) of different diameters transport water in parallel as seen in the transverse section of stems, branches or roots and thus the K_h coefficient in Hagen-Poiseuille formula is written as follows:

$$K_h = \left(\frac{\pi\rho}{128\eta}\right)\sum_{i=1}^{n}\left(d_i^{\,4}\right) \tag{3}$$

Considering ϱ constant at given temperature and η negligible in case of xylem sap, means that at least for theoretical considerations the hydraulic conductivity depends mainly on the diameter of tracheary elements being proportional to the fourth power of the capillary inside diameter or radius.

One of the problems in investigations focused on the vulnerability of xylem hydraulic conductivity failure and especially on the relationship between vulnerability to cavitation and vessel diameter is to find some kind of measure of average vessel diameter and vulnerability. The problem of appropriate measure of diameter arises also when comparison among taxa is in the scope of studies. As the Hagen-Poiseuille law tells, the hydraulic conductivity increases with the fourth power of conduit radius or diameter and thus wide vessels contribute disproportionally to the stem hydraulic conductivity, which was show e.g. for *Gnetum microcarpum* [13]. One of the ways to calculate the mean diameter is to measure 200-400 conduit diameters (D) at random and rank them from largest to smallest value after rising to the fourth power [12]. The D^4 values are summed for all conduits and then summed again until reaching 95% of the total stem conductance. The mean diameter D_{95} is calculated for number of conduits responsible for about 95% of the total stem conductance.

The mean hydraulic diameter is another estimator, based on the artificial methods of hydraulically weighting diameters distributions [12]. Mean hydraulic diameter D_H defined as:

$$D_H = \left(\frac{\sum D^4}{N}\right)^{1/4} \tag{4}$$

where N is the number of conduits and D their measured inner diameters, gives the diameter of vessels to be if total conductivity have to be the same as for sampled stem, providing the same number of conduits.

Hydraulically weighted diameter defined as:

$$D_S = \frac{\sum D^5}{\sum D^4} \tag{5}$$

has been proposed by Sperry et al. [105]. It says which vessels contribute to the greatest part in water flow and is a statistic that weights large vessels. Indeed, hydraulic diameter (D_s) has been frequently used, but some authors suggest [12] that $D_{h,}$ defined as $[(\Sigma D^4)/N]^{1/4}$ is a better estimator, which is more meaningful measure in terms of the physics of water transport.

The index of hydraulic conductivity has been proposed by Zajączkowski [14] to compare conductivities of different tress. It relays on the sum of radii raised to the fourth power, and combines two parameters: sums of fourth power of inside vessels radii and indirectly on the vessels density as first the hydraulic conductivity is calculated per limited part of ring perimeter comprising the measured vessels but then, based on the ration of this section to whole ring, recalculated for entire ring.

The principle of hydraulic conductivity measurements also arises from the Hagen-Poiseuille formula and has been proposed and empirically tested for tree shoots [15, 16]. Briefly, in case of branch, its segment of a given length has to be submitted to a water pressure difference dP. This induces a flux which is measured with suitable device. Knowing the flux, dP and the length of the sample hydraulic conductivity can be calculated. Although the method seems simple, several precautions should be taken.

Techniques that have been first developed for study of xylem hydraulics are based on the pressure chamber, xylem pressure probe and psychrometric methods. Since the 90-ties new techniques has been applied to study xylem hydraulic and these include centrifugal force methods for measuring vulnerability curves [17], freezing-stage scanning electron microscopy (cryo-SEM) [18], and magnetic resonance imaging (MRI) methods for visually assessing embolized against functional xylem vessels [19, 20]. The pressure chamber and psychrometric methods, which are chronologically older have been successfully tested against these new techniques [21, 4]. A very promising approach to the description of some special issues concerning xylem hydraulics e.g. embolism has been developed recently based on imaging techniques that use the synchrotron X-ray light source and enable monitoring of single vessels under different conditions using 2D X-ray technique [22] and 3D X-ray technique [23].

The pressure chamber is relatively quick way for estimating the water potential Ψ_w of large pieces of tissues, such as leaves and small shoots. This method was pioneered by Henry Dixon at the beginning of the twentieth century, but come into widespread use only after the instrument design was improved and its practical use demonstrated [24]. In this technique, the organ to be measured is excised from the plant and is partly sealed in a pressure chamber. Before excision, the water column in the xylem is under tension. Excision of the organ breaks the water column (i.e. its tension is relieved allowing its Ψ_p to rise to zero), water is pulled rapidly from the xylem into the surrounding living cells by osmosis. The cut surface consequently appears dull and dry. To make a measurement, the chamber is pressurized with compressed gas until the distribution of water between the living cells and the xylem conduits is returned to its initial, pre-excision, state. This can be detected visually by observing when the water returns to the open ends of the xylem conduits that can be seen in the cut surface. The pressure needed to bring the water back to its initial distribution is called the *balance pressure* and is readily detected by the change in the appearance of the cut surface, which becomes wet and shiny when this pressure is attained. The pressure chamber is often described as a tool to measure the tension in the xylem. However, this is only strictly true for measurements made on a non-transpiring leaf or shoot (for example, one that has been previously enclosed in a plastic bag). When there is no transpiration, the water potential of the leaf cells and the water potential in the xylem will come into equilibrium. The balancing pressure measured on such a non-transpiring shoot is equal in magnitude but opposite in sign to the pressure in the xylem (Ψ_p). Because the water potential of our non-transpiring leaf is equal to the water potential of the xylem, one can calculate the water potential of the leaf by adding together Ψ_p and Ψ_s of the xylem, provided one collects a sample of xylem sap for determination of Ψ_s. Luckily Ψ_s of the xylem is usually small (>–0.1 MPa) compared to typical midday tensions in the xylem (Ψ_p of –1 to –2 MPa). Thus, correction for the Ψ_s of the xylem sap is frequently omitted. Because the pressure chamber method does not require delicate instrumentation or temperature control, it has been used extensively also under field conditions [25] and modified by investigators in relation to the aim of the study. For example Knob et al. [26] enclosed the shoot or root system in a vacuum chamber with the proximal end protruding and supplied with perfusing solution. Flow through the xylem was induced by chamber vacuum. Hydraulic conductance was determined from the slope of the flow rate versus pressure relationship. Xylem embolism was quantified from the increase in hydraulic conductance following high pressure (100 kPa) perfusion of solution through the plant. The authors provided examples of the application of the method to cavitation studies in the cold desert shrub *Artemisia tridentata.*

The pressure probe device was developed by Ernest Steudle, Ulrich Zimmermann, and their colleagues in Germany [27] for measurements of hydrostatic pressure Ψ_p in the cells of higher plants. This instrument is similar to a miniature syringe and consists of a glass microcapillary tube which is pulled to a fine point and inserted into a cell. The microcapillary is filled with silicone oil, a relatively incompressible fluid that can be readily distinguished from cell sap under a microscope. When the tip of the microcapillary is first inserted into the cell, cell sap begins to flow into the capillary because of the initial low pressure of that region. Investigators can observe such movement of sap under the microscope and counteract it by pushing on the plunger of the device, thus building up a pressure. In such fashion the boundary between the oil and the cell sap can be pushed back to the tip of the microcapillary. When the boundary is returned to the tip and is held in a constant position, the initial volume of the cell is restored and the pressure inside the cell is exactly balanced by the pressure in the capillary. This pressure is measured by a pressure sensor in the device. Thus the hydrostatic pressure of individual cells may be measured directly. This method has been used to measure Ψ_p and other parameters of water relations in cells of both excised and intact tissues of a variety of plant species [28]. The pressure probe has also been adapted to measure positive and negative values of Ψ_p in the xylem [29]. However, technical problems with cavitation limit the measurement of negative Ψ_p by this technique.

Psychrometry (gr. *psychein* means to cool) is based on the fact that the vapor pressure of water is lowered as its water potential is reduced. Psychrometers measure the water vapor pressure of a solution or plant sample, on the basis of the principle that evaporation of water from a surface cools the surface. In one of psychrometric techniques, known as *isopiestic psychrometry,* measurement is made by placing a piece of tissue sealed inside a small chamber with a temperature sensor in contact with a small droplet of a standard solution of known solute concentration (known Ψ_s and thus known Ψ_w). If the tissue has a lower water potential than that of the droplet, water evaporates from the droplet, diffuses through the air, and is absorbed by the tissue. This slight evaporation of water cools the drop. The larger the difference in water potential between the tissue and the droplet, the higher the rate of water transfer and hence the cooler the droplet. If the standard solution has a lower water potential than that of the sample to be measured, water will diffuse from the tissue to the droplet, causing warming of the droplet. Measuring the change in temperature of the droplet for several solutions of known Ψ_w makes it possible to calculate the water potential of a solution for which the net movement of water between the droplet and the tissue would be zero signifying that the droplet and the tissue have the same water potential. Psychrometers can be used to measure the water potentials of both excised and intact plant tissue and the Ψ_s of solutions. There are many variations in psychrometric technique [3].

Ultrasonic acoustic emission (UAE) is a non- to minimally-invasive tool to measure xylem embolism in plant organs. Its is based on the fact that when a conduit embolizes, a small amount of energy is released. This energy is detectable with ultrasonic sensors. A close correspondence between UAE and percent loss of conductivity in stems and leaves of many species has been demonstrated. UAE can be used *in situ* to assess embolism in plants growing in the field or greenhouse or can be used for generating hydraulic vulnerability curves of plant organs on the lab bench. The UAE methods can also be used for the construction of hydraulic vulnerability curves of wood or woody organs (e.g., branches). Samples are attached to UAE sensors and are intermittently weighed while collecting UAE data. Measurements of relative water content in similar samples as they dry, along with water potential using screen-cage psychrometer. Once the relationship between relative water content and water potential has been established, the weights of the UAE samples can be converted into water potentials for the construction of the vulnerability curve. Acoustic emissions have been reported in the literature as proxies for embolism events [30, 31].

Nuclear magnetic resonance (NMR) imaging (or MRI – magnetic resonance imaging) proofed non-invasive, attractive method for providing spatially and temporally resolved quantitative information on water transport at different length scales (over membranes, cell to cell, and long distance) in intact plants. This technique has been applied for visually assessing embolized against functional xylem vessels [19,20].

The potential of synchrotron X-ray light source imaging was recently highlighted in studies monitoring the status of single vessels while refilling after embolisation. Kim and Lee [22] applied 2D X-ray imaging technique for *in vivo* monitoring sap dynamics in xylem conduits in rice leaves. Brodersen et al. [23] used the 3D High Resolution Computed Tomography (HRCT) for *in vivo* monitoring of embolism repair. This technique allowed visualization of water droplets exiting from vasicentric cells and entering embolized vessels of *Vitis vinifera.* The droplets expansion over time and refilling of vessels while forcing the dissolution of entrapped gases could be observed with HRCT. This method provides access to live plant tissue at a spatial and temporal resolution. Unlike NMR imaging, cryo-SEM, acoustic emissions, and other methods used to measure embolism spread and repair, HRCT provides quantitative data for individual vessels paired with a 3D visualization of refilling mechanism.

4. Interaction between plant hydraulics and water stress

4.1. Structural adaptations

The vascular plants as the elements of Soil-Plant-Atmosphere Continuum (SPAC), grow on the edge between two environments, and are subject to water stress at irregular intervals. The knowledge of the mechanism how trees can adapt their water transport system to the specific conditions of the environment is not complete. However, it is known that when the safety of water transport is studied the biometric features of tracheary elements i.e. diameter, length, density) should be considered. Studies on tracheary elements at subcellular level are carried out as well.

Considering structural adaptation of wood to environmental settings Carlquist [32] proposed the concepts of vulnerability (V) and mesomorphy (M) indices. Vulnerability means the possibility of cavitation to occur in vessels. The V index is obtained by dividing the mean vessel diameter by the number of vessels per mm^2 transection. The low V value indicates a great "redundancy" of vessels, what means high capability of withstanding water stress or freezing. The mesomorphy index is derived by multiplying the V index by mean vessel element length. The conditions that favor better stem growth are expected to result in longer vessel elements. Analyzing the wood of florulas from southwestern Australia, Carlquist [32] found positive correlations between the V and M values and the environmental water supply in the plants studied. Mesophytes were found to have low V and high M while xerophytes possessed high V and low M values. It means that wood of plants growing in extreme habitats has high densities of very narrow vessels that are considered safe hydrosystems adapted to drought conditions [33]. On the other hand, wood of plants grown in mesomorphic environments has low densities of wide vessels which are more efficient water conductors according to Hagen-Poiseuille formula. The safety of water transport is markedly decreased as a result of increase in vessel diameter and vessel length. On the other hand, the efficientcy of water conduction is substantially increased with increasing vessel diameter and length. As was mentioned above, in ideal capillaries the water transport is proportional to the fourth power of the diameter [34]. It means that at a given pressure gradient, the relative volume of water moving *via* capillaries of diameters 1, 2 and 4 are 1, 16 and 256, respectively. To emphasize how safety is linked with efficiency of water transport, a narrow vessels tree as maple is compared with wide vessel one as oak. The first tree has diffuse porous wood with vessels diameter of about 75 μm and length of 30 cm while oak is a ring porous tree with mean early wood vessel diameter of about 300 μm and length of 10 m. Because conductivity is proportional to four power of tracheary elements diameter and taking into account vessel diameter and length, the damage done (e.g. air ambolism) to one early wood vessels in the oak is 256 x 33 = 8448 times more serious than in the maple. These numbers illustrate the contrasting needs of safety and efficiency. It means that tracheids in conifers and diffuse porous woods are conductive system of greater safety (resistance to spreading of air embolisms in the water columns) while ring porous woods with large early vessels are more efficient conductors but less resistant to cavitation. Additionally, in conifers, air embolisms if occur, can be localized within individual cells whereas in dicots vessel members it can spread from one vessel member into an entire vessel.

Another structural adaptation to eventual water stress is the ability to form groupings of vessels [35]. Grouping of vessels is defined as vessels in contact, but not merely close and is regarded as a way to provide alternate conduits whereby water can be carried in the same pathways in case one or several vessels in a group are inactivated by air embolisms. Vessels' grouping occurs to various degrees in different taxa. The degree of grouping is connected to the probability of seriousness of vessel failure by air embolisms because of either drought or frost and shows relationship to the tracheids presence in the wood. Generally in those families or genera that had tracheids in addition to the vessels, vessels are not grouped. In the families or genera with fiber-tracheids or libriform fibers in addition to vessels, the vessels form groupings to a lesser or greater extent. In the last case the genera with the larger vessel groupings seemed to be from drier localities. Species with very large vessels at the beginning of growth rings (e.g. ring porous species) tend to have little grouping in the earlywood vessels but more grouping in latewood vessels. This phenomenon is held to relate to enhanced safety of latewood vessels, since earlywood vessels have little safety and the latewood is thereby the wood portion where safety mechanisms are concentrated [36].

In studies concerning water transport of trees in respect to the risk of bubble formation due to water stress in SPAC system the geometry of intervessel pits should not be neglected. According to pit area hypothesis, during periods of water stress, the vessels with smaller diameter are supposed to diminish the risk of cavitation because of their smaller pit area per vessel [37]. The correlation between the total pit membrane area and the cavitation vulnerability would be explained by the increased chance on an exceptionally large pit membrane pore which is prone to air-seeding [38]. The studies concerning structure of the bordered pits have demonstrated that pit membranes are responsible for at least 50% of the hydraulic resistance in the wood [39]. Therefore changes in the thickness and pit porosity have to significant impact on the total hydraulic resistance in the plant. The longer and wider tracheary elements and more porous pit membrane, the lower is its resistance to water flow. As a consequence the hydraulic conductivity could be substantially increased by the little increasing the cross-sectional lumen area of the conduits and the bordered pits, but increased the conduits diameter greatly decreases the safety of water flow against the cavitations [12]. Drought induced cavitations propagate by air seeding at interconduit pit membrane [40]. On the other hand, frost induced cavitations take place when xylem sap freezes and dissolved gases form air bubbles in the wider tracheary elements capture larger bubbles in the ice that are more likely to trigger cavitation during thawing.

As was mentioned above, the size of tracheary elements is subjected to change both with the age of tree and as a result of tree reaction to environmental signals, water stress including. Tree response depends on its plasticity and ability to adapt. In the literature we can find information that in response to water deficits trees resistance changes at different levels (tissue, organ and individual) and the adjustment of trees hydraulic conductance in respect to drought could realize in three independent ways. It means that trees could increase or decrease their conductance or no changes will occur in water transport [41]. The increase of the hydraulic conductance in trees means that, on the level of wood structure, the size of the tracheary elements should increase too.

The differences in xylem hydraulic architecture reflects not only size or age-related trends but also differences in the way trees adapt to environment, and can provide information about plasticity of a species under changing environmental conditions. Tulik et al. [42] studied the hydraulic conductivity expressed both as radial weighed diameter (Figure 5) and as an index of the theoretical hydraulic conductivity in declining ash trees. It seems that more prone to the decline were trees producing smaller vessels over the course of their life, which in consequence have had negative effect on the efficiency of water transport.

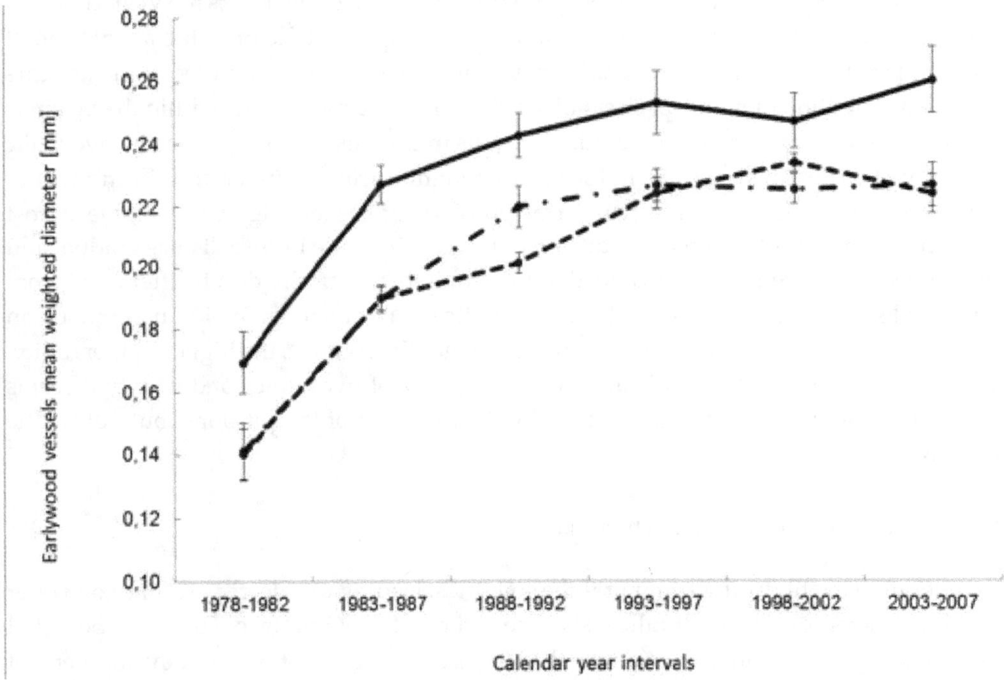

Figure 5. V ariation in earlywood vessels mean weighted diameters at breast height of European ash trees (*Fraxinus excelsior* L.), representing different health conditions at the time of sampling, over the calendar years 1978–2007. Calculations are based on at least twenty vessels diameter measurements within each annual ring and records averaged for three trees representing the group of healthy (solid line), weakened (dashed line) or dead trees (dotted-dashed line), respectively, and for 5-y time intervals. Vertical bars denote standard errors.

However, Levanič et al. [43] published somewhat contradictory results. The authors observed increased mortality of oak trees after the drainage events and concluded such results as an effect of the decrease of hydraulic conductivity by trees. Nonetheless water stress is mentioned as one among vital, possible factors responsible for the initiation and progress of the decline process of in trees.

4.2. Dysfunction due to seasonal embolism

Our understanding of the mechanism of sap ascent is based on the cohesion-tension theory ascribed to Dixon [44] which states that transpiration is the driving force pulling water from the soil to the leaves and that the water forms continuous column under negative pressure in the xylem. Under such conditions water is in a metastable state i.e. maintained in liquid phase, while below its vapour pressure. Any break of this column drains the conduit of water, disrupts further water flow and results in dysfunction of the vascular transport pathway. A break in water column is called cavitation and occurs when a gas filled void of sufficient radius forms in water under tension. Cavitation is then the abrupt change from liquid under tension to water vapour. Within hours water is withdrawn from the conduit, vapor expands to fill the entire lumen, then air diffuses and the conduit becomes embolized (i.e. air – blocked). When water moves through the SPAC under a negative pressure, cavitation often occurs in xylem conduits [30, 45, 17, 4] Vulnerability of xylem to cavitation is an important factor in the adaptation of plants to the [46]. Cavitation of xylem has been detected in stems, leaves and roots, and appears to limit the distribution of plant species [46]. Cavitation of water columns within the xylem is deleterious for plant water relations because it results in embolism thereby reducing hydraulic conductivity [47]. Embolism can be induced by drought during summer and during winter when freeze-thaw events occur. Temperate ring porous species seem highly vulnerable to frost induced embolism [48, 49] whereas for diffuse porous the increase in embolism is gradual and requires repeated freeze-thawing events [50, 16]. Conifers practically don't suffer at all from winter embolism. These studies as well as systematic data of the degree of winter embolism [51] strongly support the traditional view that the larger the conduit, the higher vulnerability. However in case of summer embolism and resulting loss of hydraulic conductivity it seems that the vulnerability is directly determined by the diameter of the pit pores but not by the diameter of conduits.

4.3. Summer embolism and its mechanism

Trees undergo drought during summer if the soil dries, leading to a decrease of the soil water potential and increase of the hydraulic resistance at the soil-root interface. The water potential of leaves and xylem negative pressure will decrease also, promoting the development of embolism and loss of hydraulic conductivity. Reduction in xylem hydraulic conductivity can impair photosynthesis and productivity as the consequence of lower rates of carbon fixation due to stomata closure, which in turn prevents further cavitation and desiccation of leaf tissue. Under prolonged drought conditions day-by-day accumulation of embolism would finally lead to failure of xylem water transport and plant death. Thus the mechanisms to limit or avoid cavitation seem of great importance in respect to plant survival in changing environment conditions.

Among potential mechanisms of drought induced xylem cavitation the air-seeding hypothesis is favorised by experimental evidence [52] whereas in case of winter embolism the frost-thaw explanation applies [53]. The air-seeding hypothesis states that cavitation occurs when air ouside (in atmosphere or adjacent air-filled conduit) a water-filled conduit is aspired into the element through pores of the pits in the walls. Xylem sap is typically under negative pressure and hence in a metastable liquid state. It is vulnerable to cavitation, which is the abrupt phase change from metastable liquid to vapor. Cavitation is triggered by a nucleating agent, which in plants appears to be gas bubbles over a critical size. Cavitation creates a vapor bubble which relaxes the surrounding negative sap pressure to atmospheric pressure. As long as the sap in adjacent conduits remains under negative pressure, water is pulled from the cavitated vessel and the vapor bubble expands to fill the cavitated conduit. The gas phase would expand beyond the conduit if it were no inter-connecting pits which arrest the gas-water meniscus and seal off the vapor-filled conduit. As air diffuses into the vapor from surrounding tissue, the gas pressure rises to atmospheric and the conduit is fully "embolized".

4.4. Restoration of hydraulic conductivity after cavitaion

Thinking about the mechanisms by which plants can restore hydraulic conductivity after cavitation at least two different possibilities emerge. First is to produce new xylem conduits and any plant that possesses the capacity for secondary growth can simply produce new xylem to replace the cavitated conduits. This mechanism, requiring carbon investment for growth processes, common for maintaining hydraulic conductivity in trees and shrubs, is not available to plants lacking secondary growth. Second possibility is embolism reversal through refililing with water gas-filled conduits. There is also third strategy i.e. embolism avoidance, requiring a tight control of xylem pressure, which will be discussed in the section 5. Plants adopt one or more of these strategies, which are not alternative to one another. In woody plants there is evidence for all three strategies to occur, ensuring the conductivity restoration at different time and plant body organization scale.

The disappearance of embolism means that the air should dissolve into the surrounding sap. According to the model of hydraulic conductivity recovery [54] a prerequisite for this is a positive or close to positive Ψ persisting for adequate time period. This raises the questions on how much pressure is needed to force gas into solution? for how long must this pressure operate? and how plants can generate positive pressure? For air to dissolve from a buble into liquid sap, the gas in the bubble has to be at a pressure in excess to atmospheric. Only if the difference between the gas pressure in the bubble and the pressure in the liquid surrounding the bubble is lower than the capillary pressure (resulting from surface tension), the gas will dissolve. The minimum xylem pressure (P_x) that must be exceeded for reversal of embolism to occur (P_{xr}) can be calculated on the basis of the Young–Laplace equation:

$$P_{xr} = P_{gas} - \left(\frac{2T}{r}\right) \quad (6)$$

where P_{gas} equals the sum of the partial pressure of gases filling the embolized conduit, T is the surface tension of water and r is the radius of the bubble (which can be roughly assumed to equal that of the conduit). Hence, the pressure required for spontaneous embolism dissolution is predicted to be higher for large conduits than for narrow ones. However, in contrast to the theory, intact plants can reverse embolism even when P_x in still functioning conduits is lower than P_{xr}.

The pressure required for refilling depends upon the composition of the gas contained within the embolized conduit. Immediately following a cavitation event, the embolus will mainly consist of water vapor, which has an absolute vapor pressure of 2.3 kPa at 20°C (which is –99 kPa relative to atmospheric pressure). At this moment, the pressure need only exceed this value to force these molecules back into solution. However, over time, cavitated conduits become air-filled as dissolved gases in the surrounding liquid phase come out of solution. Forcing an air bubble back into solution requires pressure greater than 100 kPa (1 atm).

Water vapor dissolves easily and refilling of cavitated conduits containing only water vapor should occur almost instantaneously once the vapor pressure of water is exceeded. In contrast, forcing air into solution is more difficult as the dissolving gas molecules locally increase their concentration in the liquid surrounding the embolus. The movement of more gas into solution is held in check by the rate at which the newly dissolved gases diffuse away from the gas–liquid boundary. Thus, the time needed to force an air-embolus into solution depends upon both the magnitude of the applied pressure and the diffusional limitations imposed by the surrounding environment [54].

It seems that plants may generate the positive pressures necessary for conduit refilling within the roots, throughout the stem, or locally, within the cavitated conduit. The mechanisms by which this occurs are fairly well understood in the case of root pressure but rather poorly understood in case of stems and local pressurization.

4.5. Generation of positive pressure in respect to refilling of cavitated conduits

The raise of the hydrostatic pressure throughout plant vascular system by actively loading solutes into root stele has been demonstrated in many species. Osmotically achieved root pressures of several hundred kPa were able to force emboli into solution or to push the air out of hydathodes or open vessel ends [55, 56]. The occurrence of this process seems to be limited to the conditions when the soil is saturated and transpiration is low and is much better documented in herbs and vines [56, 57, 58, 59] than for trees [61, 105]. Diurnal refilling owing to root pressure has been demonstrated in some crop species [56, 59], but it is uncertain whether trees employ it for daily recovery of hydraulic conductivity.

Drawing of water into stems has been demonstrated in some temperate plant species when temperatures dropped near or below freezing. Upon thawing, the positive pressure built up by this influx of water refilled embolism induced over the winter by both freezing and dehydration [16, 60]. The precise mechanism for this is not fully understood and the process occurs only in spring before leaf flush when transpiration is minimal and the soil saturated. The pressures generated throughout the stem, can co-occur with some degree of root pressure [16, 61].

Local refilling, relays on pressurization that occurs solely within an embolized conduit, and eventually could allow plants to recover conductivity during periods of active transpiration and therefore adjust to changing conditions throughout the day. This possibility is the most ecologically interesting mechanism, and there is convincing experimental evidence for the occurrence of local pressurization in woody plants.

4.6. The evidence and possible mechanism for conduit local refilling

There is growing evidence that once vessels or tracheids become embolized, at least some tree species are able to restore them to function by refilling with sap. This process has been documented with regards to reversal of winter-time embolism through spring root pressures [16, 58, 62]. Failure of this refilling can lead to crown dieback. However for a long time it was assumed that there is no recovery of embolism during drought. More recently, it was demonstrated that the refilling occurs under conditions of soil drought and in the absence of any positive bulk xylem pressure [59, 63, 64].

Hacke and Sperry [63] tested the ability of juvenile laurel (*Laurus nobilis*) and maple ash (*Acer negundo*) plants to refill embolized xylem vessels under conditions of soil drought when xylem sap pressure was substantially negative, thus seemingly violating the expected condition that pressure must rise to near atmospheric for refilling. Intact potted plants were dried to a stem water potential corresponding with approximately 80% loss of hydraulic conductivity (PLC) in shoots. Then plants were re-watered and kept at a less negative target water potential for 1–48 h. The water potential was measured continuously with stem psychrometers. Rewatered laurel held at the target water potential for 1 h showed no evidence for refilling unless water potential was within a few tenths of a MPa of zero. In contrast, re-watered laurel held for 24 and 48 h at water potentials well below zero showed a significant reduction in PLC. The recovery was highly variable, complete in some stem segments, and scarcely evident in others. Embolism repair was accompanied by a significant but moderate decrease in the osmotic potential of the bulk xylem to -67 kPa in recovering plants versus -31 kPa in controls. In contrast, embolized and re-watered maple ash plants held for 24 h at target water potential of -0 9 and -0 3 MPa showed no embolism reversal. Although for maple ash no evidence was found of novel refilling following relief of soil drought under the negative pressures tested, this species may show seasonal refilling in association with near-atmospheric or positive xylem pressures as do other members of the genus [16, 65]. Evidence for local refilling was also demonstrated for ring-porous oak (*Quercus gambelii* Nutt.) in contrast to the diffuse-porous maple (*Acer grandidentatum* Nutt.), which branches showed virtually no cavitation under investigated conditions [64]. Large earlywood vessels of 2–3-year-old oak stems cavitated extensively on a daily basis, resulting in a more than 80% reduction in hydraulic conductivity. Concurrently, oak vessels refilled on a daily basis, despite negative xylem pressure in the transpiration stream, indicating active pressurization of embolized vessels.

The precise mechanism allowing plants to refill embolised conduits under negative pressure is still not fully deciphered. The up to date known hypotheses fall into one of two categories: requiring that the embolised conduit is isolated from the transpiration stream during the refilling process and then reconnected once refilling has occurred or lacking such a requirement. There is also much more data available on the refilling in dicots than in conifers and the below presented experimental evidences concern vessels refilling first of all.

Holbrook and Zwieniecki [66] have proposed the 'pit valve' hypothesis, which requires the conduit to be isolated from the transpiration stream for local refilling. The xylem parenchyma cells adjacent to an embolized vessel secrete solutes that create an osmotic gradient for water movement into the embolized conduit. Among kinds of secreted solutes taken into consideration are inorganic ions [46], sugars [67, 68, 69] proteins [70] and polysaccharides [46]. The water is prevented from draining into the neighbouring transpiration stream by persistent air pockets trapped by capillary forces in the pit chambers between the embolized and functional conduits. The pits act as valves allowing a build up of positive Ψ_{PX} in the embolized vessel. In this way, the Ψ_{PX} within the embolized conduit rises above the -2T/r limit and dissolves the gas. The valves have been shown to work in six species [71], and metabolic poisons apparently reduce the refilling activity [72]. However, for the process to be completed, the air pockets in the pit chambers must all dissolve simultaneously to re-connect the refilled vessel at positive Ψ_{PX} with the functional vessels at negative Ψ_{PX}. If an air pocket in one pit chamber dissolved before the rest, the remaining air bubbles would expand to reembolize the vessel. After some mathematical models based upon geometry and surface chemistry of bordered pits [73] at least theoretically such coordination could be achieved but how this might occur in nature remains unclear. Recently Kim and Lee [22] applied synchrotron 2D X-ray imaging technique for *in vivo* monitoring sap dynamics in xylem conduits in rice leaves and demonstrated that perforation plates of rice leaf xylem offers a substantial resistance, requiring a threshold water pressure to be exceeded before the sap can move from a refilling conduit to the next one still embolized. This observation is in accordance with the idea of the hydraulic isolation of embolized conduits [66, 71].

Other authors postulated that the driving force for water movement into hydraulically isolated conduits is an increase in tissue pressure in the cortex [7, 67, 74], which induces centripetal water flow refilling embolized conduits. An input of solutes into the refilling conduit is not predicted by this hypothesis called also the reverse osmotic hypothesis [75, 76]. According to the proposed mechanism, vessel associated cells would lower their own osmotic potential through starch-to-sugar hydrolysis. This is supposed to cause water to flow into vessel associated cells thus raising their turgor pressure. This would generate a 'tissue pressure' that would be contained within the stem by tissues external to vessel associated cell. Tissue pressure would squeeze water from other living cells and squeezed water would flow under positive pressure to the embolized conduits and refill them. A number of objections have been raised against this hypothesis, the main being that parenchyma cells would have to take up water through mass flow to raise and maintain cell turgor and at the same time deliver water through mass flow to refill conduits [46, 77].

The 'pit membrane osmosis' hypothesis [63] assumes the refilling vessel remains hydraulically connected to the transpiration stream. The living xylem contact cells are to release solutes into a cavitated vessel, thereby locally lowering Ψ_p. The solutes are to be large enough to be held back in the refilling vessel by interconduit pit membranes, which therefore act as osmotic membranes. The high solute concentration in the refilling conduit attracts water both from parenchyma cells and from the transpiration stream, generating a positive Ψ_{PX} in the embolized vessel analogous to turgor pressure in a living cell.

Convincing evidence for the role of living cells as sources of water needed for refilling was demonstrated using the potential of 3D High Resolution Computed Tomography (HRCT) for *in vivo* monitoring of embolism repair [23]. Entering of water droplets derived from vasicentric cells into embolized vessels of *Vitis vinifera,* their expansion over time and refill of vessels while forcing the dissolution of entrapped gases, was visualized using this technique. The typical refilling time was further characterized and calculated to range between about 2 and 17 h for vessels with a diameter of 20 and 150 μm, respectively. Moreover, embolism reversal was observed at substantially negative stem water potentials ranging between −0.45 and −0.75 MPa. Droplets entered from many entry points oriented from the rays, as predicted by a refilling mechanism driven by solute pumping into vessels trough pits from adjacent parenchyma [63, 78, 79]. The authors presented an integrated model for vessel refilling that incorporates the phloem-loading and solute-driven refilling mechanisms already proposed [46, 63, 66, 69, 78, 80, 81] with amendment for successful and unsuccessful refilling.

The knowledge of embolism repair occurring while the bulk of xylem is under tension, has significantly improved during the last decade and some of the still obscure facts regarding embolism refilling have been highlighted in successive review papers e.g. [81, 82, 83]. As summarized by Nardini et al. [83] embolism repair under tension is a physiological process that requires fine coordination of (a) biochemical and morphological features of interconduit pits and perforation plates to assure hydraulic isolation of refilling conduits, (b) changes in vessel associated cells sugar metabolism to provide the necessary driving force for delivering water into embolized conduits, and (c) source-to-sink solute and water transport from phloem to the refilling conduit.

4.7. Vulnerability to embolism

Specialization during trees evolution has led to increase the dimensions of the tracheary elements. Increased diameter of tracheary elements increases the efficiency of water transport. However, on the other hand it decreases the safety of water transport, because wider conducts seem to be more vulnerable to embolism [84]. It means that trees living in unfavorable conditions of environment have usually to take a risk: the water transport lees efficient but safer or contrary, transport of water much more efficient but saddled with risk.

A plot of xylem tension causing 50% loss of hydraulic conductivity Ψ_{50} versus mean diameter D_{95} of vessels that account for 95% of the hydraulic conductance compiled for sixty species [85] demonstrates only weak correlation, as the regression accounts for only 22% of the variation. It is then not possible to predict the vulnerability to summer embolism of a species by measuring the mean conduit diameter. There is also evidence that vulnerability among trees from one genus [48] can be about as large as between diverse species of angiosperms for example. There is, however, a correlation between vulnerability curves and the drought tolerance between trees growing on sites with different water supply: the arid-sites species like *Quercus ilex* and *Q. suber* are less vulnerable than mesic-sites ones like *Q. robur* and *Q. petraea*. Seasonal change in percentage loss of hydraulic conductivity due to embolism in petioles and twigs of three European oak species shows also that several months of drought were necessary to induce a significant degree of embolism [86].

Several studies have suggested that resistance to drought-induced embolism is proportional to the ability of pits to prevent air-seeding [87, 88]. Carlquist [89] suggested that the structure of interconduit pits is a compromise between maintaining the wall strength and resistance to air entry of the inter-conduit wall without sacryfing hydraulic conductivity of the pit. Thus inter-conduit pits act, as check valves permitting water flow between conduits but inhibiting leaking of air into the transpiration stream [44].The great variety of pit shape, size, membrane structure, border configuration, vesture presence and pit-field pattern is found in vascular plants and presumably reflect functional significance of pits. However to meet the requirements of all functions seems not possible without some trade-offs.

In the terms of the type of valve action inter-conduit pits can be rank among capillary sealing, torus-margo and vestured pits. In all cases the nonlignified pit membrane (compound primary wall) spans the pits chamber, which opens to the lumen of each conduit by an aperture. Schematic presentations of most common in dicots capillary sealing inter-conduits pits and typical for conifers torus-margo pits are on Figure 6.

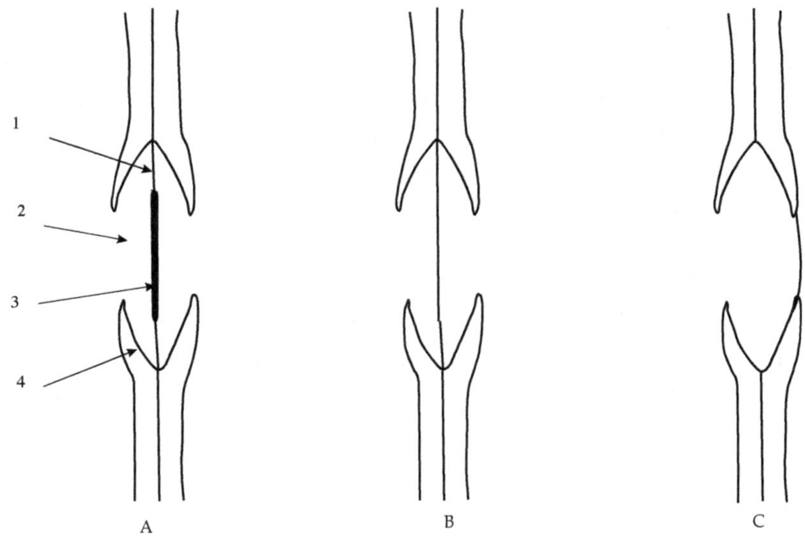

Figure 6. Schematic drawings of bordered pits in tangential view. Simplified structure of: A –bordered pit in a coniferous tree with torus in the central position allowing the water transport between tracheids, B – bordered pit in dicotyledonous tree with no torus but with uniformly small pores (not shown) in the pit membrane, C – bordered torus-margo pit when the torus closes the pit aperture. Pit elements: 1 - margo strands, 2 - pit aperture, 3 – torus, 4 – border.

Capillary sealing pits are common in angiosperms. They have uniformly small pores in their pit membranes with high air-seeding pressure, which minimizes the risk of air entry [90]. Small pores in the membrane possess however, low hydraulic conductivity. It is likely that prior to air-seeding the membrane is aspirated and distorted as the air-seeding pressure of these pores is enough to deflect the membrane to the pit border. The border of the pit supports the wall and the pit membrane against the bending force between air- and water-filled conduits. Thus pits with small or absent borders presumably indicate weak and cavitation-prone conduits but with high conductivity. Two possible ways the air seeding occurs are pointed out: irreversible plastic yielding of the membrane (creep) and reversible elastic stretching. The repeated events of membrane creep is the cause of "cavitation fatigue", which renders the conduit more prone to subsequent cavitation [91]. The investigations carried out on 27 angiosperms species with circular bordered pits with homogenous pit membranes [92] gave the evidence for the assumption that the pit chamber itself does not weaken the wall [89] but pit aperture does. The strengthening of pitted walls is achieved by thickening of the wall but in respect to the thickness-to-span ratio which translates into higher wood density. More cavitation–resistant woods turn denser i.e. more expensive to grow than cavitations susceptible woods. It is in line with earlier observations that vessels from plants of arid habitats tend to have thicker walls [89]. More air-seed resistant pits required denser pit membranes with narrower pores, resulting in lower pit conductivity. Pores in that cases were numerous nonetheless because of nonlinear pore conductivity vs. pore diameter relationship it does not compensate smaller size.

Torus-margo pits have emerged in conifers in Jura, persist in many extant conifers, *Ginkgo*, *Ephedra* and their convergent evolution occurred in certain angiosperms. The central part of the pit membrane – torus is here substantially thicker and surrounded by a very porous margo. The air seal formation occurs through the aspiration of the torus over the pit aperture driven by capillary forces at the margo pores. Air seeding occurs eventually when the torus edge is pulled through the pit aperture [93]. Much of the membrane in this type of pits is nonconductive torus, but pores in surrounding margo can be relatively large ensuring equal or even higher total conductivity than a capillary sealing pit with more small pores.

Delzon et al. [94] examined the relationship between cavitation resistance and bordered pit structure and function in 40 coniferous species. Xylem pressure inducing 50% loss of hydraulic conductance varied widely among species, from -2.9 to -11.3 MPa. The valve effect of the pit membrane, measured as a function of margo flexibility and torus overlap, explained more variation in cavitation resistance than simple anatomical traits such as pit membrane, pit aperture or torus size. Highly cavitation resistant species exhibited both a high flexibility of the margo and a large overlap between the torus and the pit aperture, allowing the torus to tightly seal the pit aperture. These results support the hypothesis of seal capillary-seeding as the most likely mode of air-seeding, and suggest that the adhesion of the torus to the pit border may be the main determinant of cavitation resistance in conifers.

Occurrence of vestured pit is limited to some gnetophytes and angiosperms [95, 96]. Vesturing most commonly appears like small wartlike lobes within pit cavities but may also be present on pitting-free parts of vessel walls. Vestures are lignified and compressively strong. Thus, after Zweypfenning [97] vestures could function as mechanism that prevents or minimizes the deflection and the risk of rupture of the pit membrane, in case of pressure drop between adjacent vessel elements. The cavitation fatigue of the pit membrane would be minimized as well. The air seeding occurs trough membrane pores and there is no distortion of the membrane prior to air-seeding. However, vestured pits are not universal in dicotyledons nor of dry areas either cold areas and the benefit of vestures may only apply to a limited range of air seeding pressures [98].

Important feature of the xylem structure is its connectivity i.e. the fact that conduits are interconnected to form a network [53, 12]. Investigations on the spatial arrangement of conduits [99, 100] gave the support to define vascular system as a network integrating all main parts of the plant's body, i.e. roots, branches and leaves. According to the model presented by Loepfe et al. [101] the maximum hydraulic conductivity and vulnerability to embolism increase with the connectivity of the xylem network. This can be explained by the fact that connectivity determines the fraction of all the potential paths or conduits actually available for both water transport and spread of embolisms. In the xylem, connectivity corresponds to the average number of different neighbor conduits to which a conduit is connected. According to the air seeding hypothesis, embolism propagates from an air-filled conduit to a functional one through the porous membrane that connects them. Thus the first condition for a conduit to be embolised is that it is connected to an air-filled conduit. The conduit will be more vulnerable to embolism the more connections it has, since more connected conduits will be more likely to be connected to an already air-filled conduit. At the tissue level high connectivity would facilitate the spread of emboli and therefore increase the vulnerability to drought induced embolism. An increase in conduit length, pit pore size or fraction of contact area occupied by pores resulted in an increment of maximal hydraulic conductivity in the presented model. The effect of conduit length on conductivity was not only through changing the number of pit membranes that water has to cross but also through a modification of the overall connectivity of the system. Earlier works also contributed to a prediction that if the gain of increased diameter of the conduit has to be fully realized in terms of conductivity it length should increase as well [102, 92]. However, there is a concept of 'saturating vessel length' above which there is no further increase in vessel conductivity and vessels longer than the saturating length contribute nothing to hydraulic conductivity being no mere that ways of embolism and disease spread [103, 92].

5. Short and long-term adjustments to water stress at different levels of woody plant organization

The continuity of water columns from soil throughout the plant linked to the evaporative flux at leaf cells is described as SPAC – soil-plant-atmosphere continuum. Maintenance of water column continuity is crucial to ensure water supply to leaves. Thus there is strong association between water transport capacity and the carbon gain by leaves which is due to the need of displaying large surface of hydrated cells in desiccating environment i.e. air. Any shortage in water supply in relation to requirements of the leaves results in water deficit and plant stress. Thus it is likely that leaf –level homeostasis in water status is maintained through ontogeny in conditions of varied and changing soil and atmosphere, possibly involving whole-plant changes in the resistance to liquid water flow [41]. In homeohydric plants i.e plants that can keep their internal water content relatively constant, regardless of the external environment, stomata exist to regulate plant water status by adjusting transpiration rate. Homehydric plants (i.e. ferns, gymnosperms, angiosperms) are further classified as either isohydric which have tight stomatal control and anisohydric with control less strict. In the initial stages of drought or for short-term drought conditions the response is always at the leaf level but differs between species between two theoretical extremes concerning perfectly isohydric plants that will close stomata reducing transpiration and simultaneously carbon gain to maintain the predrought leaf water status and perfect anisohydric plants that would keep stomata more open reducing leaf potential just enough to maintain leaf transpiration. An excellent exemplification of this differential behavior are the results of Tandea and Sperry [64] work on cavitation resistance, vessel refilling, transport capacity and water status in ring-porous oak (*Quercus gambelii* Nutt.) and diffuse-porous maple (*Acer grandidentatum* Nutt.) that co-dominated summer-dry foothills in the western Rocky Mountains of the USA. It has been demonstrated that the large earlywood vessels of ring-porous trees can be extraordinarily vulnerable to cavitation making it necessary that these trees maintain a consistent and favorable water status. Native embolism measurements, dye perfusions and balance pressure exudation patterns indicated that the large earlywood vessels of 2–3-year-old oak stems cavitated extensively on a daily basis as predicted from laboratory vulnerability curves, resulting in a more than 80% reduction in hydraulic conductivity. Maple branches showed virtually no cavitation. Oak vessels refilled on a daily basis, despite negative xylem pressure in the transpiration stream, indicating active pressurization of embolized vessels. Conductivity and whole-tree water use in oak were between about one-half and two-thirds that in maple on a stem-area basis; but were similar or greater on a leaf-area basis. Oak maintained steady and modest negative xylem pressure potentials during the growing season despite little rainfall, indicating isohydric water status and reliance on deep soil water. Maple was markedly anisohydric and developed more negative pressure potentials during drought, suggesting use of shallower soilwater. Although ring porosity may have evolved as a mechanism for coping with winter freezing, this study suggests that it also has major consequences for xylem function during the growing season.

Common short-term respose to drouht is reduction in hydraulic conductivity of both xylary and extra-xylary pathways of water movement [104]. This reduction occurs not only as the consequence of xylem cavitation development but includes also interrupton of water flow at the soil-root interface, suberinization of root epidermal cells and down-regulation of aquaporin expression.

According to the model presented by Maseda and Fernández [41] the response to prolonged drought depends mainly on the type of leaf-level response and will include reduction of leaf area in anisohydric plant whereas isohydric ones should be able to keep it only little reduced. In the time scale close to an individual life span the adjustment to prevailing soil and atmosphere conditions are expected through developmental plasticity.

6. Conclusions

Facing a potentially lethal loss of hydraulic conductivity while draught periods vascular plants have developed structural and physiological mechanisms some of which can be categorized as embolism avoidance, embolism reversal and embolised conduits replacement through production of new xylem. Embolism avoidance is likely the first to act at the initial stages of draught; it requires a tight control of xylem pressure which plants can execute through stomatal control. Embolism reversal relays on the refilling with water of gas-filled conduits once cavitation event occurred. It can be driven by over atmospheric root/stem pressures originating in specific conditions at define period of the year or by local pressurization, while the neighboring conduits are under tension. There is evidence for daily cycles of cavitation and refilling which is of great ecological importance in respect to photosynthetic capacity restoration. Production of new xylem conduits in aim to replace cavitated ones, requires carbon investment for growth processes, is common for maintaining hydraulic conductivity in plant that possesses the capacity for secondary growth e.g. trees and shrubs. Plants adopt one or more of these strategies, which are not alternative to one another. In woody plants there is evidence for all three strategies to occur, ensuring the conductivity restoration at different time and plant body organization scale. A number of ecophysiological features will determine whether a tree species can capture and maintain site dominance. Xylem anatomy and chemistry is among these strategic features and the compromises needed for mechanical support, hydraulic conduction, light interception and gas exchange are elements of life strategy. Experimental observations and theoretical modeling revealed that xylem embolism repair is a physiological process requiring fine coordination of sugar metabolism, solute and water transport and biochemical and morphological traits of conduits. In the light of these findings the image of xylem as an inert, vulnerable network of nonliving, pipe-like cells only passively transporting water by tension driving force seem inadequate. Xylem should rather be looked as a finely regulated water transport system tightly depending on plant life processes for optimal functioning and maintenance, especially under environmental stress conditions.

Acknowledgements

This publication was supported by the Polish National Science Centre through grant Number NN 309 1086 40.

Author details

Katarzyna Marciszewska* and Mirela Tulik

*Address all correspondence to: katarzyna.marciszewska@wl.sggw.pl

Department of Forest Botany, Warsaw University of Life Sciences – WULS-SGGW, Warsaw, Poland

References

[1] Zimmermann MH. Hydraulic architecture of some diffuse-porous trees. Canadian Journal of Botany 1978;56 2286-2295.

[2] Dixon HH, Joly J. On the ascent of sap. Philosophical Transactions of the Royal Society London 1894; Series B 186 563-76.

[3] Kramer PJ, Boyer JS. Water relations of plants and soil. San Diego: Academic Press; 1995.

[4] Tyree MT. The cohesion-tension theory of sap ascent: Current controversies. Journal of Experimental Botany 1997;48 1753-1765.

[5] Steudle E. The cohesion-tension mechanism and the acquisition of water by plant roots. Annual Reviews 2001;52 47-875.

[6] Zimmermann U, Schneider H, Wegner LH, Haase A. Transley review. Water ascent in tall trees: does evolution of land plants rely on a highly metastable state? New Phytologist 2004;162 575-615.

[7] Canny M J. A new theory for the ascent of sap – cohesion supported by tissue pressure. Annals of Botany 1995;75 343-357.

[8] Plumb RC, Bridgman WB. Ascent of sap in trees. Science 1972;179 1129-1131.

[9] Zimmermann D, Westhoff M, Zimmermann G, Geßner P, Gessner A, Wegner LH, Rokitta M, Ache P, Schneider H, Vásquez JA, Kruck W, Shirley S, Jakob P, Hedrich R, Bentrup FW, Bamberg E, Zimmermann U. Foliar water supply of tall trees: evidence for mucilage-facilitated moisture uptake from the atmosphere and the impact on pressure bomb measurements. Protoplasma 2007;232 11-34.

[10] Tyree MT, Sperry JS. Vulnerability of xylem to cavitation and embolism. Annual Review of Plant Physiology and Molecular Biology 1989;40 19-38.

[11] Tulik M, Marciszewska K. Plant and soil as hydraulic systems. In: Lakshmanan E. (ed.) Hydraulic conductivity. Issues, determination and applications. Rijeka: InTech; 2011. p. 85-102.

[12] Tyree MT, Zimmermann MH. Xylem structure and the ascent of sap. 2nd ed. Berlin: Springer-Verlag; 2002.

[13] Fisher JB, Ewers FW. Vessel dimension in Liana and tree species *Gnetum* (Gnetales). American Journal of Botany 1995;82 1350-1357.

[14] Zajączkowski S. Changes in cambial activity and secondary xylem structure of *Pinus sylvestris* L. associated with aging. In: Paschalis P, Zajączkowski Z, (eds.) Biodiversity Protection of Bialowieza Primeval Forests. Warsaw: Fundacja Rozwój SGGW; 1996. p. 8-21.

[15] Sperry JS, Donnelly JR, Tyree MT. A method for measuring hydraulic conductivity and embolism in xylem. Plant Cell & Environment 1988a;11 35-40.

[16] Sperry JS, Donnelly JR, Tyree MT. Seasonal occurrence of xylem embolism in sugar maple *(Acer saccharum)*. American Journal of Botany 1988 b;75 1212-1218.

[17] Pockman WT, Sperry JS, O'Leary JW. Sustained and significant negative water pressure in xylem. Nature 1995;378 715-716.

[18] Cochard H, Bodet C, Amegilo T, Cruiziat P. Cryo-scaning electron microscopy observations of vessel content during transpiration in walnut petioles: Facts or artifacts? Plant Physiology 2000;124 1191-1202.

[19] Holbrook NM, Ahrens ET, Burns MJ, Zwieniecki MA. *In vivo* observation of cavitation and embolism repair using magnetic resonance imaging. Plant Physiology 2001;126 27-31.

[20] Kaufmann I, Schulze-Till T, Schneider HU, Zimmermann U, Jakob P, Wegner LH. Functional repair of embolized vessels in maize roots after temporal drought stress, as demonstrated by magnetic resonance imaging. New Phytologist 2009;184 245-256.

[21] Holbrook NM, Burns MJ, Field CB. Negative xylem pressures in plants: A test of the balancing pressure technique. Science 1995;270 1193-1194.

[22] Kim HK, Lee SJ. Synchrotron X-ray imaging for nondestructive monitoring of sap flow dynamics through xylem vessel elements in rice leaves. New Phytologist 2010;188 1085-1098.

[23] Brodersen CR, McElrone AJ, Choat B,, Matthews MA, Shackel KA. The dynamics of embolism repair in xylem: in vivo visualizations using High Resolution Computed Tomography. Plant Physiology 2010; 154 1088-1095.

[24] Scholander PF, Bradstreet ED, Hemmingsen EA, Hammel HT. Sap Pressure in Vascular Plants. Negative hydrostatic pressure can be measured in plants. Science 1965 Apr16:148:339-346. DOI: 10.1126/science.148.3668.339

[25] Tyree MT, Hammel HT. The measurement of the turgor pressure and the water relations of plants by the pressure-bomb technique. Journal of Experimental Botany 1972;23 267-282.

[26] Kolb KJ, Sperry JS, Lamont BB. A method for measuring xylem hydraulic conductance and embolism in entire root and shoot systems. Journal of Experimental Botany 1996;47(304) 1805-1810.

[27] Hüsken D, Steudle E, Zimmermann U. Pressure probe technique for measuring water relations in higher plants. Plant Physiology 1978; 61 158-63.

[28] Steudle E. Pressure probe techniques: basic principles and application to studies of water and solute relations at the cell, tissue and organ level. In: Smith JAC, Griffith H, (eds.) Water deficits: plant responses from cell to community. Oxford: Bios Scientific Publishers; 1993. p. 5-35.

[29] Heydt H, Steudle E. Measurement of negative root pressure in the xylem of excised roots. Effects on water and solute relations. Planta 1991;184 389-396.

[30] Tyree MT, Dixon MA. Cavitation events in *Thuja occidentalis* L.? Ultrasonic acoustic emissions from the sapwood can be measured. Plant Physiology 1983;72 1094-1099.

[31] Mayr S, Rosner S. Cavitation in dehydrating xylem of *Picea abies*: energy properties of ultrasonic emissions reflect tracheid dimensions. Tree Physiology 2011;31 59-67.

[32] Carlquist S. Ecological factors in wood evolution: a floristic approach. American Jornal of Botany 1977;64(7) 887-896.

[33] Baas P, Carlquist S. A comparison of the ecological wood anatomy of the floras of Southern California and Israel. IAWA Bulletin 1985;6(4) 349-353.

[34] Zimmermann MH. Xylem structure and ascent of sap. Berlin: Springer-Verlag; 1983.

[35] Carlquist S. Vessel grouping in dicotyledon wood: significance and relationship to imperforate tracheary elements. Aliso 1984;10(4) 505-525.

[36] Utsumi Y, Sano Y, Funada R, Fujikawa S, Ohtani J. The progression of cavitation in earlywood vessels of *Fraxinus mandshurica* var *japonica* during freezing and thawing. Plant Physiology 1999;121 897-904.

[37] Wheeler JK, Sperry J, Hacke UW, Hoang N. Inter-vessel pitting and cavitation in woody Rosaceae and other vesseled plants: a basis for safety versus efficiency tradeoff in xylem transport. Plant Cell & Environment 2005;28 800-812.

[38] Hargrave KR, Kolb KJ, Ewers FW, Davis SD. Conduit diameter and drought-induced embolism in *Salvia mellifera* Greene (Labiatae). New Phytologist 1994;126 695-705.

[39] Hacke UG., Sperry JS., Wheeler JK., Castro L. Scaling of angiosperm xylem structure with safety and efficiency. Tree Physiology 2006;26 689-701.

[40] Hacke UG., Sperry JS. Functional and ecological xylem anatomy. Perspectives in Plant Ecology. Evolution and Systematic 2001;4 97-115.

[41] Maseda PH, Fernández RJ. Stay wet or else: three ways in which plants can adjust hydraulically to their environment. Journal of Experimental Botany 2006;57(15) 3963-3977.

[42] Tulik M, Marciszewska K, Adamczyk J. Diminished vessel diameter as a possible factor in the decline of European ash (*Fraxinus excelsior* L.). Annals of Forest Science 2010;67 103-110.

[43] Levanič T, Čater M, McDowell NG. Associations between growth, wood anatomy, carbon isotope discrimination and mortality in a *Quercus robur* forest. Tree Physiology 2011;31 298-308.

[44] Dixon HH. Transpiration and the Ascent of Sap in Plants. London: Mac Millan; 1914.

[45] Oertli, JJ. Effect of cavitation on the status of water in plants. In: Borghetti M, Grace J Raschi A, (eds) Water Transport in Plants under Climatic Stress. Cambridge: Cambridge University Press; 1993. p. 27-40.

[46] Tyree MT, Sallo S, Nardini A, Gullo MAL, Mosca R. Refilling of embolized vessels in young stems of laurel. Do we need a new paradigm? Plant Physiology 1999;120 11-21.

[47] Jackson GE, Grace J. Field measurements of xylem cavitation: are acoustic emissions useful? Journal of Experimental Botany 1996; 304 1643-1650.

[48] Tyree MT, Cochard H. Summer and winter embolism in oak: impact on water relations. Annales des Sciences Forestières 1996;53 173-180.

[49] Cochard H, Pfeiffer M, Le Gall K, Garnier A. Developmental control of xylem hydraulic resistances and vulnerability to embolism in *Fraxinus excelsior* L.: impact on water relations. Journal of Experimental Botany 1997;48 655-663.

[50] Hacke U, Sauter JJ. Xylem dysfunction during winter and recovery of hydraulic conductivity in diffuse-porous and ring-porous trees. Oecologia 1996;105 435—439.

[51] Wang J, Ives NE, Lechowicz MJ. The relation of foliar phenology to xylem embolism in trees. Functional Ecology 1992;6(4) 469-475.

[52] Sperry JS, Tyree MT. Mechanism of water stress-induced xylem embolism. Plant Physiology 1988;88 581-587.

[53] Cruiziat P, Cochard H, Améglio T. Hydraulic architecture of trees: main concepts and results. Annals of Forest Science 2002;59 723-725.

[54] Yang S, Tyree M.T. 1992. A theortical model of hydraulic conductivity recovery from embolism with comparison to experimental data on *Acer saccharum*. Plant Cell & Environment 1992;15 633-643.

[55] Sperry JS, Holbrook NM, Zimmermann MH, Tyree MT. Spring filling of xylem vessels in wild grapevine. *Plant Physiology* 1987;83 414-417.

[56] Fisher J B, Angles G, Ewers F W, Lopez-Portillo J. Survey of root pressure in tropical vines and woody species. International Journal of Plant Sciences 1997;158 44-50.

[57] Tyree MT, Fiscus EL, Wullschleger SD, and Dixon MA. Detection of xylem cavitation in corn under field conditions. Plant Physiology 1986;82 597-599.

[58] Sperry JS, Holbrok NM, Zimmermann MH, Tyree MT. Spring filling of xylem vessels in wild grapevine. Plant Physiology 1987;83 414-417.

[59] Stiller V, Lafitte HR, Sperry JS. Hydraulic properties of rice and the response of gas exchange to water stress. Plant Physiology 2003;132 1698-1706.

[60] Marvin J W and Greene M T. Temperature-induced sap flow in excised stems of *Acer*. Plant Physiology 1951;26 565-580.

[61] Ewers FW, Ameglio T, Cochard H, Beaujard F, Martignac M, Vandame M, Bodet C, Cruizat P. Seasonal variation in xylem pressure of walnut trees: root and stem pressures. Tree Physiology 2001;21 1123-1132.

[62] Sperry JS, Sullivan JEM. Xylem embolism in response to freeze-thaw cycles and water stress in ring-porous, diffuse-porous, and conifer species. Plant Physiology 1992;100 605-613.

[63] Hacke UG, Sperry JS. Limits to xylem refilling under negative pressure in *Laurus nobilis* and *Acer negundo*. Plant Cell & Environment. 2003;26 303--311.

[64] Taneda H, Sperry JS. A case-study of water transport in co-occurring ring- versus diffuse-porous trees: contrasts in water-status, conducting capacity, cavitation and vessel refilling. Tree Physiology 2008;28 1641-1651.

[65] Hacke U, Sauter JJ. Xylem dysfunction during winter and recovery of hydraulic conductivity in diffuse-porous and ring-porous trees. Oecologia 1996;105 435–439.

[66] Holbrook NM, Zwieniecki MA. Embolism repair and xylem tension: do we need a miracle? Plant Physiology 1999;120 7-10.

[67] Bucci SJ, Scholz FG, Goldstein G, Meinzer FC, and Sternberg LDL. Dynamic changes in hydraulic conductivity in petioles of two savanna tree species: factors and mechanisms contributing to the refilling of embolized vessels. *Plant Cell and Environment* 2003;26 1633-1645.

[68] Salleo S, Lo Gullo MA, Trifilò P, Nardini A. New evidence for a role of vessel associated cells and phloem in the rapid xylem refilling of cavitated stems of *Laurus nobilis* L. Plant Cell & Environment 2004;27 1065-1076.

[69] Améglio T, Decourteix M, Alves G, Valentin V, Sakr S, Julien JL, Petel G, Guilliot A, Lacointe A. Temperature effects on xylem sap osmolarity in walnut trees: evidence for a vitalistic model of winter embolism repair. Tree Physiology 2004;24 785-793.

[70] Neumann PR, Weissman R, Stefano G, Mancuso S. Accumulation of xylem transported protein at pit membranes and associated reductions in hydraulic conductance. Journal of Experimental Botany 2010; 6 1711-1717.

[71] Zwieniecki MA, Holbrook NM. Bordered pit structure and vessel wall surface properties. Implications for embolism repair. Plant Physiology 2000;123 1015-1020.

[72] Zwieniecki MA, Hutyra L, Thompson MV, Holbrook NM. Dynamic changes in petiole specific conductivity in red maple (*Acer rubrum* L.), tulip tree (*Liriodendron tulipifera* L.) and northern fox grape (*Vitis labrusca* L.). Plant Cell & Environment 2000;23 407-414.

[73] Konrad W, Roth-Nebelsick A. The significance of pit shape for hydraulic isolation of embolized conduits of vascular plants during novel refilling. Journal of Biological Physics 2005;31 57-71.

[74] Domec JC, Scholz FG, Bucci SJ, Meinzer FC, Goldstein G, and Villalobos-Vega R. Diurnal and seasonal variation in root xylem embolism in neotropical savanna woody

species: impact on stomatal control of plant water status. Plant Cell & Environment 2006;29 26-35.

[75] Canny MJ. Vessel content during transpiration – embolism and refilling. American Journal of Botany 1997b;84(9) 1223-1230.

[76] Canny MJ. Applications of the compensating pressure theory of water transport. American Journal of Botany 1998;85 897-909.

[77] Comstock JP. Why Canny's theory doesn't hold water. American Journal of Botany 1999;86 1077-1081.

[78] Salleo S, Trifilò P, Lo Gullo M. Phloem as a possible major determinant of rapid cavitation reversal in stems of *Laurus nobilis* (laurel). Functional Plant Biology 2006;33 1063-1074.

[79] Salleo S, Trifilò P, Esposito S, Nardini A, Lo Gullo M. Starch-to-sugar conversion in wood parenchyma of field-growing *Laurus nobilis* plants: a component of the signal pathway for embolism repair? Functional Plant Biology 2009;36 815-825.

[80] Sauter J, Wisniewski M, Witt W. Interrelationships between ultrastructure, sugar levels, and frost hardiness of ray parenchyma cells during frost acclimation and deacclimation in poplar (*Populus* x *canadensis* Moench 'robusta') wood. Journal of Plant Physiology 1996;149 451-461.

[81] Zwieniecki MA, Holbrook NM. Confronting Maxwell's demon: biophysics of xylem embolism repair. Trends in Plant Science 2009;14 530-534.

[82] Clearwater MJ, Goldstein G. Embolism repair and long distance water transport. In: Holbrook NM, Zwieniecki MA. (eds.)Vascular Transport in Plants. Burlington: Elsevier Academic Press; 2005. p.375-400.

[83] Nardini A, Lo Gullo M, Salleo S. Refilling emolized xylem conduits: Is it a matter of phloem unloading? Plant Science 2011;180 604-611.

[84] Tyree MT, Sperry JS. Do woody plants operate near the point of catastrophic xylem dysfunction caused by dynamic water stress? Plant Physiology 1988;88 574-580.

[85] Tyree MT, Davis SD, Cochard H. Biophysical perspectives of xylem evolution: is there a tradeoff of hydraulic efficiency for vulnerability to dysfunction? IAWA Journal 1994;15 335-346.

[86] Cochard H, Breda N, Granier A, Aussenac G. Vulnerability to air embolism of three European oak species (*Quercus petraea* (Matt) Liebl, *Q. pubescens* Willd, *Q. robur* L.). Annales des Sciences Forestières 1992;49 225-233.

[87] Jarbeau JA, Ewers FW, Davis SD. The mechanism of water-stress-induced embolism in two species of chaparral shrubs. Plant Cell & Environment 1995;18 189-196.

[88] Hacke UG, Sperry JS, Pittermann J. Analysis of circular bordered pit function. II. Gymnosperm tracheids with torusmargo pit membranes. American Journal of Botany 2004;91 386–400.

[89] Carlquist S. Comparative wood anatomy. Berlin: Springer; 1988.

[90] Crombie DS, Hipkins MF, Milburn JA. Gas penetration of pit membranes in the xylem of *Rhododendron* as the cause of acoustically detectable sap cavitation. Australian Journal of Plant Physiology 1985;12 445-454.

[91] Hacke UG, Stiller V, Sperry JS, Pittermann J, McCulloh KA. Cavitation fatigue: embolism and refilling cycles can weaken the cavitation resistance of xylem. Plant Physiology 2001;125 779-786.

[92] Sperry JS, Hacke UG. Analysis of circular bordered pit function. I. Angiosperm vessel with homogenous pit membranes. American Journal of Botany 2004;91(3) 369-385.

[93] Sperry JS, Tyree MT. Water-stress-induced xylem embolism in three species of conifers. Plant Cell & Environment 1990;13 427-436.

[94] Delzon S, Douthe C, Sala A, Cruiziat H. Mechanism of water-stress induced cavitation in conifers: bordered pit structure and function support the hypothesis of seal capillary-seeding.p Plant Cell & Environment 2010; DOI: 1-1110.1111/j. 1365-3040.2010.02208.xc

[95] Carlquist S. Wood and bark anatomy of *Gnetum gnemon* L. Botanical Journal of Linneanian Society 1994;116 203-221.

[96] Carlquist S. Comapartive wood anatomy. Systematic, ecological and evolutionary aspects of Dicotyledon wood. Berlin; Heidelberg: Springer; 2001.

[97] Zweypfenning RCVJ. An hypothesis on the function of vestured pits. IAWA Bulletin 1978;1 13-15.

[98] Sperry JS. Evolution of water transport and xylem structure. International Journal of Plant Sciences 2003;164(3) 115-127.

[99] Burggraaf PD. Some observations on the course of the vessels in the wood of *Fraxinus excelsior* L. Acta Botanica Neerlandica 1972;21 32-47.

[100] Kitin PB, Fujii T, Abe H, Funada R. Anatomy of the vessel network within and between tree rings of *Fraxinus lanuginosa* (Oleaceae), American Journal of Botany 2004;91 779-788.

[101] Leopfe L, Martinez-Vilalta JM, Pinol J, Mencuccini M. The relevance of xylem network structure for plant hydraulic efficiency and safety. Journal of Theoretical Biology 2007;247 788-803.

[102] Schulte PJ, Gibson AC. Hydraulic conductance and tracheid anatomy in six species of extant seed plants. Canadian Journal of Botany 1988;66 1073-1079.

[103] Comstock JP, Sperry JS. Tansley rereview no. 119. Some theoretical considerations of optimal conduit length for water transport in plants. New Phytologist 2000;148 195-218.

[104] Mencuccini M. The ecological significance of long-distance water transport: short-term regulation, long term acclimation and the hydraulic costs of stature across plant life forms. Plant Cell & Environment 2003;26 163-182.

[105] Sperry JS, Nicholas KL, Sullivan JEM, Eastlack SE. Xylem embolism in ring-porous, diffuse-porous, and coniferous trees of northern Utah and interior Alaska. Ecology 1994;75 1736-1752.

[106] Hejnowicz Z. Anatomia i histogeneza roślin naczyniowych. Warszawa: Wydawnictwo Naukowe PWN; 2002.

4

Detection of Permeable Bodies: From Laboratory Measurements to Seismic Measurements

Jean Luc Mari and Béatrice Yven

1. Introduction

The seismic reflection method has the advantage of providing a picture of the subsurface in three dimensions (3D) with a regular grid. In high resolution seismic surveys, the size of the grid cell is about tens of meters for horizontal distances, and of several meters for vertical distances. The classical approach of seismic processing leads to obtain migrated seismic sections in 2D or migrated seismic blocks in 3D. It is indispensable to have a good velocity model to carry out the migration process. The migrated sections can then be transformed into acoustic impedance sections if well data, mainly acoustic and density logs, are available. The procedures used to obtain acoustic impedance sections are often referred to as Model-based seismic inversions which require an a priori impedance model (obtained from well data) which is iteratively refined so as to give a synthetic seismic section to match the seismic section to be inverted. The impedance model is the fruit of a processing sequence which is usually done in time. 3D seismic impedance blocks and logging data are often used for geological model building in time. The geological model must be then converted from time to depth, thanks to the computation of a time to depth conversion model.

The final impedance model can be converted into porosity by using an empirical relationship between porosity and acoustic impedance established at well locations. To model porosities, an other option is to use porosity at wells location and interpolate between the wells by means of kriging. Partly due to the small number of wells, this outcome is really smooth and usually does not seem geologically consistent. More dense information can be integrated in order to improve the estimation of porosity. As porosity is linked to acoustic impedance, it is relevant to use dense seismic acoustic impedance information. So a collocated cokriging of porosity integrating the seismic information is performed by using the normalized acoustic impedance as the secondary variable [1]. 3D cube makes possible to provide 3D imaging of the connectivity of the porous bodies [2]. Core analysis is usually carried out to establish porosity vs. permeability laws [3].

It has been shown that it is possible to extract new attributes from seismic sections, leading to a better understanding of the distribution of the porous and permeable bodies [4]. The attributes are also used to detect the impermeable layers. The methodology is based on laboratory experiments which have shown that a formation permeability indicator can be obtained via the computation of four input data: P-wave frequency, attenuation, porosity and specific surface. The procedure has been firstly conducted in acoustic logging to estimate permeability of porous layers and to detect water inflows [5]. In seismics, the processing is performed in order to measure these parameters. The analytic signal is used to compute the instantaneous frequency and attenuation (Q factor). The porosity and specific surface are computed from seismic impedances obtained by acoustic inversion of the migrated seismic sections. The input parameters are used to compute a new index named Ik-Seis factor. (Indicator (I) of permeability (k) from acoustic or seismic (Seis) data).The chapter is written mainly from two journal articles: "Detection of porous and permeable formations: from laboratory measurements to seismic measurements" [4] and "Characterization of geological formations by physical parameters obtained through full waveform acoustic logging" [5].

In a first step we show how the methodology has been transposed from laboratory measurements to geophysical data. Then we illustrate the potential of the proposed procedure via field examples: acoustic logging, 2D seismic line with acoustic impedance inversion after migration, 3D seismic section with elastic impedance inversion after migration and depth conversion.

2. From laboratory experiments to geophysical data

We present a short review of the laboratory measurements conducted by P. Morlier and J.P. Sarda [6]. The possibilities of using laboratory results for field geophysical applications are then discussed.

2.1. Laboratory measurements

Laboratory experiments [6] have shown that the attenuation of a clean formation can be expressed in terms of three structural parameters: porosity, permeability and specific surface. Both theoretical and experimental studies have identified the relationship between acoustic attenuation and petrophysical parameters:

$$\delta=\left(C.S/\varphi\right).\left(2\pi.k.f.\rho_f/\mu\right)^{1/3} \tag{1}$$

with:

δ : attenuation (dB/cm), f : frequency (Hz), ρf : fluid density, μ : fluid viscosity (centipoise) φ: porosity, S : Specific surface (cm^2/cm^3), C : calibration coefficient, k : permeability (mD).

Figure 1 is an example of laboratory measurements on sandstone core plugs. The upper part of the figure shows the results obtained on cores with constant specific surface, the lower part on cores with variable specific surface, the specific surface being estimated on the basis of the average pore radius measurement. It is necessary for computing the permeability from eq 1, to measure the attenuation of the formation and calculate the effective specific surface of the formation.

Fabricius et al. [7] have found that the specific surface with respect to grain volume (Sg) is apparently independent from porosity. In an attempt to remove the porosity effect on Vp/Vs and mimic a reflected φ vs log (Sg) trend, they propose to use the following relationship between porosity φ, Vp/Vs and Sg

$$\log(Sg.m) = a.\varphi + b.(Vp/Vs) + c \,\text{with}\, Sg = S/(1-\varphi) \tag{2}$$

where it should be observed that Sg is multiplied by m to make Sg dimensionless. To establish eq 2, Fabricius et al. [7] have looked at ultra sonic data, porosity, and permeability of 114 carbonate core plugs.

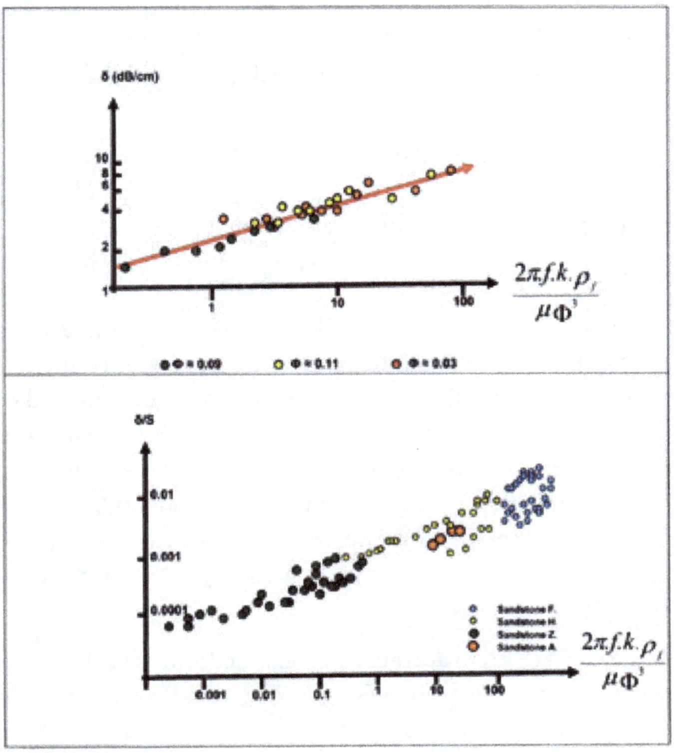

Figure 1. Relationship between attenuation and petro-physical parameters [6]. Laboratory measurements on cores with constant specific surface (top) and with variable specific surface (bottom).

An ideal porous medium made up of spheres (porosity 30%) of any size from cm (pisolites), mm (oolites) to a fraction of mm has an inverse trending specific surface that is the surface of solid exposed to the intergranular void. Fine grained well rounded sands or sandstones have a higher specific surface because the number of grains per unit volume is high. Clay content either as a coating of grains or as lumps or lenses enhances the specific surface up to a limit of effectiveness as long as pore connections prevail. The mineralogical nature of the grains, either silica (quartz or any siliceous rock debris) or carbonate is of little effect on the specific surface as long as secondary changes for instance diagenesis (for instance dolomitization), dissolution (karstification) or fracturation did not occur. Any process that simplifies the mineral surfaces exposed to the pore space tends to decrease the specific surface, whereas fractures and fissures, often occurring as networks or swarms, create new specific surfaces but very little added porosity.

Such processes remain minimal in tectonically quiet basins (Paris basin) where and when connate waters remain in equilibrium with the reservoir minerals. In subsurface exploration when no exposures more existing wells (logs, core and cutting information's) are available, it is difficult to assess the degree of transformation that a granular sedimentary reservoir formation has undergone through the elapsed geologic time.

2.2. To geophysical measurements

In practice, the parameter Ik-Seis (Indicator (I) of permeability (k) from acoustic or seismic (Seis) data) computed from equation 1 is proportional to permeability k.

$$\text{Ik-Seis} = (\varphi.\delta/S)^3/f = (\varphi/SQ)^3/f \quad (3)$$

with f : P-wave frequency, Q quality factor, δ : attenuation,

S : specific surface, φ : porosity

The natural rocks are never perfectly elastic. The viscoelastic media always exhibit a wave amplitude decline as a function of time, independently of geometrical effects. From a wave propagation point of view, this results in attenuation on the one hand, and on the other hand in a dispersion of the propagation velocity, i.e. velocity depends on wave frequency. In order to conduct calculations and measurements on these parameters, it is necessary to build a model, the best known and used being the constant Q model. Q, called quality factor, is inversely proportional to the attenuation often designated by δ or α. In the case where Q is constant with frequency, which is true in the domain of frequencies used with seismic, we have: $\alpha = \pi f/Q$, f being the frequency.

Wave propagation modelling in a saturated porous medium reveals the existence of three propagation waves: two P-waves and one S-wave. The two P-waves (P_1-wave and P_2-wave) have two different propagation velocities V_{p1} and V_{p2} ($V_{p1} > V_{p2}$)and two different characteristic particle motions corresponding to a movement in which overall and fluid displacements are in phase for the P_1-wave (fast P-wave) and out of phase for the P_2-wave (slow P-wave) [8]. The laboratory experiments conducted by Plona [9] have confirmed the Biot 's theory [8]. Frequency is very important. Two major domains, separated by a critical frequency must be distinguished. Above the critical frequency, it is possible to estimate a permeability knowing that the calculated permeability is only an approach of hydraulic permeability. As shown in figure 2, above 2 kHz, permeability has an influence on velocities and attenuations [10]. The attenuation may reach a maximum for frequency in the order of 10 kHz, which is the domain of full waveform acoustic logging. It is the reason why several authors [2, 5-7, 10-12] attempted to predict permeability from acoustic data. The historical focus has been on predicting permeability from P-wave velocity and attenuation.

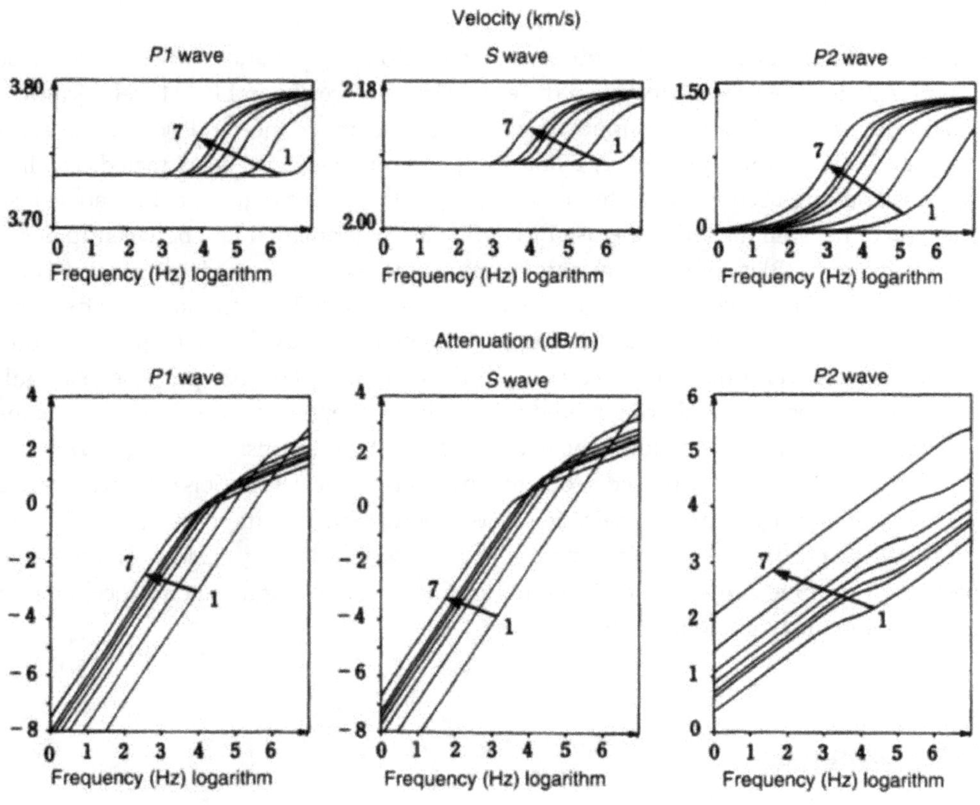

Figure 2. Theoretical sensitivities of velocity and attenuation as a function of permeability, for a 19%-porosity sandstone (Schmitt, 1986). Curves 1 to 7 were calculated for a permeability of 2, 32, 200, 500, 1000, 1500 and 5000 milli-Darcys. P_1-wave and P_2-wave are respectively the fast and the slow P-waves.

The transmission of an acoustic wave through geological formations is used for formation characterisation, in the acoustic frequency domain (ranging between 1 and 25 kHz). Acoustic logging allows the measurement of the propagation velocities and frequencies of the different waves (P-wave, S-wave and Stoneley wave) which are recorded by an acoustic tool. The analysis of the acoustic waves recorded simultaneously on both receivers of the acoustic tool is used to compute additional logs, defined as acoustic attributes, useful for the characterization of the formation, such as amplitude, shape index, wavelength and attenuation logs. Singular value decomposition (SVD) filtering method can be used to attenuate the noise, to measure the attenuation and to extract the acoustic wavelets [4].

Above the critical frequency, in the domain of acoustic frequencies, the Ik-Seis factor can be seen as a pseudo- permeability log after calibration on core data or hydraulic tests. Below the critical frequency (low frequency approximation), in the domain of seismic frequencies, the Ik-Seis factor can only be seen as a relative indicator which varies from 0 for less porous and permeable bodies to 1 for more porous and permeable bodies. In the same way, the specific surface is a pseudo specific surface which varies from 0 for less shaly bodies to 1 for more shaly bodies.

The seismic data must be inverted in order to obtain seismic impedance sections. If an elastic inversion is done, it is possible to obtain the elastic impedances Ip and Is. At well location, it is usually possible to obtain cross plots between acoustic impedance and porosity φ and to define a law between the two. Usually a linear or polynomial law can be extracted. The Ip, Is and φ quantities are used to compute the seismic specific surface (equation 2). If an acoustic inversion is done, well logs must be used to define an experimental law between Ip and Is. The analytic signal [13] is computed in order to extract, from the migrated seismic section, the variation of the seismic frequency and the Q factor versus time. The instantaneous frequency gives the frequency variation versus time and the envelop decrease leads to an estimation of the Q-factor. However a high signal to noise ratio is required. For that purpose, a singular value decomposition (SVD) filtering method [13] is used to enhance the coherent reflections and to attenuate the noise. Whatever the geophysical method (acoustic logging or reflection seismic surveying), equation 3 is used to obtain the Ik-Seis factor. The Ik-Seis factor can be used to detect in acoustic logging or on seismic sections permeable and impermeable bodies. For that purpose, we need to compute four quantities : P-wave frequency f, Q factor or attenuation δ, specific surface S and porosity φ. More information concerning the data processing and analysis is given in [4] and in the field examples.

3. Fields examples

2D and 3D seismic data were recorded in France at the boundary of the Meuse and Haute-Marne departments in the vicinity of the Andra Center (National radioactive waste management Agency). The acoustic data were recorded in well "EST 431" located on the Ribeaucourt township, in the national forest of Montiers-sur-Saulx, 8 km North-North-West of the Andra Center. Figure 3 indicates the location of 2D and 3D seismic surveys and of the selected seismic lines: line "07EST09" from the 2D survey and the in line 405 from the 3D survey. Figure 3 also indicates the location of the well " EST 431".

In a first step, we present and discuss the results obtained in acoustic logging. In a second step we describe those obtained with seismic data (2D then 3D).

3.1. Geological context

One of the drilling platforms, located in the centre of the studied zone, was used to study formations ranging from Oxfordian to Trias. The analysis presented here concerns borehole "EST 431" and covers the Oxfordian formation. The objective of this borehole is to complement the geological and hydrogeological knowledge of this formation. This formation consists essentially of limestone deposited in a vast sedimentary platform. The limestone facies, which vary from one borehole to another, are generally bio-detritic with reef constructions. The base of the Kimmeridgian shale was observed at –258.3 m (100 m ASL) and the base of the Oxfordian limestone at - 544.3 m (-186 m ASL).In this formation, porosity ranges between 5 and 20% and "porous horizons" of kilometric extension have been identified. As far as hydrogeology is concerned the observed water inflows are usually located in high porosity zones [14]. During the drilling, water inflows were detected at – 368 m and - 440 m. At the end of the drilling, the well was left in its natural water.

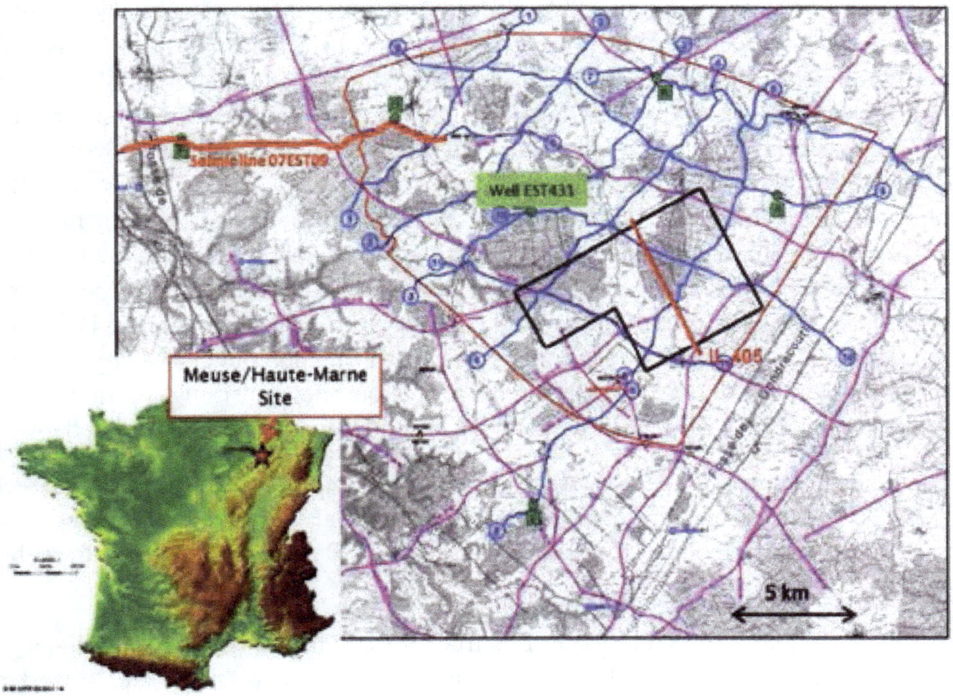

Figure 3. Location map Well EST431 and seismic surveys. The selected seismic lines are the line 07EST09 from the 2D survey and the in line 405 from the 3D survey.

3.2. Acoustic logging

The acoustic tool used for the field experiment described in this paper is a flexible monopole tool with two pairs of receivers: a pair of near receivers (1 and 1.25 m offsets) and a pair of far receivers (3 and 3.25 m offsets). The data have been recorded through the far offset configuration. The sampling depth interval is 10 cm. The sampling time interval is 5 microseconds. The length of recording is 5 ms.

The acoustic log has been run in the Oxfordian carbonate formation, in the 333 – 510 m depth interval. Figure 4 (top left) shows the 3 m constant offset section, opposite the geological description. On the acoustic section, the refracted P-waves appear in the 0.6 – 1.2 ms time interval, the converted refracted shear waves in the 1.2 – 2 ms time interval, and the Stoneley wave in the 2 – 2.4 ms time interval.

On the acoustic section, we can differentiate: an event at 345 m showing a very strong attenuation of all the waves; an interval showing a very strong slowing down of the P and S waves (363 – 375 m); a relatively homogeneous mid-level interval (375 – 400 m); a level which stands out because of its strong variations in P, S and Stoneley velocities (400 – 455 m); a very homogeneous zone below 455 m with easily identifiable P and S waves, and an image of alteration between 501 and 507 m.

The processing of the acoustic data has been described in detail in [5]. Figure 4 is a display of acoustic logs: P-wave velocity (VP), P-wave frequency, acoustic porosity, P-wave attenuation, Sg specific surface and acoustic permeability. For the carbonate formation, a "V_P - V_S " cross plot led us to define a linear relationship between the two logs and to compute a shear velocity model. The experimental linear relationship computed as a regression line between "V_P and V_S" is $V_S = 0.37\ V_P + 879$. The correlation coefficient between the V_S measurements and the V_S values given by the linear equation is equal to 0.93. For a clean formation, if the matrix and fluid velocities are known, an acoustic porosity log can be computed from the acoustic V_p velocities by using the formula given by Wyllie [15] expressed in velocities. As the maximum P-wave velocity is 6150 m/s at a depth of 464.5 m, the matrix velocity value has been chosen at 6300 m/s, and the fluid velocity at 1500 m/s, since the formation fluid is water. The acoustic porosity log, valid only in the clean part of the formation, shows a strong correlation (correlation coefficient : 0.86) with a NMR porosity log (not displayed here) recorded in the well.

The attenuation (expressed in dB/m) of the formation is computed from the first eigensection (obtained by SVD) of the refracted P-wave acoustic signal recorded by the two adjacent receivers of the acoustic tool. The results obtained by the SVD processing procedure are shown in figure 5.

Figure 4 (bottom right) shows the *Sg* specific surface log and the acoustic permeability log (Ik-Seis) calculated from eq. 2 and 3. The fluid viscosity μ and density ρ_f have been assumed to be constant (μ = 1 centipoise, ρ_f = 1 g/cm^3). The Ik-Seis log detects three permeable zones at 368 m, between 400 and 440 m, and 506 m. The permeable zone located at 506 m corresponds to a high value of conductivity and is characterized by a low porosity (6 %), a 10 dB/m attenuation, but a significant decrease of the P-wave frequency and of the specific surface. During drilling, water inflows have been detected at – 368 m and - 440 m. At the end of the drilling, the well was left in its natural water. The hydraulic tests and conductivity measurements conducted later on did not confirm the inflow at 368 m seen during the drilling, but have validated the 400 – 440 m and 506 m permeable zones detected by the acoustic logging.

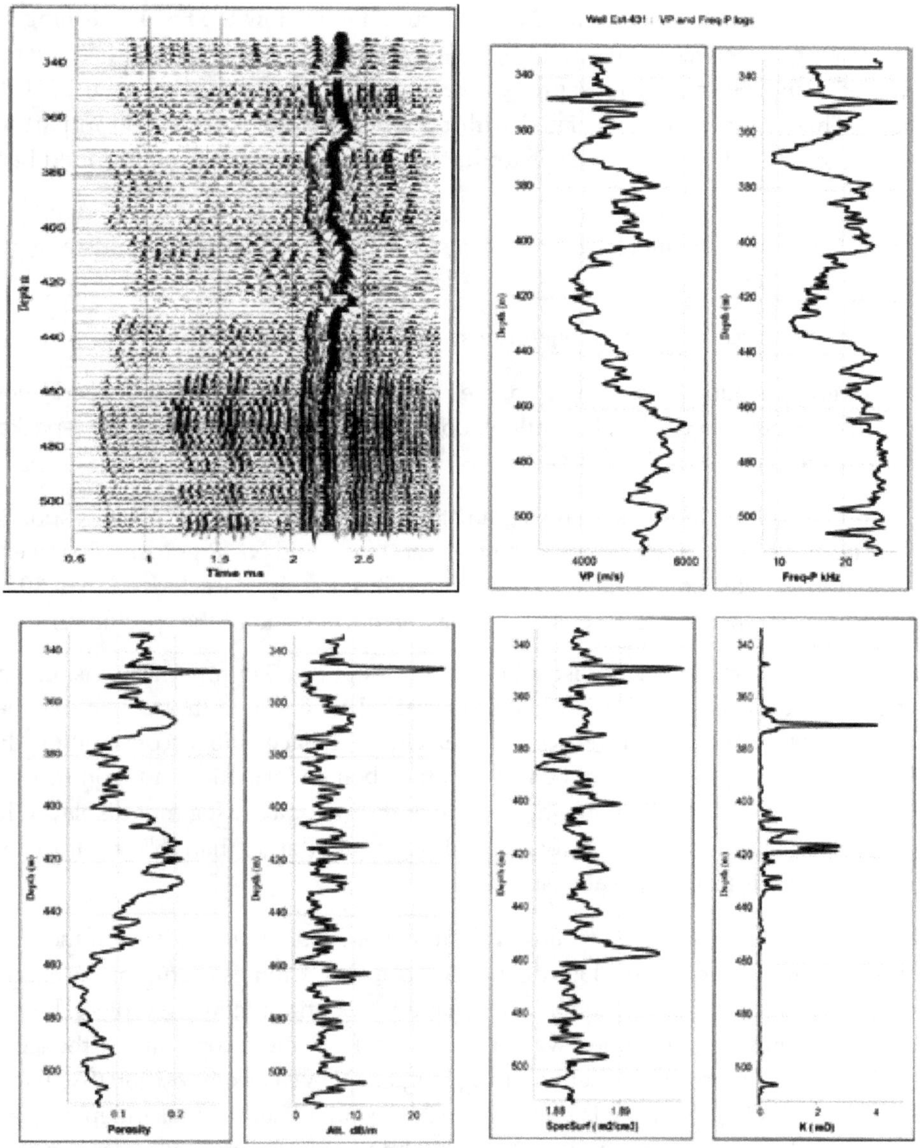

Figure 4. Permeability estimation from acoustic logs [5] Top : acoustic section and P-velocity and frequency logs Bottom from left to right: porosity and attenuation logs; specific surface and predicted permeability (Ik-Seis) logs.

A short pumping test was conducted between the 4th and 7th of March, 2008. This test, associated with four geochemical logs, highlighted five productive zones between - 297 and - 507 m. The overall productivity is weak with a 15 L/min flow below 23.5 m drawdown. In general, there is a good correlation between the zones identified through the analysis of the geochemical logs, the natural gamma-ray log and the NMR porosity log. All the identified production zones correspond to low clay content zones.

The mean overall transmissivity of the Oxfordian formation below the EST431 drilling pad is $7.2\ 10^{-06}\ m^2/s$, ranging between $5\ 10^{-06}$ and $1\ 10^{-05}\ m^2/s$. A total of five inflows have been identified between – 297.5 and -506 m. The most productive inflow (- 506 m) ranges between $2.5\ 10^{-06}$ and $5\ 10^{-06}\ m^2/s$. These inflows are associated with pore porosity and correspond in part to the porous horizons (HP) described in the Oxfordian formation below the Underground Laboratory:

- -297.5 to -301 m inflow, alternation of bioclastic and oolites packstone/mudstone, which corresponds to HP7,
- -328.5 m inflow, coral reefs packstone/grainstone, which corresponds to HP6,
- -413 m inflow, carbonated sand with oolites and oncolites, and numerous coral polyps, which corresponds to HP4, -439 m inflow, oolites and coral polyps grainstone-packstone interface, which corresponds to HP3,
- -506 m inflow, oolites and coral polyps grainstone-packstone interface, no correspondence with the porous horizons. The acoustic porosity and free fluid NMR porosity do not exceed 6%. The inflow, detected by the acoustic permeability log, is characterized by a 10 dB/m attenuation, a significant decrease of the P-wave frequency and of the specific surface.

The logs reveal a strong acoustic discontinuity at a depth of 345 m, clearly visible on the attenuation. The acoustic discontinuity is also revealed by a strong increase of the specific surface, a significant decrease of the acoustic velocities. The acoustic discontinuity is due to the presence of a thin shaly layer in the Oxfordian carbonate formation. It is confirmed by a change in the borehole diameter and a high value of the gamma ray log (not displayed here). Since at that depth, the acoustic porosity log has high values (larger than 25%, figure 4 bottom left), the shaly layer is probably water saturated.

The analysis of the acoustic waves recorded simultaneously on both receivers of the acoustic tool is used to compute additional logs defined as acoustic attributes useful for the characterization of the formation, such as amplitude, shape index and attenuation logs. The results obtained are optimum if the studied wave is extracted from the records and if the signal to noise ratio is high. We show the benefit of using Singular Value Decomposition (SVD) for that purpose [13]. The SVD processing is done on the 2 constant offset sections independently, in a 5 traces (N=5) depth running window. After flattening of each constant offset section with the picked times of the refracted wave, the refracted wave signal space is given by the first eigensection obtained by SVD:

$$\underline{\underline{r}}^{sig} = \lambda_1 \underline{u}_1 \underline{v}_1^T \tag{4}$$

$\underline{v}_1$ is the first singular vector giving the time dependence, hence named normalized wavelet, $\underline{u}_1$ is the first singular vector giving the amplitude in depth, therefore called propagation vector and λ_1 the associated eigenvalue. The amplitude variation of the refracted wavelet over the depth interval is $\lambda_1 \underline{u}_1$.

Figure 5 (top) shows the normalized wavelet ($\underline{v}_1$) and the associated amplitude ($\lambda_1\underline{u}_1$) log versus depth, for the two constant offset sections associated with the two receivers (R1 and R2) of the acoustic tool. Figure 5 (bottom left) also shows the refracted wave signal space versus depth for the two receivers. The amplitude logs have been used to compute the attenuation log (figure 4, bottom left) expressed in dB/m. The correlation coefficient (figure 5, bottom right) between the two normalized wavelets has been computed at each depth. We can notice some anomalies at local depth (358, 390, 460, 492 and 503 m) and a significant decrease of the correlation coefficient in the 400 – 440 m depth interval. The interval corresponds to the porous and permeable zone detected by the Ik-Seis factor (figure 5, bottom right). It is therefore suggested that changes in phase or distortion of the acoustic signal is linked to propagation through a porous and permeable zone. The distortions can be measured by a shape index attribute. To measure the shape variation, an acoustic attribute, named Ic, independent of the energy of the source, has been introduced [10]. The Ic parameter is given by the following equation:

$$Ic = \left(\left(A_2 + A_3\right) / A_1\right)^n \tag{5}$$

where A_1, A_2 and A_3 are the amplitudes of the first three arches, respectively, of the studied signal and n an exponent.

The shape index is computed from formula 5 with an exponent value of 3 (n = 3). Figure 5 (top) allows the comparison between the two shape index logs (Ic-R1 and Ic-R2).

The shape index logs highlight anomalic zones, in the 350 – 380 m depth interval with a maximum value at 358 m, and in the 400 – 440 m depth interval with a peak at 415 m which corresponds to the porous layer HP4. In order to reduce the noise to extract the common component of the two shape index logs (Ic-R1 and Ic-R2), the geometric mean of the two has been computed. The resulting shape index log Ic is compared with the Ik-Seis log and the correlation coefficient log (Figure 5, bottom right). The comparison shows a good coherence between the Ic log and the correlation coefficient log. The permeable and porous layers in the 400-440 m depth interval and the inflow at 506 m are seen both by the shape index log and by the Ik-Seis log. Shape index and correlation coefficient logs can be used as a quick look method to detect permeable bodies and inflows which must be confirmed by the Ik-Seis log.

3.3. Seismic surveying: 2D seismic line with acoustic impedance inversion after migration

The 2D seismic line was recorded in 2007 (see location map, figure 3). The 2D design is a split dip spread composed of 240 traces. The distance between 2 traces is 25 m. The source is a vibroseis source generating a signal in the 14 - 140 Hz frequency bandwidth. The bin size is 12.5 m. The nominal fold is 120.

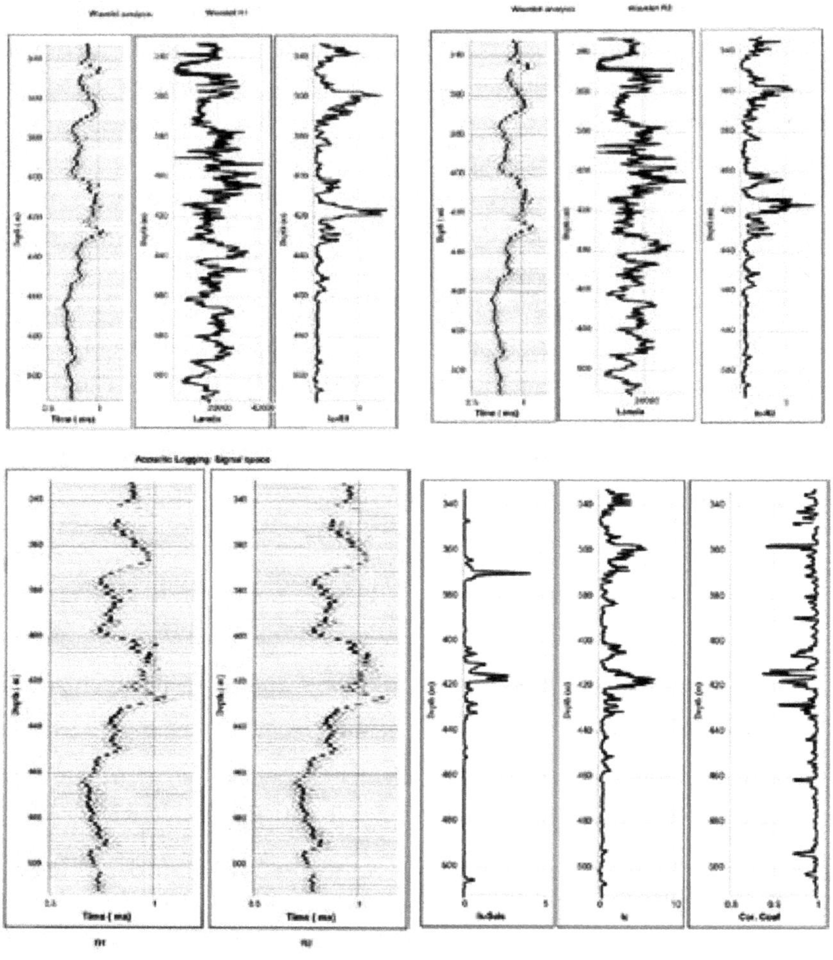

Figure 5. Analysis of refracted waves in acoustic logging by SVD [4]. Top : SVD analysis (Wavelet, amplitude, shape index) on receivers R1 and R2. Bottom from left to right: signal space (R1, R2); predicted permeability (Ik-Seis), shape index and correlation coefficient logs.

3.3.1. Seismic procedure

A conventional seismic sequence was applied to the data set. It includes amplitude recovery, deconvolution and wave separation, static corrections, velocity analysis, CMP-stacking and time migration.

However, the amplitude recovery has been done in two main steps: spherical divergence compensation by using the Newman's law [16], and attenuation compensation. Besides amplitude changes at each interface, the signal undergoes a general decay as a function of the propagation distance from the source (decaying as 1/r) and with its transmission through the interface. The application of a gain scalar that is a function of t can compensate for these effects. The commonly used gain function is:

$$G_s(t) = t.\, V^2_{rms}(t) / V_1 \tag{6}$$

where t is propagation time, V_1 is the propagation velocity of sound in the first layer and $V_{rms}(t)$ is the rms velocity at time t.

In order to be able to apply the Newman's law, the velocity model given by the rms velocity model must be estimated by velocity analysis. An a priori gain scalar t.V is applied to each shot point. The deconvolution is then performed by spectrum equalization in the 10-130 Hz frequency band. The wave separation by frequency – wavenumber filter (f-k filter) is then done to cancel the direct, refracted and surface waves and to enhance the reflected waves. The data are sorted in CMP gathers and the velocity analysis is performed. The knowledge of the velocity model allows the computation of the gain function $G_s(t)$. The initial gain function is retrieved and replaced by the function $G_s(t)$. After such a processing, a residual decay of the amplitude of the stacked traces has been noticed. The residual decay observed on the envelope of the stacked trace has been used to extract a residual compensation law $G_r(t)$, after a strong smoothing in time of the envelope. The amplitude recovery residual law has been then approximated locally by an exponential law $e^{\alpha t}$, where t is propagation time and α attenuation factor. Figure 6 (top, left) shows the migrated trace at CMP 600 after amplitude recovery by Newman's law, the amplitude recovery function which is used to compensate the attenuation, and the seismic trace after compensation of attenuation.

The migrated section has been filtered by SVD in order to enhance the signal to noise ratio before computing seismic attributes (instantaneous frequency and envelope). Before SVD filtering, the migrated section has been shifted in time to flatten a reference seismic horizon, noted S1. In that case, the SVD processing is done on the migrated section, in a 5 traces (N=5) CMP running window and the signal space is composed of the two first eigensections in order to take into account the local dips which can been present on the time migrated section. Figure 6 (top, right) shows the amplitude recovery function $G_r(t)$, the instantaneous frequency trace f(t) and the associated Q function. The Q factor at time t is obtained from the following equation:

$$Q(t) = \pi f(t)/\alpha \quad \text{with } \alpha = \text{Ln}\left(G_r(t)\right)/t \tag{7}$$

After migration, a model-based stratigraphic inversion (a priori impedance model obtained from well data) provides a 2D impedance model section. The 2D impedance model section has also been shifted in time to flatten the seismic horizon S1. At well locations, the logs of velocity V_p, density ϱ and porosity φ have been used to define laws between porosity and acoustic impedance (φ vs Ip) and between V_p and I_p. The porosity vs impedance cross-plot, displayed in figure 6 (bottom right) was used to define a linear law between the two. The porosity law obtained in the Oxfordian limestone and the relation between V_s and V_p ($V_s = 0.37\, V_p + 879$)

defined at well EST 431 allow the computation of both the pseudo specific surface and Ik-Seis functions, thanks to equations 2 and 3 (figure 6, bottom left). The processing is done on each migrated CMP independently in order to build the Ik-Seis section. The results are shown in the vicinity of CMP 600 in figure 7 and for the total line in figure 8.

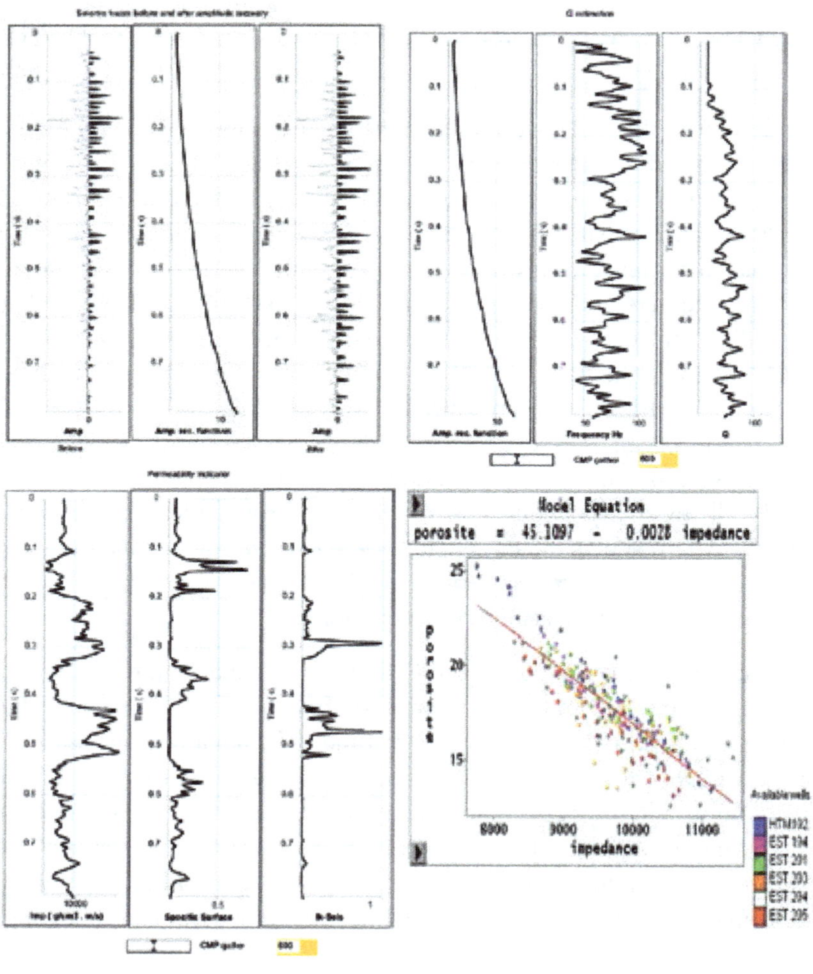

Figure 6. Seismic analysis at CMP 600. [4] Top from left to right: seismic trace before and after amplitude recovery, amplitude recovery function, instantaneous frequency trace and Q seismic trace. Bottom from left to right: acoustic impedance trace, Specific surface and Ik-Seis traces, porosity – acoustic impedance relationship.

3.3.2. Seismic analysis at CMP 600

Figure 6 (bottom left) displays, at location of CMP 600, the acoustic impedance trace, the pseudo specific surface seismic trace with its associated Ik-Seis trace. The specific surface seismic trace clearly shows the main geological units: the Kimmeridgian marls (between 0.1 and 0.2 s), the Oxfordian carbonates (between 0.2 and 0.33 s), the Callovo Oxfordian claystone (between 0.33 and 0.42 s), the Dogger carbonates (between 0.42 and 0.53 s) and the Toarcian claystone after 0.53 s. The shaly units have a high pseudo specific surface and a low value of the Ik-Seis factor. Figure 7 shows the migrated section and the associated Ik-Seis section in the vicinity of the CMP 600. As far as hydrogeology is concerned the observed water inflows are usually located in high porosity zones located in the lower part of the Oxfordian limestone, as it can be seen on the Ik-Seis section between 0.28 and 0.33 s (Figure 7, left). The Ik-Seis section also shows the distribution of the permeable bodies in the Dogger formation (Figure 7, right).

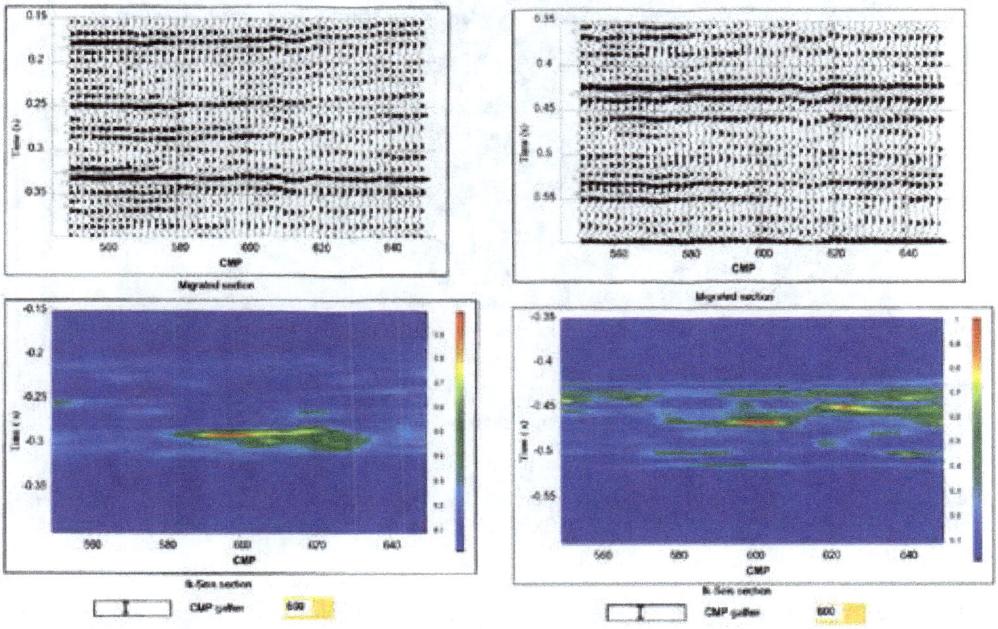

Figure 7. Migrated and Ik-Seis sections at the vicinity of the CMP 600. [4] Left: 150 – 400 ms time interval (Oxfordian carbonate formation), Right: 350 – 600 ms time interval (Dogger carbonate formation).

3.3.3. Seismic line analysis

Figure 8 shows the distribution of the permeable bodies in the carbonate formations via the Ik-seis factor.

Several carbonated shelves have been emplaced during the middle and upper Jurassic, according to second order stratigraphic sequences. The Callovo-Oxfordian claystone is deposited over the Bathonian platform and is in turn overlaid by the Oxfordian platform.

The contact between the Dogger formation and the Callovo-Oxfordian claystone is sharp, associated to retrogradation condensed facies. The Dogger carbonate formation displays discontinuous thin porous layers in the Figure 7, but the correlation with the two known thin porous layers cannot be assumed due to the seismic vertical resolution.

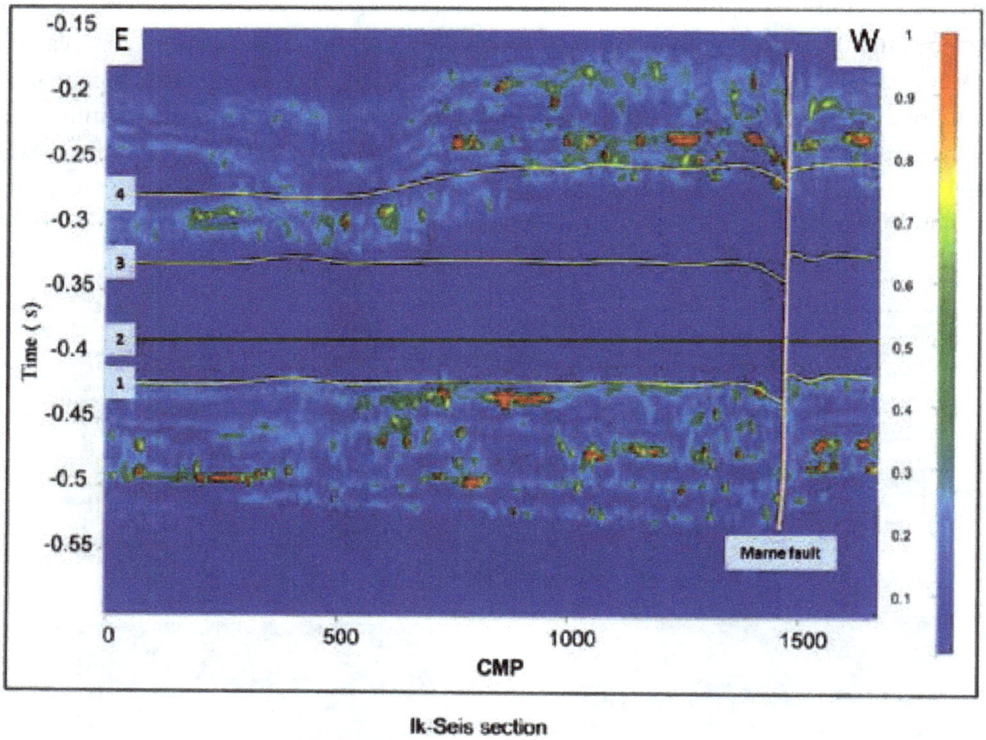

Figure 8. Ik-Seis section [4] 1: base of the Callovo-Oxfordian claystone; 2: carbonated layer used for flattening (S1); 3: top of the Callovo-Oxfordian claystone; Interval 3-4: Oxfordian limestone

The Callovo-Oxfordian claystone displays very low values of the Ik-Seis factor, with no significant lateral variation in good agreement with the known lithology. A carbonated layer, corresponding to a sequence boundary, has been used for flattening. This layer is a known marker-bed in the Paris basin (RIO: repère inférieur oolithique).

The value of the factor in the lower part of the Oxfordian limestone (around 0.30 s) is low in the interval CMP 850 - 1600 according to the transition from high energy inner ramp carbonated facies toward outer ramp marly facies in the west direction. The porosity decreases drastically in the western facies.

3.4. Seismic surveying: 3D seismic section with elastic impedance inversion after migration and depth conversion

The 3D seismic survey was recorded in 2010. The 3D design is a cross spread. The active spread is composed of 12 receiver lines with 120 stations each. The source lines are perpendicular to the receiver lines. The receiver and source line spacing's are respectively 80 m and 120 m. The receiver and source point spacing's are 20 m. The source is a vibroseis source generating a signal in the 14 - 140 Hz frequency bandwidth. The bin size is 10 x 10 m^2. The nominal fold is 60.

A conventional seismic sequence was applied to the data set. It includes amplitude recovery, deconvolution and wave separation, static corrections, velocity analysis and pre-stack time migration. The pre-stack time migration allows a model-based inversion (a priori impedance model obtained from well data) which provides a 3D elastic impedance block. No well being located in the 3D area, additional 2D lines have been recorded to calibrate 3D data on 3 wells situated outside of the 3D area. After migration, the seismic block is depth converted using consistent velocity model thanks to Bayesian kriging [17]. The velocity model is also used for estimating a Ik-Seis model in depth.

3.4.1. Time to depth conversion

Geostatistical approach for time-to-depth conversion of seismic horizons is often used in many geo-modelling projects. From a geostatistical point of view, the time-to-depth conversion of seimic horizons is a classical estimation problem involving one or more secondary variables.The converted depth and associated uncertainty can be estimated using a kriging method which can be constrained by the well markers, velocity model and interpreted horizons. For the multilayer case, the kriging estimator should take into account all the relationships between horizons determined by the velocity model associated to each layer.

The more appropriate kriging method for this problem is the Bayesian Kriging (BK) [17 -19]. Bayesian approach provides an excellent estimator which is more general than the traditional kriging with external drift(s) and fits very well to the needs for time-to-depth conversion of seismic horizons. The advantage of BK as estimator comparing to the others consists in the fact that we can manage simultaneously the uncertainty on the trend velocity model and the local uncertainty defined by the uncertainty of interpreted time maps and local fluctuations of interval velocities.

The input information's for BK are:

- Two-way-time (TWT) maps for interpreted horizons
- Well markers for each horizon
- Prior velocity model and associated uncertainty for each layer
- Local uncertainty definition for each time map (picking uncertainty, and spatial variogram definition)
- Local uncertainty definition of interval velocity for each layer (local velocity fluctuations around the trend model, and spatial variogram definition)

As any Kriging based estimator, the Bayesian Kriging provides as results:

- The estimated variable (estimated depth for each horizon)
- Variance of estimation (associated uncertainty of estimated depth).

The use of BK in depth conversion has the advantage to combine the prior knowledge of the velocity model with a certain degree of uncertainty and the well data. All sources of uncertainty (velocity and time) are integrated in a consistent way in a unique probabilistic model used for estimation or simulation.

For each selected horizon, the Bayesian Kriging provides its estimated depth Z associated with its time t. The " Z versus t " data set is interpolated in the whole space (3D block) at the time sampling rate (1 ms) in order to obtain a time to depth conversion model.

The time to depth conversion procedure is illustrated via the In line 405 extracted from the 3D seismic block (see location map, figure 3).

10 seismic horizons numbered from 1 to 10 have been picked in time and depth converted. The 10 seismic horizons are:

1. Top of Kimmeridgian White Limestones
2. Top of Porous Horizon HP4
3. Top of Lower Oxfordian (Top of target interval)
4. Top of Upper Callovian (RIO)
5. Top of Carbonated Dogger (Base of target interval)
6. Base of Argillaceous limestone and Longwy marls
7. Base of Carbonated Dogger
8. Top of Domerian
9. Base of Lias (base of Gryphees limestone)
10. Top of Beaumont dolomite

Figure 9 shows the acoustic impedance section (I_p) in time and the picked times of the 10 seismic horizons, the time- to – depth conversion model and the depth conversion of the 10 horizons. The time-to-depth conversion model can be used to convert in depth any type of seismic sections (amplitude, velocity, acoustic impedance, Ik-Seis factor).

3.4.2. Ik-Seis section in depth

The Ik-Seis section has been computed in time following the procedure described for the 2D line (figure 8) and then depth converted using the time to depth conversion model shown in figure 9.

Figure 10 shows the Ik-Seis sections in time (a) and in depth (b). The Ik-Seis sections allow the identification of a porous and permeable horizon (HP4) in the Oxfordian limestone (depth : 500 m, CMP : 100). They show the distribution of porous layers in the Dogger formation (between 800 m and 1000 m depth). The contact between the Dogger carbo nate formation and Callovo-Oxfordian argillite is clearly marked. The Callovo-Oxfordian claystone displays very low values of the Ik-Seis factor. The results obtained confirm the observations done on the 2D line (figure 8).

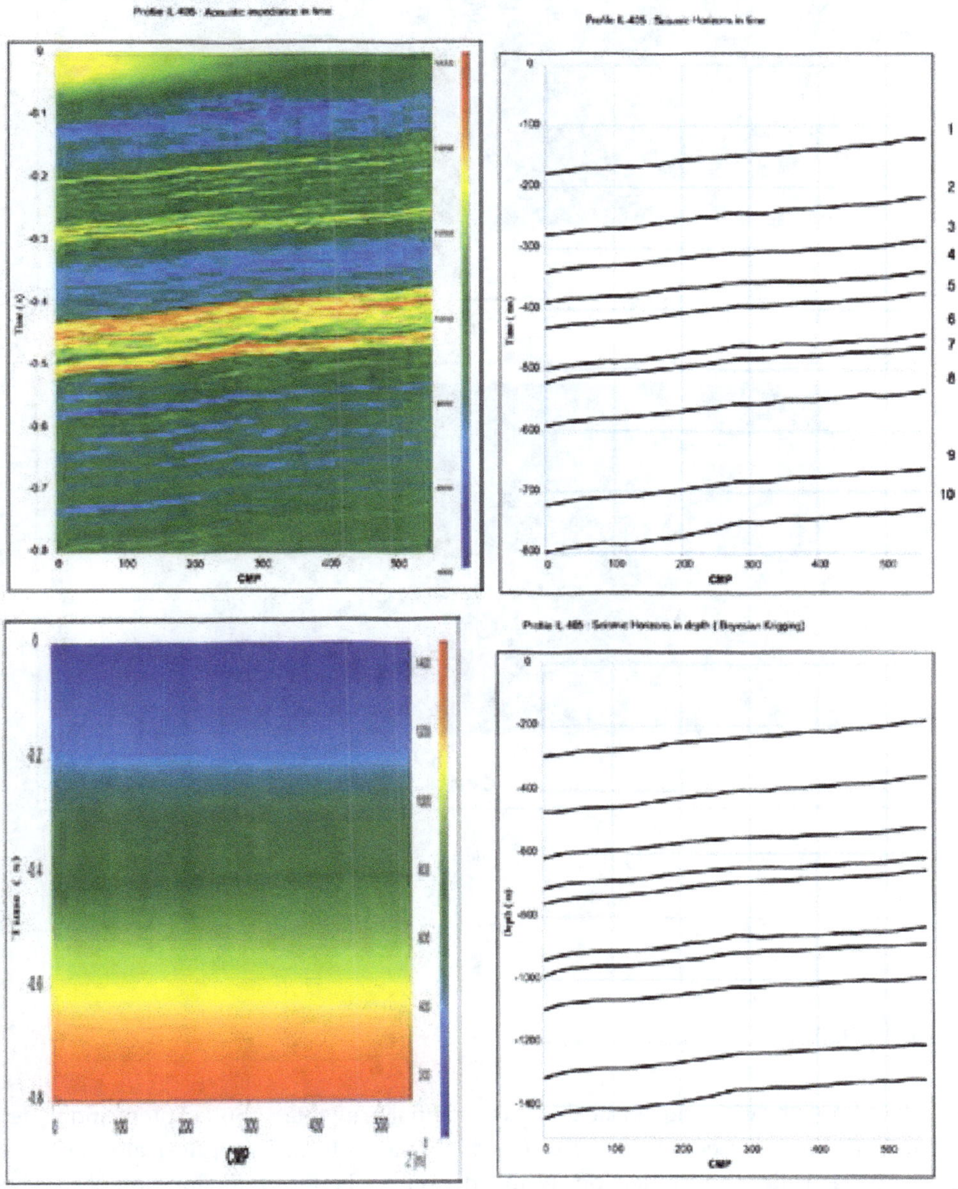

Figure 9. Time to depth conversion. Top : acoustic impedance section in time and the picked times of the 10 seismic horizons, Bottom: time- to – depth conversion model and the depth conversion of the 10 horizons.

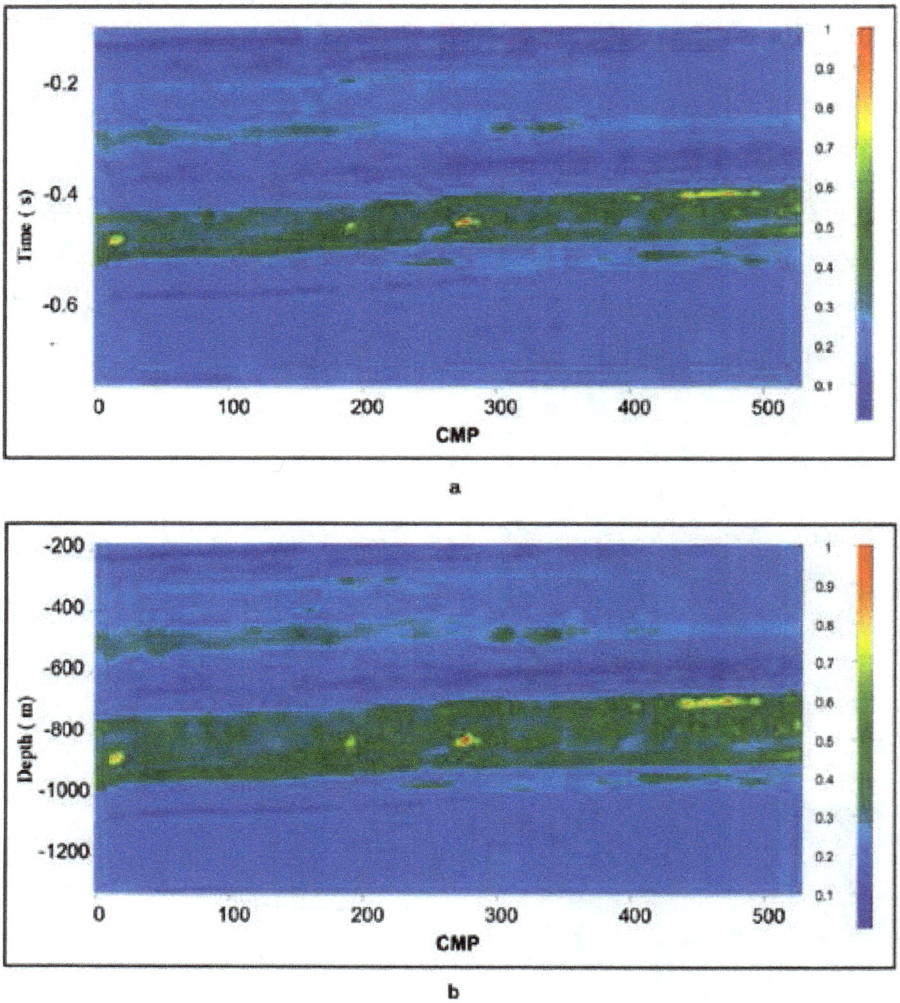

Figure 10. Ik-Seis sections a : in time, b: in depth.

4. Conclusion

Knowledge about porosity and permeability is essential to evaluate fluid content and to detect fluid flow. We have presented a procedure which allows to use geophysical data (i.e. full waveform acoustic data and reflection seismic data) for a better understanding of the distribution of the porous and permeable bodies. The methodology is based on laboratory experiments which have shown that a formation permeability indicator, named Ik-Seis factor, can be obtained via the computation of 4 quantities: P-wave frequency, attenuation, porosity and specific surface.

The ability to use laboratory results for field geophysical applications has been discussed. The frequency content is important. Regarding frequencies above 2 kHz, permeability has an influence on velocities and attenuations. The attenuation may reach a maximum for frequency in the order of 10 kHz, this being the domain of full waveform acoustic logs.

Consequently, the procedure has been firstly conducted in acoustic logging to estimate the permeability of porous layers and to detect water inflows. Full waveform acoustic data were recorded in an Oxfordian carbonate formation. The Ik-Seis factor computed in the acoustic frequency domain (ranging between 10 and 25 kHz) has detected permeable zones, both associated with high porosity (20 %) but also with low porosity (6 %). The hydraulic tests and conductivity measurements conducted later on have validated the permeable zones detected by acoustic logging. The benefit of using the SVD method to evaluate the signal to noise ratio, to compute the attenuation log and to extract the acoustic wavelets from the acoustic data has been shown. It has also been observed that the correlation coefficient computed between acoustic wavelets recorded by two adjacent receivers of an acoustic tool significantly decreases in porous and permeable zones. It is therefore suggested that changes in phase or distortion of the acoustic signal is linked to propagation through a porous and permeable zone. The distortions can be measured by a shape index attribute. After calibration on core data or hydraulic tests, the Ik-Seis could be seen as a pseudo acoustic permeability log.

In seismic, after signal to noise ratio enhancement by the SVD method, processing is carried out in order to measure the needed parameters (frequency, attenuation, impedance) to compute the Ik-Seis factor. The analytic signal is used to compute the instantaneous frequency and attenuation (Q factor). The porosity and specific surface are computed from seismic impedances obtained by acoustic inversion of the migrated seismic sections. The Ik-Seis factor should only be used as a relative indicator which varies from 0 for less porous and permeable bodies to 1 for more porous and permeable bodies.

Our results suggest that it is possible to extract a significant Ik-Seis factor from seismic sections. This factor leads to a better understanding of the distribution of the porous and permeable bodies. The potential of the proposed procedure has been demonstrated via a 2D seismic profile and a seismic in-line extracted from a 3D block. We have demonstrated the benefit of combining time–to-depth conversion of seismic horizons by Bayesian kriging, consistent seismic velocity model and acoustic impedance in time for building a 3D geological model in depth. The field data example illustrates the potential of the proposed depth conversion procedure for estimating a Ik-Seis model in depth.

Acknowledgements

We thank Andra for permission to use the data presented in the field examples. We thank Jean Dellenbach, Jean Paul Sarda and Daniel Guillemot for their valuable help and advices. We thank Arben Shtuka (Seisquare) for very useful discussions on various occasions, specifically for his experience in the use of geostatistical methods for time to depth conversion.

Author details

Jean Luc Mari[1] and Béatrice Yven[2]

1 IFPEN, France

2 ANDRA, France

References

[1] Bourges M, Mari J.L., Jeannée N., A practical review of geostatistical processing applied to geophysical data: methods and applications, Geophysical Prospecting, 2012, 60, 400-412, DOI: 10.1111/j.1365-2478.2011.00992.x

[2] Mari J.L., Delay F., Contribution of seismic and acoustic methods to reservoir model building, in Lakshmanan Elango (ed.) Hydraulic Conductivity - Issues, Determination and Applications, InTech, 2011; 329–354, ISBN 978-953-307-288-3. Available from http://www.intechopen.com/articles/show/title/contribution-of-seismic-and-acoustic-methods-to-reservoir-model-building.

[3] Zinszner B., Pellerin F.M., A geoscientist's guide to petrophysics, ISBN 978-2-7108-0899-2, Paris, Editions Technip, 2007

[4] Mari J.L, Guillemot D., Detection of porous and permeable formations: from laboratory measurements to seismic measurements, Oil & Gas Science an Technology _Rev. IFP Energies nouvelles, 2012, DOI: 10.2516/ogst/2012209.

[5] Mari J.L., Gaudiani P., Delay J., Characterization of geological formations by physical parameters obtained through full waveform acoustic logging., J. Phys. Chem. Earth, 2011; DOI:10.1016/jpce.2011.07.11.

[6] Morlier P., Sarda J.P., Atténuation des ondes élastiques dans les roches poreuses saturées, Revue de l'Institut Français du Pétrole, 1971 ;26 (9), p. 731-755.

[7] Fabricius I. L., Baechle G., Eberli G.P., Weger R., Estimating permeability of carbonate rocks from porosity and Vp/Vs, Geophysics, 2007; 72(5), 185-191, DOI: 10.1190/1.2756081.

[8] Biot, M.A.,Theory of propagation of elastic waves in a fluid-saturated porous solid, The Journal of the Acoustical Society of America, 1956; 28(2), 168-191.

[9] Plona T.J., Observation of a second bulk compressionnal wave in a porous medium at ultrasonic frequencies, Appl. Phys. Lett., 1980; 36, 259-261.

[10] Lebreton F., Morlier P, A permeability acoustic logging, Bulletin of the International Association of Engineering Geology, 1983; 1, 101-105.

[11] Klimentos T., McCann C., Relationships among compressional wave attenuation, porosity, clay content, and permeability in sandstones, Geophysics, 1990; 55(8), 998-1014.

[12] Yamamoto T., Imaging permeability structure within the highly permeable carbonate earth: inverse theory and experiment, Geophysics, 2003; 68(4), 1189-1201, DOI: 10.1190/1.1598103.

[13] Glangeaud F., Mari, J.L., Signal Processing in Geosciences, 2000. [CD-ROM] Paris: Editions Technip, ISBN 2-7108-0768-8; 2000.

[14] Delay J, Rebours H, Vinsot A, Robin P., Scientific investigation in deep wells for Nuclear waste Disposal studies at the Meuse/Haute-Marne underground research laboratory, North Eastern France, J. Physics and Chemistry of the Earth, 2007; 32,42-57.

[15] Wyllie M.R., Gregory R.J., Gardner H.F., Elastic wave velocities in heterogeneous and porous media, Geophysics,1956; 21(1),41-70.

[16] Newman P., Divergence effects in a layered earth, Geophysics, 1973; 38, 377-406.

[17] Sandjivy L., and Shtuka A., Depth conversion and associated uncertainties using consistent velocity model: a probabilistic unified model based on Bayesian approach, proceedings of the 11th International Congress of the Brazilian Geophysical Society, August 24-28 2009, Salvador, Brazil.

[18] Abrahamsen P., Bayesian Kriging for Seismic Depth conversion of Multi-layer Reservoir, in A. Soares (ed.) Geostatistics Troia 92, 1993; 385-398.

[19] Omre, H. and Halvorsen, K. B., 'The Bayesian bridge between simple and universal kriging'. Math. Geol., 1989; 21(7), 767-786.

5

Role of Hydraulic Conductivity Uncertainties in Modeling Water Flow through Forest Watersheds

Marie-France Jutras and Paul A. Arp

1. Introduction

Soil hydraulic conductivities at saturation (K_{sat}) are highly variable in space and time. For example, K_{sat} varies along vertical and lateral flow paths depending on directional changes in soil texture, density, and structure [1, 2, 3]. Temporal changes are caused by changes in soil structure and bulk density (D_b) in response to, e.g., (i) gradual soil formation processes, and (ii) operationally induced soil compaction or de-compaction due to various land-uses [4]. Changes in weather and climate also affect K_{sat} through freezing and thawing [5, 6], swelling and shrinking [7], extent of rooting and related organic matter build-up [8]. This chapter explores how changes in hydraulic conductivity may affect modelled rates of water flow through forested watersheds, with flows referring to infiltration, percolation, run-off, inter-flow, base flow, and stream discharge. This is done by way of sensitivity analyses centered on two well-studied watershed studies, referring to Moosepit Brook, Nova Scotia [1, 9] and Turkey Lakes, Ontario [1, 10]. Also addressed are:

K_{sat} impacts on the retention of soil water and the transmittance of the same towards streams as influenced by evapotranspiration from open conditions to forests [14];

the relationship between K_{sat} and the state of organic matter decomposition, as characterized by the von Post index from fibric (H1) to fully humified or sapric (H10) [11, 12, 13].

The sensitivity analysis is based on using the forest hydrology model ForHyM2 [1, 15] to determine how scenario-set K_{sat} variations affect soil water retention and flow including stream discharge through the watersheds. The scenarios vary K_{sat} by changing organic matter (OM) and sand content from their actual values within the 0 to 100% per soil weight range.

2. Quantitative background

The equations used for estimating the sensitivity of K_{sat} on account of changes in soil texture, structure, density and organic matter content is given by [16], as follows:

$$\log_{10}K_{sat} = a + 7.94\log_{10}(D_p - D_b) + 1.96\ \text{SAND} \tag{1}$$

$$D_b = \frac{1.23 + (D_p - 1.23 - 0.75\ \text{SAND})(1 - \exp(-\ 0.0106\ \text{DEPTH})}{1 + 6.83\ \text{OM}} \tag{2}$$

$$\frac{1}{\text{Dp}} = \frac{\text{OM}}{\text{Dp}_{om}} + \frac{1 - \text{OM}}{\text{Dp}_{min}} \tag{3}$$

where, Dp_{om} is the particle density of OM (1.3 gcm^{-3}), Dp_{min} is the particle density of mineral soils (2.65gcm^{-3}), SAND and OM are dry soil weight fractions (fine earth fraction only), DEPTH is the mid depth of each soil layer (cm), "a" represents K_{sat} when D_p-D_b = 1 g cm^3 and SAND = 0%. Fig. 1 illustrates how variations in D_b, OM, and SAND affect K_{sat} in general.

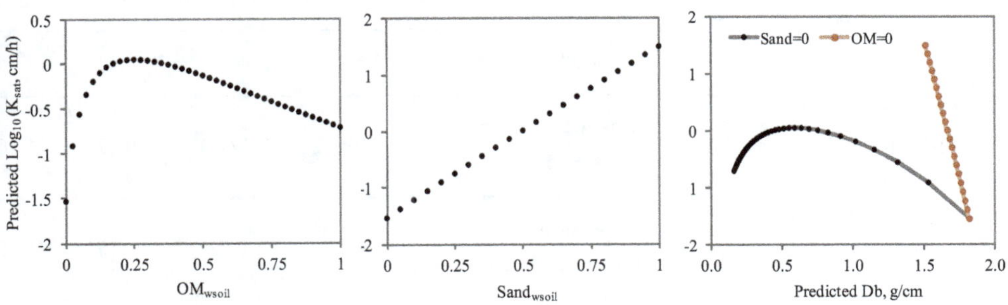

Figure 1. Left and middle: how $\log_{10}K_{sat}$ varies with increasing OM, and sand fraction. Right: Changes in $\log_{10}K_{sat}$ and Db when OM and Sand fraction = 0.

For organic soils, it is important to adjust a, D_b, and D_p in Eqs. 1 to 3 by extent of organic matter decomposition and humification [12, 11, 17, 13, 18, 19]. These adjustments are based on the von Post humification index ([11, 20], Table 1) as follows:

$$D_b = 0.035 + 0.0159\ vP\ (\text{R2} = 0.93) \tag{4}$$

$$D_{pom} = \frac{D_b}{1 - \phi} \tag{5}$$

$$\phi = 100.38 - 76.7D_b \ (R^2 = 0.99) \tag{6}$$

$$a = (2.05 \pm 0.2) - (0.046 \pm 0.004)vP^2 \ (R^2 = 0.89) \tag{7}$$

where vP is the von post index (Table 2) and ϕ is the soil porosity. Fig. 2 illustrates the relationship between the von Post adjusted $\log_{10}K_{sat}$ (Eqs. 1, 4-7) and $\log_{10}K_{sat}$ based on literature sources. Fig. 2 shows (i) a plot of actual versus best-fitted K_{sat} values (left), and actual as well as best-fitted K_{sat} values with increasing organic matter humification in peaty soils (right).

Peat Class	von Post Index	Squeeze Test: Exudate condition	$\log_{10}K_{sat}$	K_{sat} [1] cm h^{-1}	D_b [2] g cm^{-1}	D_p [2] g cm^{-1}
Fibric	H1	Water colourless	3.15	1406.48	0.05	1.44
Decomposition: none to slight; Amorphous content: low	H2	Water yellowish	2.88	756.62	0.07	1.41
	H3	Water brown, muddy; no peat	2.55	356.24	0.08	1.39
Mesic	H4	Water dark brown, muddy; no peat	2.15	141.08	0.10	1.37
Fibers still recognisable; Decomposition: moderate to strong; Amorphous content: medium	H5	Water muddy; some peat	1.66	46.20	0.11	1.36
	H6	Water dark brown; 33% peat	1.09	12.40	0.13	1.36
	H7	Any water very dark brown, 50% peat	0.43	2.72	0.15	1.35
Sapric	H8	66% peat, Water pasty	-0.32	0.48	0.16	1.34
Fibers unrecognisable; Decomposition: very strong to complete; Amorphous content: high	H9	Nearly all peat; paste uniform	-1.16	0.07	0.18	1.34
	H10	100% peat paste; no water	-2.09	0.006	0.19	1.34

[1]Eq. 1 from [16]

[2]Eqs. 4-5 from [12]

Table 1. von Post humification index, with K_{sat}, D_b and D_p for 100% OM content according to Eqs. 1 and 4 to 7; adapted from [21] and [11]

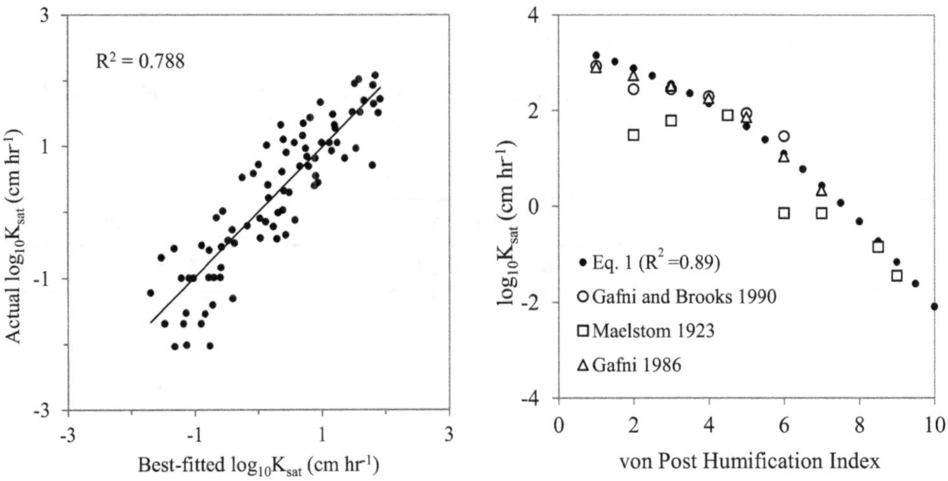

Figure 2. Best-fitted $\log_{10}K_{sat}$ versus actual data from New Brunswick and Nova Scotai, Canada, as seen in [1] and [16] (left), best-fitted $\log_{10}K_{sat}$ versus von Post humification index from literature sources (right).

3. Methods

The two study areas, Moosepit Brook and Turkey Lakes, have contrasting terrain (generally flat versus hummocky), climate (maritime versus continental), vegetation (mostly coniferous versus deciduous), and soil parent material (ablation till versus basal till) (Table 2, Fig. 3)

Eight scenarios were adopted to examine the impacts of K_{sat} variations on water flow through these locations, as follows: the actual soil conditions in terms of soil texture and organic matter (Scenario 1, Table 3), varying the soil texture sand, silt, or clay (Scenarios 2, 3, and 4), and varying the soil organic matter content (Scenarios, 5, 6, 7, and 8).

Scenario 1:

1. Actual soil texture and OM content

Scenarios 2 to 4: Changing soil texture

2. Sand = 95% sand, 1% silt, 4% clay
3. Silt = 7% sand, 87% silt, 6% clay
4. Heavy clay = 25% sand, 25% silt, 50% clay

Scenarios 5 to 8: Changing organic matter content

5. Half actual OM

6. Double actual OM
7. No OM throughout entire soil profile
8. 100% OM throughout entire soil profile using the von Post profile of FF = 3, A&B = 5, C = 9, Subsoil = 10.

Watershed characteristics	Moosepit Brook	Turkey Lakes
	Nova Scotia (NS)	Ontario (ON)
Abbreviation	MP	TL
Latitude (N)	44°28'	47°03'
Longitude (W)	65°03'	84°25'
Area (ha)	1670	1050
Elevation (m)	100-150	350-400
Slope (%)	1	8
Deciduous:coniferous	50:50	100:0
Rooting habit	Medium	Deep
Forest floor depth (cm)	5	7
Mineral soil: depth (cm); texture	50; SL	60; SilL
Subsoil: depth (cm); texture	70; LS	100; LS
Bedrock	Metamorphic greenschist slate	Metavolcanic basalt
Land Formation	Glacial till	Ablation till on basal till
Topography	Rolling	Undulating to rolling
Mean yearly temperature (°C)	7.02	4.52
Mean yearly snow depth (cm)	5	23
Mean yearly rainfall (mm)	1140	790
Model Run Years	1999-2004	1997-2004

Table 2. Site description for the Moosepit Brook and Turkey Lakes watersheds.

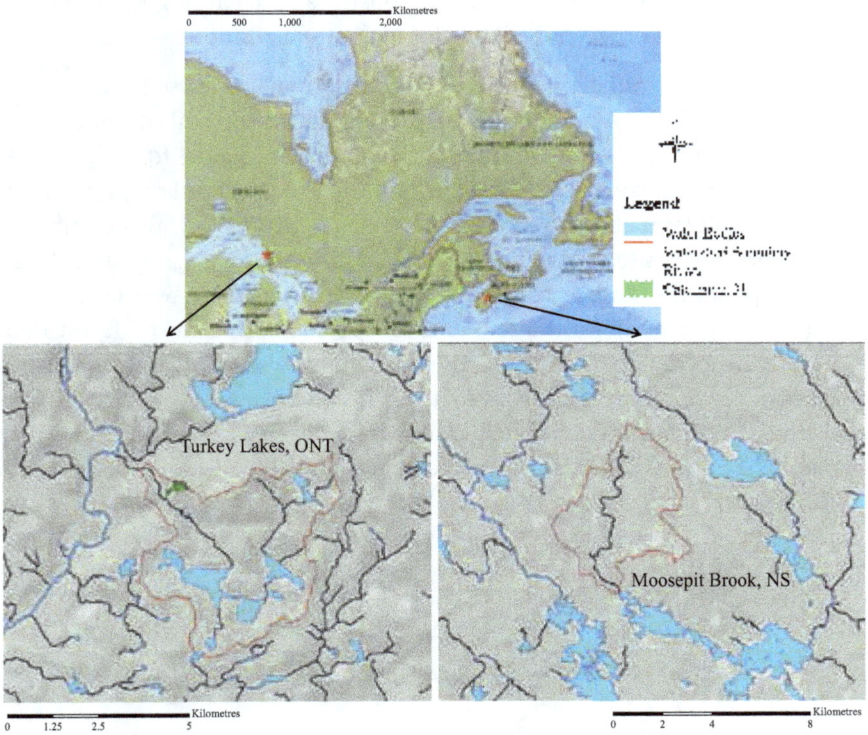

Figure 3. Locator maps for the Turkey Lakes (left) and Moosepit Brook (right) study areas.

Actual watershed inputs: Scenario 1

Horizon	Moosepit Brook, NS					Turkey Lakes, ONT				
	Depth (cm)	Sand (%)	Silt (%)	Clay (%)	OM (%)	Depth (cm)	Sand (%)	Silt (%)	Clay (%)	OM (%)
A	21	66	24	10	4.5	15	66	24	10	5
B	21	66	24	10	9	30	66	24	10	3
C	50+	82	12	6	4.5	50+	22	65	13	2

Table 3. Actual scenario soil input for Moosepit Brook and Turkey Lakes.

The sand texture percentages for scenarios 2-4 demonstrate the effects of varying texture on K_{sat} from sandy and sandy loam soils to silty and clayey soils (Fig. 4). The organic matter levels for scenarios 5-8 were chosen to demonstrate the effects of changing the organic matter on content from very small in mineral soils to fully organic soils. For the 100% organic soil condition (Scenario 8), three sub-scenarios were chosen to account for variations in forest cover from 100 % (fully forested), 50% (varying from forested to boggy) and 0% (open moss and shrub-covered bogs with no trees). This is to demonstrate how varying K_{sat} levels from high to low increase the amount of water available for evapotranspiration

Each scenario was used for initializing the ForHyM2 requirements for soil texture and organic matter by soil layer, with the A and B layers representing the top soil conditions, and the C layer representing the subsoil conditions Table 2). Layer-specific values for $D_{p,}$ D_b and K_{sat} were then generated automatically via Eqs. 1 to 3. All other site-specific input requirements for daily weather (rain, snow air temperature), slope, aspect, elevation and soil layer depths were kept the same. Scenario 1 was used to refine the Eq. 1 estimates for K_{sat}, by adjusting the K_{sat} adjustment multipliers for surface run-off, interflow (forest floor, A&B layers combined), baseflow (C layers combined), infiltration, and soil percolation from the forest floor to the topsoil, and from the topsoil to the subsoil. The calibrations were done by matching modeled with actual stream discharge a the daily level, using local weather records for daily rain, snow and air temperature as model input. Modelled snowpack depth was also calibrated using daily snowpack data. The ForHyM2 model runs were done for 1999 – 2004 for Moosepit Brook, and for 1997-2002 for Turkey Lakes.

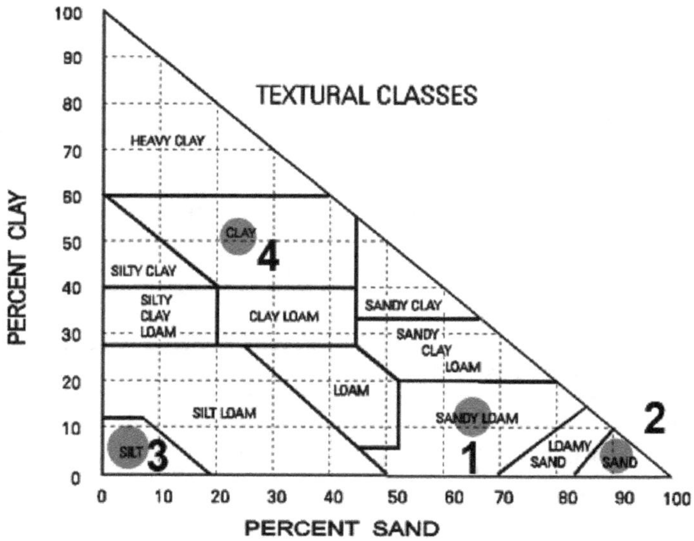

Figure 4. Mineral texture class triangle for fine soil showing texture classes for scenarios 1 - 4 (adapted from CANSIS 2000).

4. Results

The results of this analysis are shown in Tables 3 to 8 and in Figs. 5 to 12 for the Moosepit Brook and Turkey Lakes study areas. Tables 3 and 4 inform about the Scenario-based changes on Dp, Db and K_{sat} for each of the two sites by topsoil and subsoil. The Db numbers indicate that the subsoil at both locations is compacted, with K_{sat} values typically 10 to 50 times lower in the subsoil than in the topsoil. Since the soil texture is sandier at Moosepit Brook than at Turkey Lakes, K_{sat} values remain higher in the subsoil at Moosepit Brook than at Turkey Lakes. Changing the topsoil texture from the actual values changes K_{sat} by about 5x upwards, and by about 10x downwards at both locations. These K_{sat} changes are similar for the somewhat coarser subsoil at Moosepit Brook. In contrast, subsoil K_{sat} is not much affected by increasing the clay and silt content, but increases with increasing sand content towards 95% by a factor of 147

Table 5 and Fig. 5 inform about the 5-year cumulative effects of the texture and OM changes on ForHyM2-modelled run-off, forest floor interflow, topsoil interflow, baseflow and stream discharge in terms of modelled mm per study period, and also in terms of modeled flow rate percentages per stream discharge. As shown, the interflow and baseflow percentage contributions to stream discharge so compiled are very sensitive to K_{sat} as well as basin slope: for intermediate K_{sat} values, interflow would dominate the base flow contributions to stream discharge within the steeper watershed at Turkey Lakes (average slope=8%). The reverse would occur at the flatter Moosepit Brook watershed (average slope = 1%). Low subsoil permeability at Turkey Lakes would further accentuate this difference. In detail, base flow would dominate in both watersheds or at any location within the watersheds with high soil permeability and where the subsoil would not be blocked by impervious bedrock. In contrasts, locations with low overall soil permeability and low slopes would be most variable in terms of their cumulative run-off, interflow and baseflow contributions, varying from mostly baseflow to mostly interflow (Fig. 6). For example, mineral soils with high silt content (Scenario 3) would support more lateral flow in the topsoil as opposed to soils with high sand content (Scenario 2). Doubling the OM in the mineral soil (Scenario 5) would also increase baseflow, whereas reducing OM (Scenario 6) would induce the opposite. The extent of water infiltration in Scenario 4, as modeled, would be midway between Scenarios 2 and 3

Site	Scenarios	K_{sat}, cm h^{-1}		D_b, g cm^{-1}		D_p, g cm^{-1}	
		Mineral	Subsoil	Mineral	Subsoil	Mineral	Subsoil
Moosepit Brook	1: Actual	48.40	5.95	0.95	1.61	2.48	2.59
	2: Sand	162.90	29.15	0.93	1.50	2.48	2.59
	3: Silt	3.05	0.15	1.00	1.86	2.48	2.59
	4: Heavy clay	7.15	0.50	0.99	1.80	2.48	2.59
	5: Double OM	60.60	13.30	0.72	1.48	2.33	2.54
	6: Half OM	31.35	2.70	1.14	1.70	2.56	2.62
	7: No OM	12.60	0.75	1.41	1.83	2.65	2.65
Turkey Lakes	1: Actual	39.80	0.10	1.09	1.85	2.55	2.61
	2: Sand	136.25	14.70	1.06	1.54	2.55	2.61
	3: Silt	2.50	0.05	1.15	1.92	2.55	2.61
	4: Heavy clay	5.80	0.10	1.15	1.84	2.55	2.61
	5: Double OM	56.05	0.30	0.90	1.68	2.45	2.57
	6: Half OM	30.55	0.05	1.19	1.85	2.59	2.61
	7: No OM	14.50	0.00	1.39	2.06	2.65	2.65

Table 4. Results for various levels of sand and OM against K_{sat}, D_b and D_p for Moosepit Brook and Turkey Lakes.

Figs. 6 to 9 inform about the changes in daily variations in run-off, interflow and baseflow for both locations as the soil texture changes from actual to sandy, silty and clayey (Scenarios 1 to 4, respectively, Figs. 6, 7), and soil organic matter content changes actual to 0.5 and 2 x, and 100% (Scenarios 1, and 5 to 8, Figs. 8, 9). As shown, these flows would peak faster with increasing K_{sat} (increasing sand and organic matter content), and would saturate the lower soil layers more quickly with decreasing K_{sat} and decreasing pore space, or increasing bulk density. Among the scenarios, the largest textural change on the flow regime was incurred by increasing the silt content within the already compacted subsoil at Moosepit Brook. Note that organic soils with 100% sapric organic matter would also have very low interflow and baseflow rates, and would therefore lead to relative fast soil saturation as well.

Site	Scenario	Runoff (mm)	%[2]	Interflow FF (mm)	%[2]	Interflow A&B (mm)	%[2]	Base flow (mm)	%[2]	Total Discharge (mm)
Moosepit Brook	1	2.7	0.1	202.3	4.8	682.8	16.1	3341.2	79.0	4229.0
	2	2.6	0.1	202.4	4.7	512.0	11.9	3580.4	83.3	4297.0
	3	7.0	0.2	209.5	5.1	2653.0	64.4	1251.5	30.4	4121.0
	4	2.7	0.1	202.4	4.8	1609.7	38.3	2391.4	56.9	4206.0
	5	2.6	0.1	202.3	4.8	450.0	10.6	3578.0	84.5	4233.0
	6	2.7	0.1	202.4	4.8	964.0	22.8	3054.8	72.3	4224.0
	7	2.7	0.1	202.3	4.8	1675.8	39.8	2329.9	55.3	4210.7
	8	0.0	0.0	594.8	13.2	3771.1	83.4	153.7	3.4	4519.6
	8[1]	0.0	0.0	612.4	11.6	4495.9	85.3	161.0	3.1	5269.3
	8[2]	0.0	0.0	668.6	10.6	5487.3	86.7	171.1	2.7	6327.0
Turkey Lakes	1	1.1	0.0	375.9	9.1	2990.0	72.2	772.9	18.7	4139.0
	2	3.3	0.1	376.1	8.8	301.0	7.1	3588.5	84.1	4269.0
	3	114.9	2.8	521.2	12.8	2635.0	64.8	796.5	19.6	4068.0
	4	19.4	0.5	389.1	9.4	1965.0	47.3	1782.1	42.9	4156.0
	5	2.2	0.1	376.1	9.1	2314.0	55.8	1454.4	35.1	4146.0
	6	0.5	0.0	375.8	9.1	3273.0	79.2	485.6	11.7	4135.0
	7	0.0	0.0	374.8	9.1	3457.0	83.9	290.7	7.1	4122.0
	8	0.0	0.0	69.8	1.5	4250.5	92.1	292.9	6.3	4613.1
	8[1]	0.0	0.0	70.9	1.2	5463.3	93.6	305.0	5.2	5839.2
	8[2]	0.0	0.0	72.6	1.0	6609.0	94.6	305.0	4.4	6986.5

[1] 50% coverage [2] 10% coverage

[2] % values refer to the calculated percent contributions of run-off, FF interflow A&B interflow and baseflow to stream discharge.

Table 5. Lateral stream discharge by cumulative and percent runoff, interflow, and base flow for scenarios 1-8 for Moosepit Brook (1999-2004) and Turkey Lakes (1997-2004).

Assessing the waterflow through peatland locations within each of the two watersheds, and setting the state of decomposition of the peat equal to H1, H4, H7 and H10 produced the results compiled in Table 6. As shown, organic soils mostly composed of fibric to mesic peat (H1) would support deep percolation and baseflow, whereas organic soils mostly composed of humic peat (H10) would contain pooled water from the subsoil upwards to the surface, thereby encouraging surface run-off

Note also from Table 4 and 6 that the changing K_{sat} values for decomposing peat would also have strong effects on forested peatland evapotranspiration and on stream discharge: the lower K_{sat}, the higher would be the rate of water retention and subsequent forest water uptake and evapotranspiration during the growing season (Fig. 10). In contrast, the higher K_{sat}, the faster water would be lost due to quick baseflow (Fig. 11). Outside the growing season, run-off increases, as modeled and as to be expected (Fig. 10 and 11)

Site	von Post	Runoff (mm)	%	Interflow FF (mm)	%	Interflow A&B (mm)	%	Base flow (mm)	%	Total Discharge (mm)
Moosepit Brook	H1	0.00	0.0	1.09	0.0	53.00	1.1	4618.27	98.8	4672.35
	H4	0.00	0.0	20.27	0.4	222.67	4.9	4291.43	94.6	4534.38
	H7	0.00	0.0	240.34	5.9	184.86	4.5	3676.06	89.6	4101.26
	H10	3275.13	97.3	74.12	2.2	13.32	0.4	4.67	0.1	3367.25
Turkey Lakes	H1	0.00	0.0	0.08	0.0	0.84	0.0	5954.36	100.0	5955.27
	H4	0.00	0.0	1.03	0.0	7.64	0.1	5641.93	99.8	5650.61
	H7	0.00	0.0	55.87	1.2	8.18	0.2	4655.45	98.6	4719.50
	H10	3057.12	97.7	50.55	1.6	4.51	0.1	16.86	0.5	3129.03

Table 6. Lateral stream discharge by cumulative and percent runoff, interflow, and baseflow for scenario 8 to represent 100% peat surface deposits (von Post index set at H1, H4, H7, and H10 for the entire profile) underneath forest cover at each of the two locations.

The extent water retention in terms of mm per soil layer is illustrated in Fig. 12 for the two study locations as modeled for the actual soil (Scenario 1) and for organic soil conditions (100% organic matter content, Scenario 8), starting the soil moisture content at field capacity for January 1. For the slowly draining peatland scenario (Scenario 8), subsoil moisture levels would increase from field capacity towards saturation in about one year. For the well-drained upland soil conditions (Scenario 1), K_{sat} values would be sufficiently high so that soil moisture conditions would fluctuate around the field capacity, depending on season as well as rainfall and snow melt events

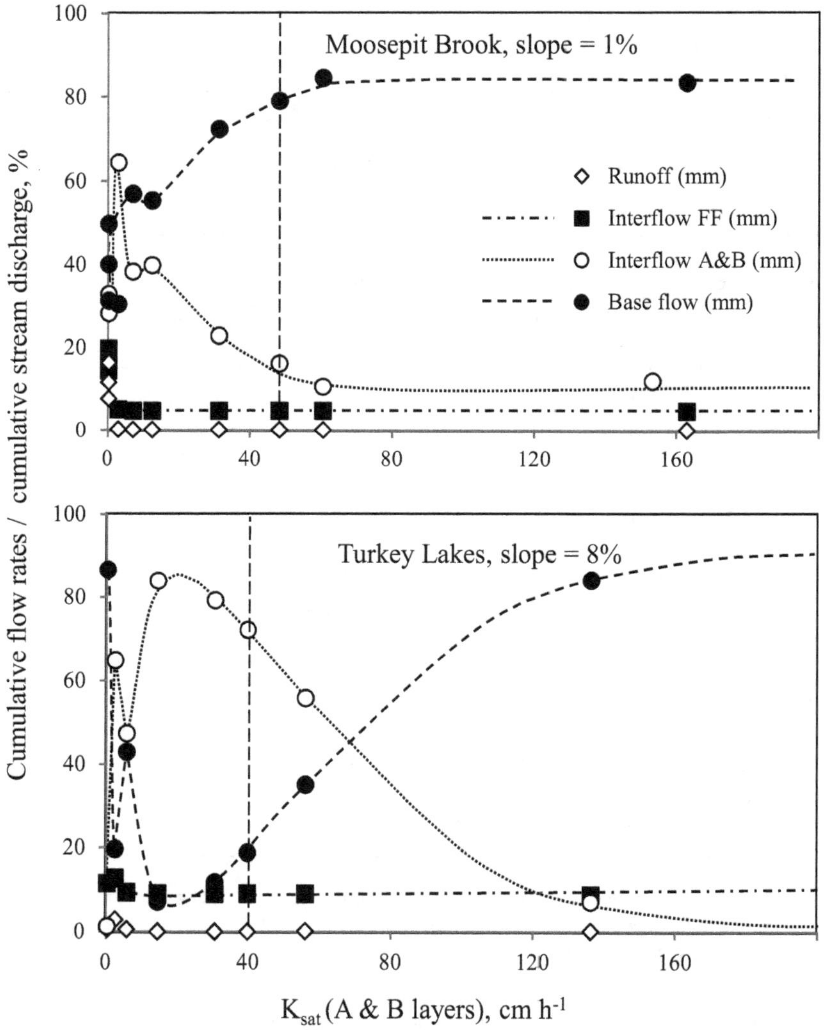

Figure 5. K_{sat} of the A&B layers by cumulative stream discharge % for Moosepit Brook (top), and Turkey Lakes (bottom) across all 8 scenarios (vertical dashed line represents the actual scenario)

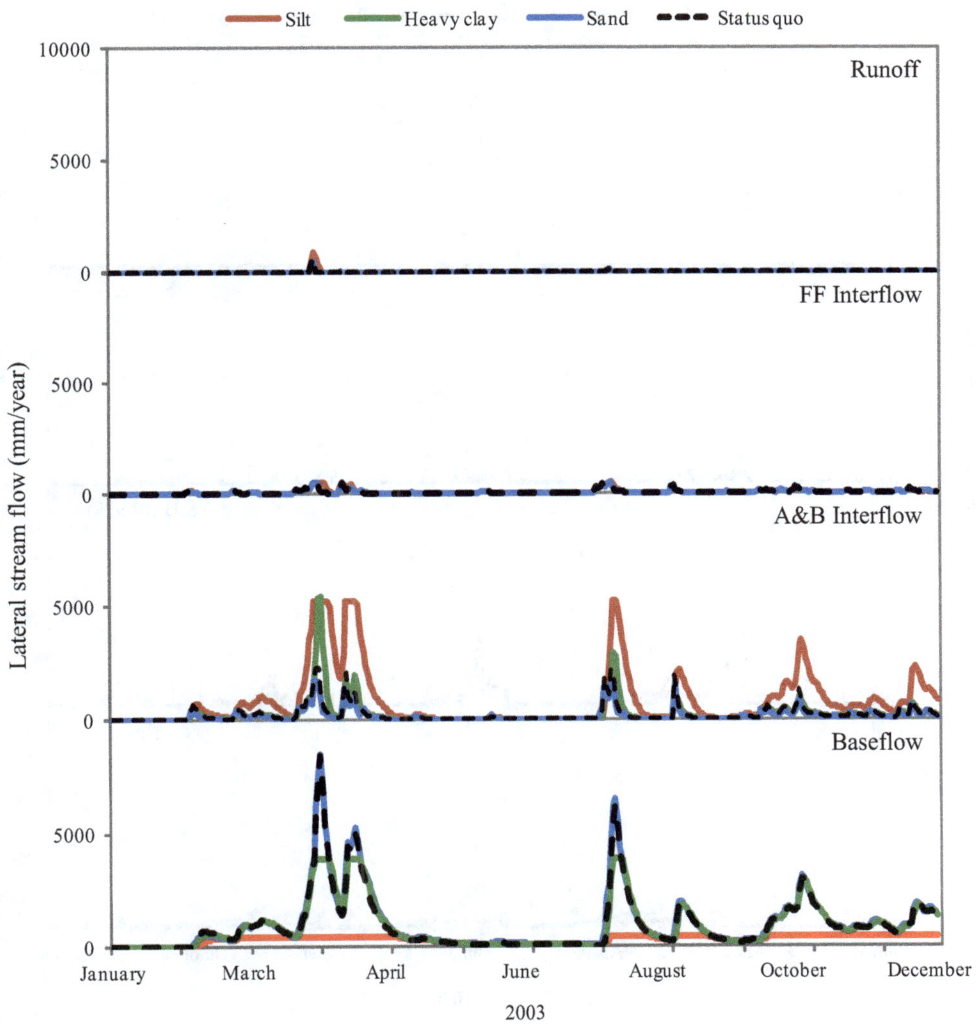

Figure 6. Run-off, forest floor and A&B interflow and base flow for Moosepit Brook, by scenario from actual to sandy, silty and clayey (Scenarios 1 to 4, respectively; 2003)

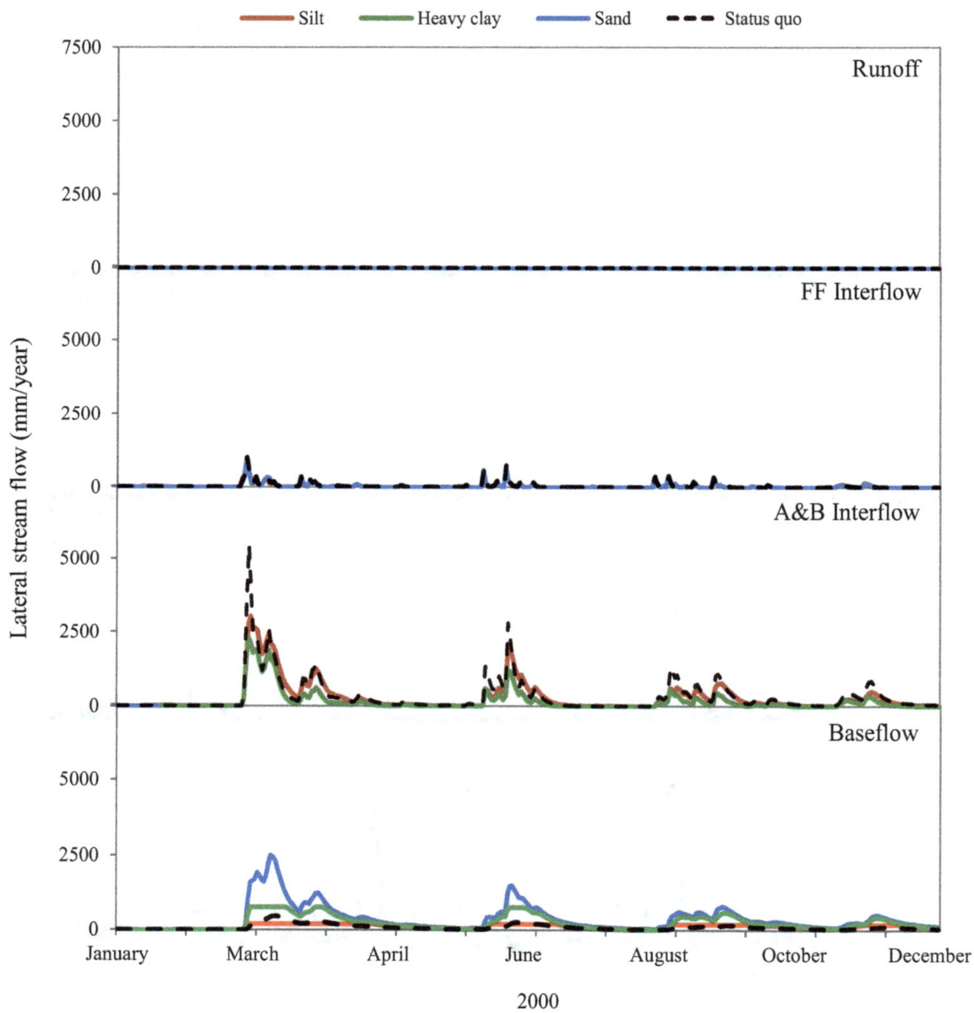

Figure 7. Run-off, forest floor and A&B interflow and base flow for Turkey lakes, by scenario from actual to sandy, silty and clayey (Scenarios 1 to 4, respectively; 2000), no runoff for any of the scenarios

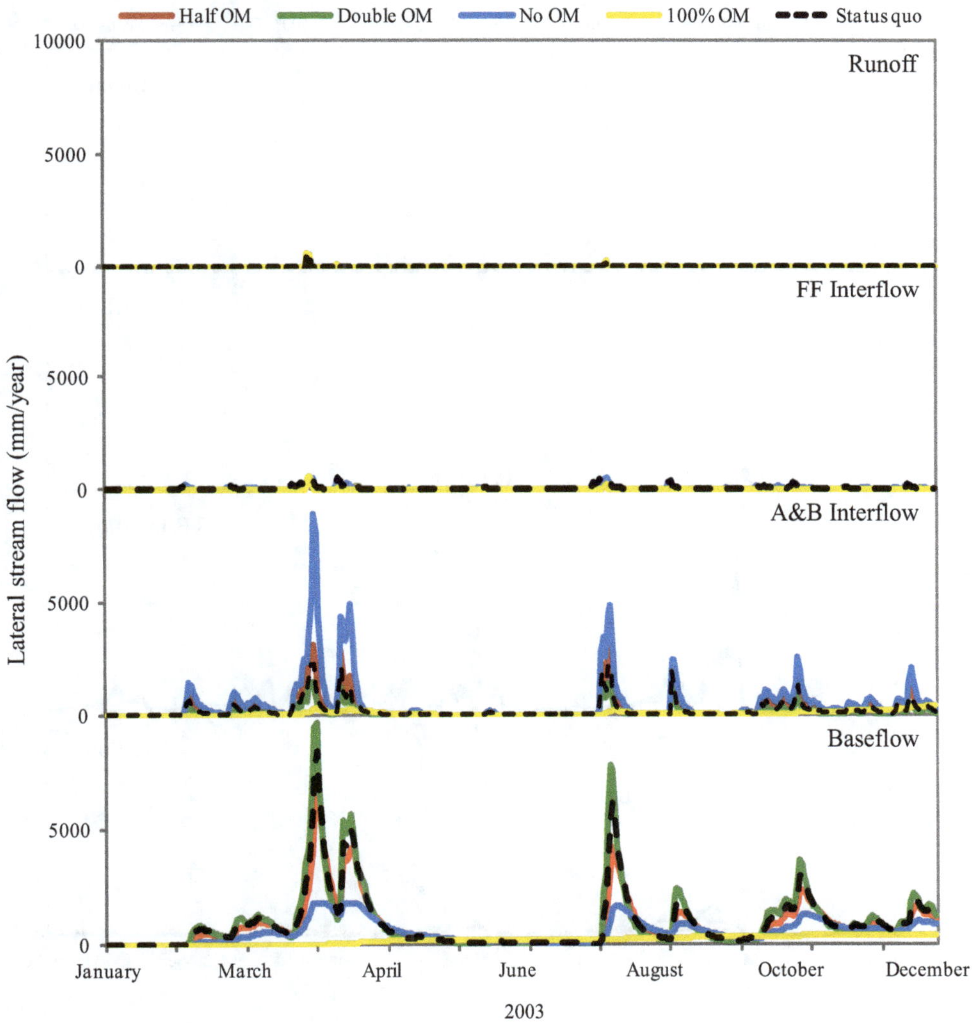

Figure 8. Run-off, forest floor and A&B interflow and base flow for Moosepit Brook, by scenario from actual to no, 0.5x and 2x actual organic matter content, and 100 % sapric organic matter (Scenarios 1 and 5 to 8, respectively; 2003)

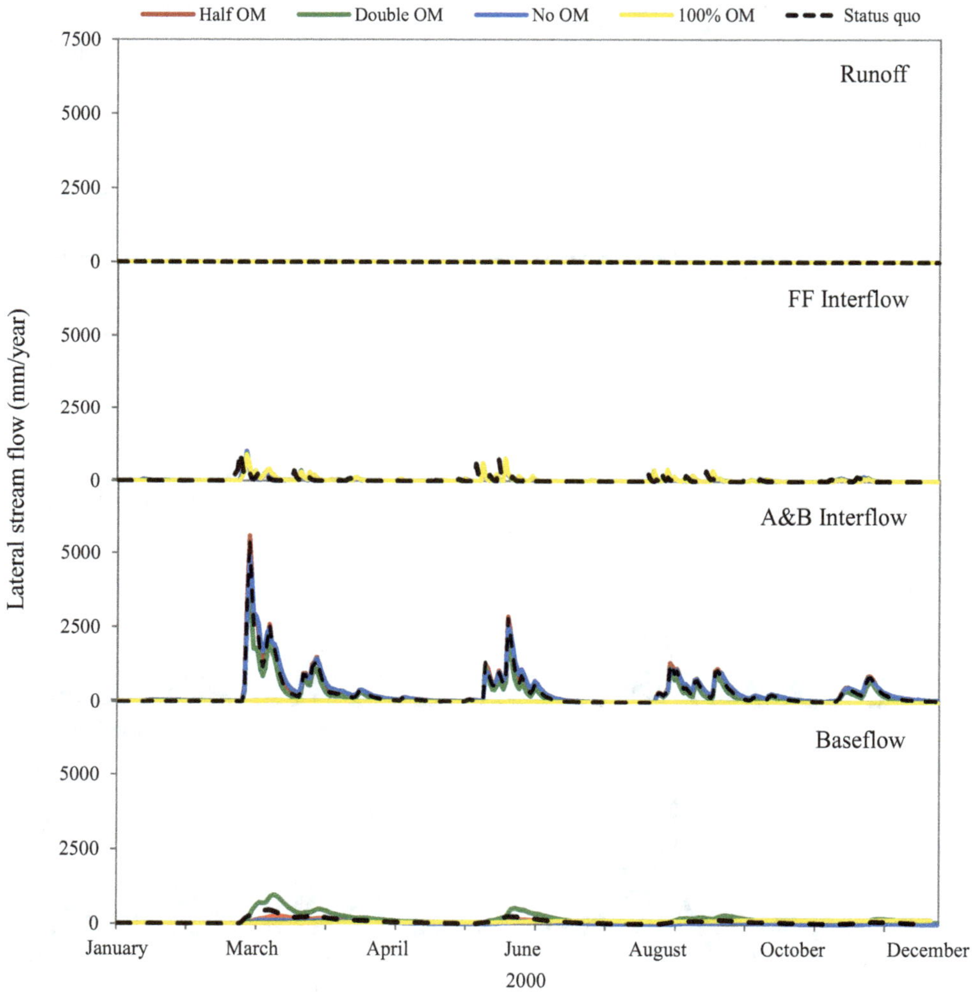

Figure 9. Run-off, forest floor and A&B interflow and base flow for Turkey Lakes, by scenario from actual to no, 0.5x and 2x actual organic matter content, and 100% sapric organic matter (Scenarios 1 and 5 to 8, respectively; 2000), no runoff for any of the scenarios.

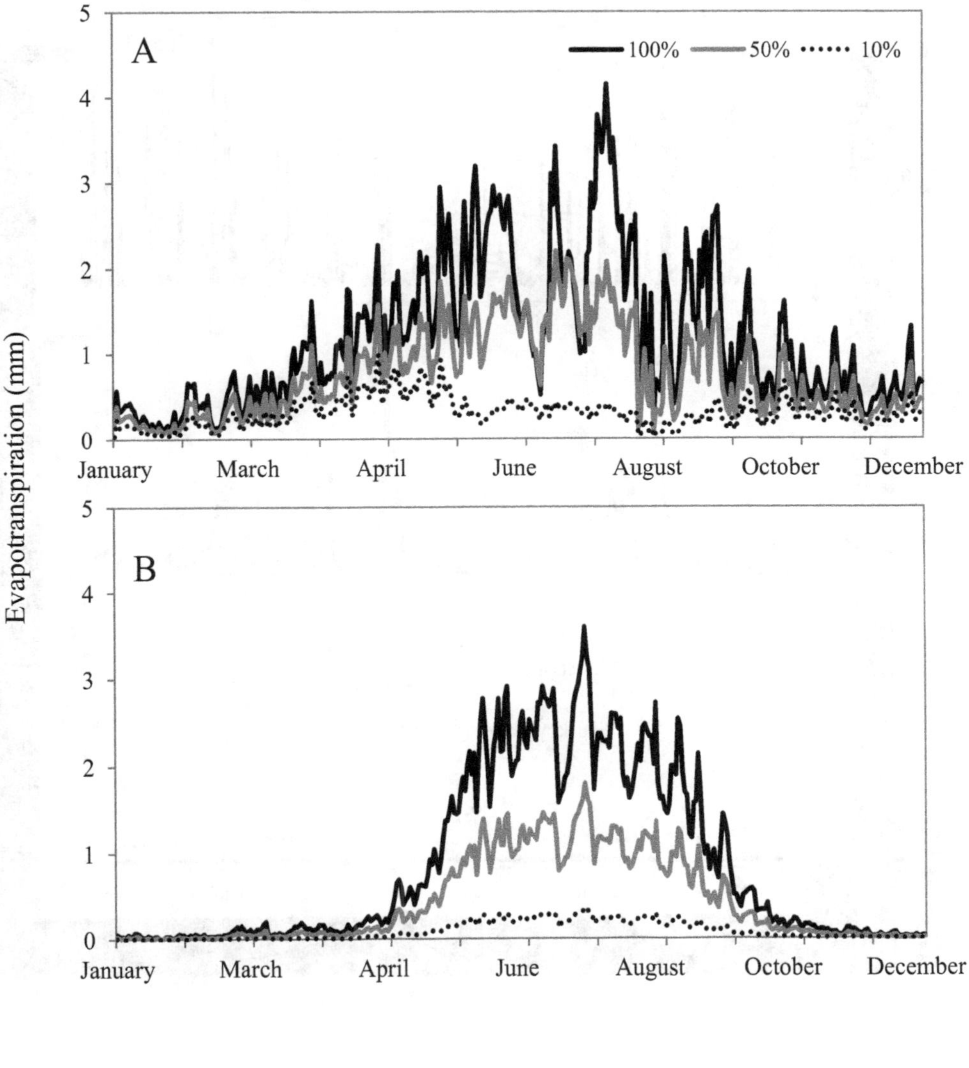

Figure 10. Evapotranspiration at Moosepit Brook (A) during 2003 and Turkey Lakes (B) during 2000, for Scenario 8 (100% OM), with actual (100% vegetation), Scenario 8^1 (50% vegetation), and Scenario 8^2 (10% vegetation). Note the difference in the extent of the growing season: wide for Moosepit Brook (maritime climate), and narrow for Turkey Lakes (continental climate).

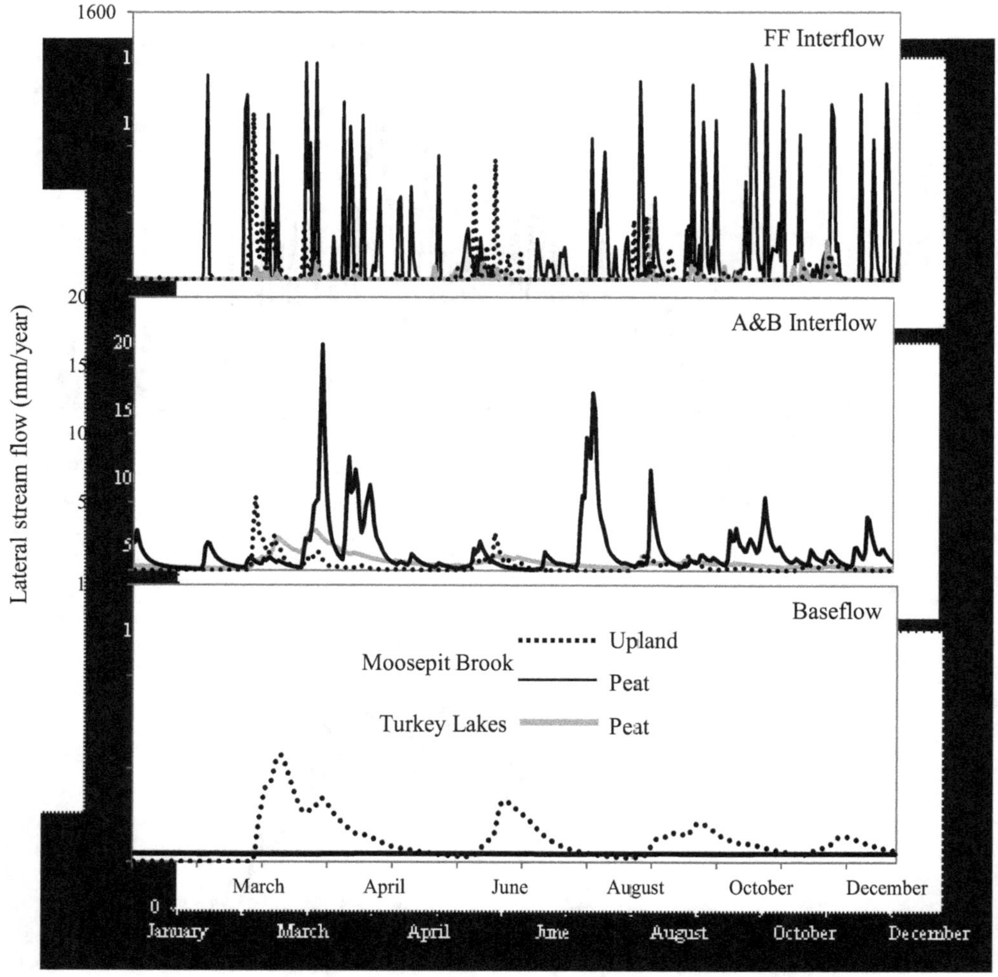

Figure 11. ForHyM2 estimated rates for daily forest floor and A&B interflow and base flow for peatland locations with a fibric - mesic – sapric layer profile at Moosepit Brook and Turkey Lakes, with 100% forest cover. Also shown: upland interflows and baseflows for the Moosepit Brook watershed (2000).

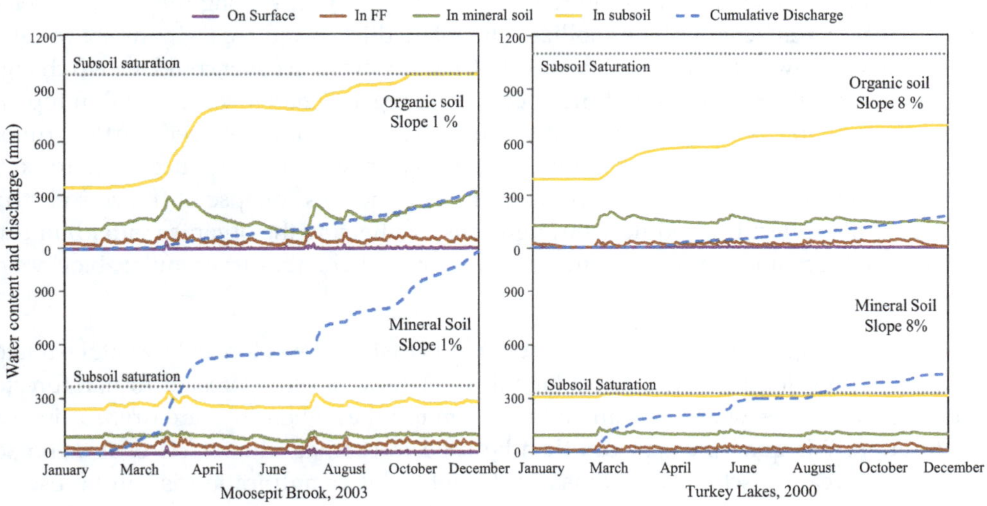

Figure 12. Soil water content on the surface, in the forest floor, in the mineral soil, and in the subsoil, as well as the cumulative discharge for Scenario 8 regarding organic soil (100% OM, top), and actual mineral soil conditions (bottom), for Moosepit Brook (2003, left) and Turkey Lakes (2000, right). Simulations start with unsaturated soil condition. Discussion

The above watershed-based K_{sat} evaluations have shown that the effective K_{sat} values for downward and lateral flow generally vary by a factor of 2 to 3 in comparison to corresponding values generated via Eqs. 1-3 [1]. As illustrated via Table 5 and subsequent figures, these variations lead to uncertainties in quantifying how water percolates through watersheds as run-off, interflow and baseflow (Fig. 5). These uncertainties also affect the flow response time, ranging generally from small delays to extended periods of flow as K_{sat} values decrease (Figs. 6 to 9). Across watersheds, however, flows tend to be well synchronized, regardless of major differences in texture, density, and organic matter content [28]. Typically, watersheds with the more compacted soils and therefore low K_{sat} values will be more peaked and will therefore be flashier than watersheds that allow deep percolation [25, 26, 27, 2]. The strongest impact of shallow to deep flow would deal with the water quality: deep water percolation during summer would lead to cooler and purer stream and seepage water with elevated pH than shallow water percolation [6]. During winter, deep percolation and persistent base flow would be warmer compared to the frost-affected surface water on poorly drained soils [6, 5]. Water flowing along the surface would also be more colored towards brown and more acidic than the more filtered and mineral-exposed water flowing at greater soil and subsoil depth [28]

While organic matter and soil density would not change drastically throughout undisturbed watersheds, such changes would occur during and after times of intense surface operations, especially under poor weather conditions. For example, forest operations during times of poor soil trafficability lead to ruts and increased soil compaction [29, 30]. In turn, soil compaction leads to lower K_{sat} values and therefore lower infiltration and hence higher surface run-off rates, thereby accelerating soil erosion and subsequent sediment transfer to streams and lakes [4]. Trails across the slopes of watersheds also affect downslope flow by compacting the soil underneath the trails, which means more water retention upslope along the trails, therefore leading to weather-effected trail destabilization, unless ditches and cross drains are installed to divert the water away from the trail beds [31]. Changes in forest cover could lead to changes in rooting space, which would – in turn – reduce the organic matter content within top and subsoils. This reduction would then alter the overall interplay between surface runoff, interflow and baseflow. Similarly, variations in climate from wet to dry (induces soil shrinking, may reduce root biomass), from frozen to non-frozen (induces collapse of frozen soil structures) would also affect K_{sat} and flow through soils by affecting the organic matter build-up, the state of soil organic matter humification, and overall changes in granular, blocky and columnar soil structures

The main advantage of the above K_{sat} formulation is that it allows for daily weather-related projections concerning downward and lateral water flow rates in forested to non-forested watersheds from times when soils are at saturation to times when soils are dry. At times of soil saturation, this quantification can then be used to estimate the effects of flow on soil stability and stream discharge. At times of drought, this quantification is can be used to estimate the effects of no flow on the remaining water reserves within soils and watersheds with and without peatland components (Fig. 12). Using ForHyM2 has the additional advantage of conducting these calculations year-round, summers through winters, based on already existing daily weather records, and extending these by way of daily, weekly, monthly or annual weather forecasts

Acknowledgements

Financial support for this research was received from Environment Canada, Alberta Sustainable Resource Department, and NSERC (Discovery Grant and CRD project grants).

Author details

Marie-France Jutras and Paul A. Arp

*Address all correspondence to: arp2@unb.ca

Faculty of Forestry and Environmental Management, University of New Brunswick, Fredericton, Canada

References

[1] Jutras MF, Arp PA. Determination of hydraulic conductivity from soil characteristics and its application for modelling stream discharge in forest catchments. In (Ed.) LE,

editor. Hydraulic Conductivity - Issues, Determination and Applications.: InTech; 2011. 189-202.

[2] Brady NC, Weil RR. The Nature and Poperties of Soils. 13th ed. Helba S, editor. Upper Saddle River: Prentice Hall; 2001.

[3] Crawford JW. The relationship between structure and the hydraulic conductivity of soil. European J Soil Sci. 1994;45: 493-502.

[4] Harr RD, Fredriksen RL, Rothacher J. Changes in streamflow following timber harvest in southwestern Oregon. Forest Service Resoure Paper PNW-249. Portland, Oregon: Pacific Northwest Research Station, USDA; 1979.

[5] Chi X. Hydrogeological assessment of stream water in forested watershed: temperature, dissolved oxygen, pH, and electrical conductivity. MSc. Thesis. Fredericton: University of New Brunswick, Forestry; 2008.

[6] Steeves MT. Pre- and post-harvest groundwater temperatures and levels, in upland forest catchments in Northern New Brunswick. MSc. F Thesis. Frederiction, NB: University of New Brunswick, Department of Forestry; 2004.

[7] Bouma J. Field measurement of soil hydraulic properties characterizing water movement through swelling clay soils. J Hydol.. 1980; 45: 149-158.

[8] Löfkvist J. Modifying soil structure using plant roots. PhD Thesis. Uppsala: Swedish University of Agricultural Sciences, Department of Soil Science; 2005.

[9] Yanni S, Keys K, Meng FR, Yin X, Clair T, Arp PA. Modelling hydrological conditions in the maritime forest region of south-west Nova Scotia. Hydrol. Processes. 2000b; 14: 195-214.

[10] Arp PA, Yin X. Predicting water fluxes through forests from monthly precipitation and mean monthly air temperature records. Can. J. For. Research. 1992; 22: 864-877.

[11] Verry ES, Boelter DH, Paivanen J, Nichols DS, Malterer T, Gafni A. Physical properties of organic soils. Chapter 5. In Kolka RK, Sebestyenm SD, Verry ESBKN, editors. Peatland biogeochemistry and watershed hydrology at the Marcell Experimental Forest. Boca Raton: CRC Press; 2011. 135-176.

[12] Päivänen J. Hydraulic conductivity and water retention in peat soils. Acta Forestalia Fennica. 1973; 129: 1-70.

[13] Letts MG, Roulet NT, Comer NT, Skarupa MR, Verseghy DL. Parametrization of peatland hydraulic properties for the Canadian land surface scheme. Atmosphere-Ocean. 2000; 38: 141-160.

[14] Boelter DH. Important physical properties of peat materials. In Proceedings, third international peat congress; 1968 August 18-23. Quebec, Canada: Department of Energy, Minds and Resources and National Research Council of Canada; 1968. 150-154.

[15] Meng FR, Bourque PA, Jewett K, Daugharty D, Arp PA. The Nashwaak experimental watershed project: analysing effects of clearcutting on soil temperature, soil moisture, snowpack, snowmelt and streamflow. Water, Air Soil Pollute. 1995; 82: 363-374.

[16] Balland V, Pollacco JAP, Arp PA. Modeling soil hydraulic properties for a wide range of soil conditions. Ecol. Model. 2008; 219: 300-316.

[17] Päivänen J. The bulk density of peat and its determination. Silva Fennica. 1969; 3: 1-19.

[18] Gafni A, Brooks KN. Hydraulic characteristics of four peatlands in minnesota. Can. J. Soil Sci. 1990; 70: 239-253.

[19] Silc T, Stanek W. Bulk density estimation of several peats in northern Ontario using the von Post humification scale. Can. J. Soil Sci. 1977; 57: p. 75.

[20] Szajdak LW, Szatylowicz J, Kolli R. Peats and peatlands, physical properties. In Gliński K, Horabik J, Lipiec J, editors. Encyclopedia of Agrophysics.: Springer; 2011. 551-555.

[21] Letts MG, Roulet NT, Comer NT, Skarupa MR, Verseghy DL. Parametrization of peatland hydraulic propoerties for the Canadian land surface scheme. Atmosphere-Ocean. 2000; 1: 141-160.

[22] Dube S, Plamondon A, Rothwell R. Watering up after clear-cutting on forested wetlands of the St. Lawrence lowland. Water Resource Research. 1995; 31: 1741-1750.

[23] Boelter DH. Hydraulic Conductivity of Peats. Soil Sci. 1965; 100: 227-231.

[24] Boelter DH, Verry E. Peatland and water in the northern lake states. General Technical Report NC-31. St. Paul, MN: Forest Service, North Central Forest Experiment Station, U.S. Dept. of Agriculture; 1977.

[25] Lull HW. Soil Compaction on Forest and Range Lands: Forest Services, U.S. Department of Agriculture; 1959.

[26] Kramer PJ, Boyer JS. Chapter 4: Soil and Water. In Water relations of plants and soils.: Academic Press, Inc; 1995. 84-114.

[27] Kozlowski TT. Soil compaction and growth of woody plants. Scand. J. For. Research. 1999; 6: 596-619.

[28] Jutras MF, Nasr M, Castonguay M, Pit C, Pomeroy J, Smith TP, et al. Dissolved organic cardon concentrations and fluxes in forest catchments and streams: DOC-3 model. Ecol. Model. 2011; doi:10.1016/j.ecolmodel.2001.03.035.

[29] McNabb DH, Startsev AD, Nguyen H. Soil wetness and traffic level effects on bulk density and air-filled porosity of compacted boreal forest soils. Soil Sci. Soc. J. America. 2001; 65: 1238-1247.

[30] Froehlich HA, McNabb DH. Minimizing soil compaction in Pacific Northwest forests. Proceedings Forest Soils and Treatment Impacts Conference. Knoxville, TN; 1984.

[31] Jamshidi R, Jaeger D, Raafatnia N, Tabari M. Influence of two ground-based skidding systems on soil compaction under different slope and gradient conditions. Int. J. For. Eng. 2008; 19: 9-16.

6

Evaluation o the Quality of the Soil from Soil Physical-Hydrical Indicators

Paulo Leonel Libardi and Flávia Carvalho Silva

1. Introduction

Due to the high demand for food, farmers have intensified agriculture seeking high productions. Hence, agriculture, amongst other activities, has been considered a highly potentially polluting activity of all the system water-soil-plant-environment. As a consequence, the intensive use of the soil causes its degradation. The accelerated waste will always exist if the farmer does not take proper measures to combat the causes related to various processes such as: chemical depleting and leaching, erosion, physical and biological degradation. The growing concern about the environment raised the concern about the quality of the soil. Ever since, several concepts of soil quality have been proposed, however, currently "soil quality" is defined as the capacity of the soil to keep biological productivity, environment quality and the vegetal and animal lives healthy on earth [1]. There has currently been a wide discussion towards environmental patterns and indicators, specially in Brazil, where there are very few studies and even a lack of systematization over the subject, since there is plenty of data that could provide support for an extensive discussion over the topic. The evaluation of the quality of the soil could be carried out by the monitoring of its features or physical, chemical and biological characteristics. In this approach, the expression "soil quality indicators" is being used, since it is the parameter or reference that best translate the conditions of a specific environment compartment. Among them, some attributes or physical indicators that might go through a few medium term changes have been recommended, such as density, porosity, aggregation and compression state. Hydrical conductivity, water retention, storage and density of water flow in the soil may also be indicators of great importance to assess the quality of the soil. Although such parameters are not frequently studied in Brazil, in foreign literature they are reported to vary according to different soil preparations and management. Detailed familiarity of the water dynamics during the development of a culture provides essential elements for the stablishment or improvement of agricultural management practices that aim to optimize the productivity. Water is a fundamental factor in the development of a culture, interfering mainly in the development of the root system and the absortion and tranference of nutrients for the plants. When a certain culture is irrigated, or when it rains, the water penetrates the soil and redistributes in its interior. If the quantity added by these processes exceeds the soil infiltration capacity, the excess moves sideways over the soil surface. The theoretical-practical knowledge of the runoff water, infiltration and evapotranspiration, the movement of the water in the interior of the soil profile, the absortion of the water by the plants, the draining, etc, is extremely relevant to obtain an increase in the productivity of the cultivars and to prevent the environmental degradation.

2. Body

When hydrical conductivity of a soil in mentioned, it is implicit that a sample of soil is being considered a porous media and as a fluid that moves in it an aqueous solution (hydrical). The methods of direct determination of hydrical conductivity of the soil may be classified in laboratory methods and field methods. In the laboratory methods, samples may be used with deformed structure or samples with undeformed structure and they are subdivided for saturated and unsaturated conditions. For the measurement of the hydrical conductivity of saturated samples a simple equipament is used, denominated permeameter, and for unsaturated samples the hydrical conductivity may be measured under the steady-state condition, in which big or small columns are used and also under transient conditions. The measurement of hydrical conductivity in the field may also be carried out under the saturation condition (below the groundwater) and under unsaturated conditions (above the groundwater). The Instantaneous Profile Method [2]; [3]; [4] is the most commonly used in the field and it is applied for unsaturated conditions in situations in which the groundwater is nonexistent or very deep; it also has the advantage of being possibly used in heterogenous soil [3]. To apply this method in the field, a portion of sufficiently large area of soil must be chosen, so that the processes in the center are not affected by the boundaries. This area must be conveniently instrumented for the measurement of the amount of water in the soil and the matric potencial, along the profile until the depth of interest. The surface of the soil is, then, mantained under water depth, in a way that the profile, until de depth of interest is as humid as possible. Reaching this condition, which may be perceived by the virtually no variation of the content of water and the matric potencial with the infiltration time along the profile, infiltration is interrupted and the surface of the soil is covered with a plastic cover to avoid evaporation and the water inlet through the surface. The water in the profile redistributes, then, by the process of internal drainage, and as this process occurs, periodical measurements of water and matric potencial are carried out, data to determine the hydric conductivity K according to the water content θ, that is, the function $K(\theta)$ [5]. The measurements of the water content for this purpose have normally been carried out through a) techinque of neutron moderation or, simply, neutron probe (for example [6], b) tensiometers, installed in several depths, together with their respective retention curves [7] and, more recently, c) time-domain reflectometry technique or, simply TDR [8]. Other techniques are or will be able to be utilized, however, for the measurement of the water content, since the thermogavimetric method, taken as a pattern, the gamma radiation attenuation techinque and new techniques that have been developed as the computerized tomography (CT), Magnetic Resonance Imaging (MRI), electromagnetic induction, etc. [9]. In Brazil, there are few studies about hydric conductivity for different preparations and soil management and, besides that, the experimental results and contradicting, implicating in detailed analysis of the physical-hydrical properties of the place where the research is being carried out, as this indicator depends on the type of soil, management systems, vegetation, etc. Empirical parameters of K adjustment function θ, allowing function K(θ) of Oxisol, under different management systems, were obtained by [10] in laboratory. These authors verified discrepancies in the conductivity behavior for different managements and depths, with the same tendency observed in the field determination (Table 1). More recently, [11] observed, in Latosol, that the average hydrical conductivity of saturated soil until 1m deep is larger in bare soil than in cultivated soil (Table 2).

Depth (m)	Woods			Growing upland rice			Irrigated lowland rice		
	a	b	r^2	a	B	r^2	a	b	r^2
0,2	53,47	-22,22	0,85	30,95	-13,78	0,79	59,56	-27,10	0,94
0,3	54,61	-21,61	0,97	48,84	-20,02	0,75	57,76	-27,47	0,95
0,4	54,30	-21,33	0,91	47,26	-18,60	0,94	46,98	-22,05	0,93
0,5	50,86	-20,98	0,92	57,32	-22,42	0,93	52,08	-22,56	0,97
0,6	35,32	-14,91	0,88	37,92	-15,71	0,91	59,33	-24,96	0,93
0,7	39,08	-19,52	0,95	40,86	-19,16	0,91	41,17	-16,96	0,97
0,8	33,38	-14,47	0,92	40,87	-17,08	0,96	43,16	-18,48	0,89
0,9	35,27	-15,65	0,98	38,35	-15,23	0,93	35,58	-15,00	0,94

Empirical parameters a and b of the equation $\ln K = a\theta + b$ e r^2 = determination coefficient.

Table 1. Empirical parameters to calculate unsaturated soil conductivity as a function of adjusted managements from data obtained in the field (1); *taken from [10].*

Depth (m)	Management	
	Bare	Cultivated
0,2	$3{,}25.10^{-3}$	$3{,}55.10^{-4}$
0,4	$2{,}07.10^{-3}$	$6{,}16.10^{-4}$
0,6	$4{,}45.10^{-3}$	$1{,}03.10^{-3}$
0,8	$2{,}79.10^{-3}$	$1{,}93.10^{-3}$
1,0	$3{,}74.10^{-3}$	$2{,}55.10^{-3}$
Average	$3{,}26.10^{-3}$ a	$1{,}30.10^{-3}$ b

Averages followed by the same letters differ from each other to 5% in probability by the Tukey test.

Table 2. Average hydrical conductivity of saturated soil (mh^{-1}) as a function of management systems and depth taken from Angelotti Neto & Fernandes (2005).

The soil, water reservoir for the plants, is affected by the management and culture practices, changing the dynamics and the retention of water in its pores. This retention of water in the soil matrix is controlled by two types of force, capillary and adsorption, which are denominated matric forces and originated the term matric potential of water in soil. The retention curve connects the amount of solution in the soil in equilibrium with the tension applied. It is also known that once the retention curve is obtained one can calculate, from it, the distribution of pores size in the soil. With the water retention curve or, simply, retention curve, the grower, during the management of water, will know exactly when and how much water he will need to add to the system of production to supply the plant properly with no loss of the root system for internal drainage. It was [12], in his work searching for an equation to quantify the movement of water in the soil under unsaturated conditions, who introduced in the soil science this relation between the content of water and the matric potencial. The methodology of retention curve determination had its beginning, apparently, as a [13] work that utilized a funnel fitted with a porous plate in the lower part with which potential matric values or the water retention in the soil were fixated and it measured the corresponding content of water after the equilibrium. For this reason, this kinf of funnel available in the market is also called Haines funnel [5] and it is used to determine of the most humid part of the retention curve (tension levels below 10 kPa). After that [14, 15, 16] developed a porous plate pressure chamber to measure higher tensions (10 to 2000 kPa) also known as Richards pressure chamber. Nowadays, both these devices are routine equipment to determine the retention curve in soil physics laboratories. Some variations of the funnel may be found (but the theoretical principle is the same) under names such as tension table, sand tank, etc. [17]. It is important to make it clear that the retention curve can be obtained through wetting or drying, resulting in the hysteresis phenomenon [18, 19]. In this context, to utilize the results practice, there is a need to specify the curve branch which is being used [20]. The complete retention curve of a soil (from 0 to 1500 kPa), besides reflecting the water behavior in the soil in terms of water availability for the plants, is also a reflex of the important physical properties of the soil such as texture, structure, pores distribution, consistence, etc. [21, 22, 23]. In this context, it is important to mention the recent work of [24] about the physical quality of a soil through the retention curve. Basically, what the author does is define a parameter S that is nothing but a slope of the retention curve (expressed as the content of water in the soil as a function of the tension logarithm of the water in the soil) in its inflection point and based on several arguments, the higher the S, in absolute values, the better the physical quality of the soil. As an illustration, in figure 1 below, the exemple introduced by Dexter himself [24] of two retention curves of the same degraded (or compressed) and non-degraded soil. Besides that, as regarded to water conduction through soils, the retention curve may be an alternative to the direct hydric conductivity of the soil K as a function of the content of water in the soil θ, that is, this function K (θ) can be theoretically calculated from water retention data more easily measured in laboratories. [25, 26, 27, 28, 29, 30].

In figure 2 retention curves of five different depths of Oxisol under cerrado are presented, tillage and plowing discs preparation. Data, obtained by [31], show, according to the authors, that, in the tension belt between 100 and 1500 kPa, particularly in the systems that involve cultivation, in many depths, the curves showed an aspect nearly rectilineal, asymptotic to the

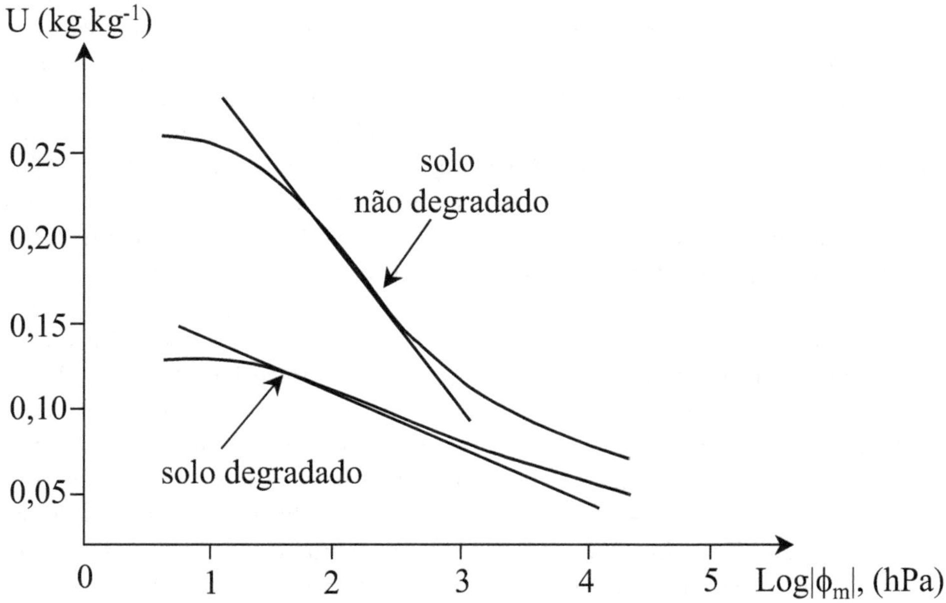

Figure 1. Retention curves of the same soil in two situations: degraded and non-degraded [24].

abscissa axis, indicating the existence of ultra-micropores with low water-storage capacity. [32] concluded, from results obtained in figure 3, that in the layer from 0,4-0,8 m, significant increase in the water retention in the soil under the irrigated system was noticed between the matric potencial of –33 to –1.500 kPa, comparing to the woods soil, due to a more expressive reduction of macropores in this soil.

Knowing how the plants utilize the water retained in the soil and how they respond to the level or storage in the profile, may be a viable solution to establish effective management strategies aiming a better possible use of water reserves in the soil by the crops. Water storage in the soil represents the water that the soil is storing in a specific layer in a specific depth. In figure 4 the results obtained by [33] are illustrated and determine the variation of water storage in the soil for three managements, between current humidity and PWP, for eight different redistribution times and five depths in determination experiment of function K(θ) by the instantaneous profile method. They concluded that until the 0,25 m layer, the under rainfed soil always retained more water than the woods soil and irrigated, having a higher variation between the woods and two other management systems.

Nitrogen leaching is extremily important because it can decrease the quantity of ammonium and nitrate in the topsoil sensibly and, consequently, reduce the nutrient availability; when in excess, leaching constitutes a potential danger of groundwater contamination by nitrate [34]. Over the last few decades, surface and underwater sources contamination with nutrients, particularly N and P, has become a significant subject for people in general,

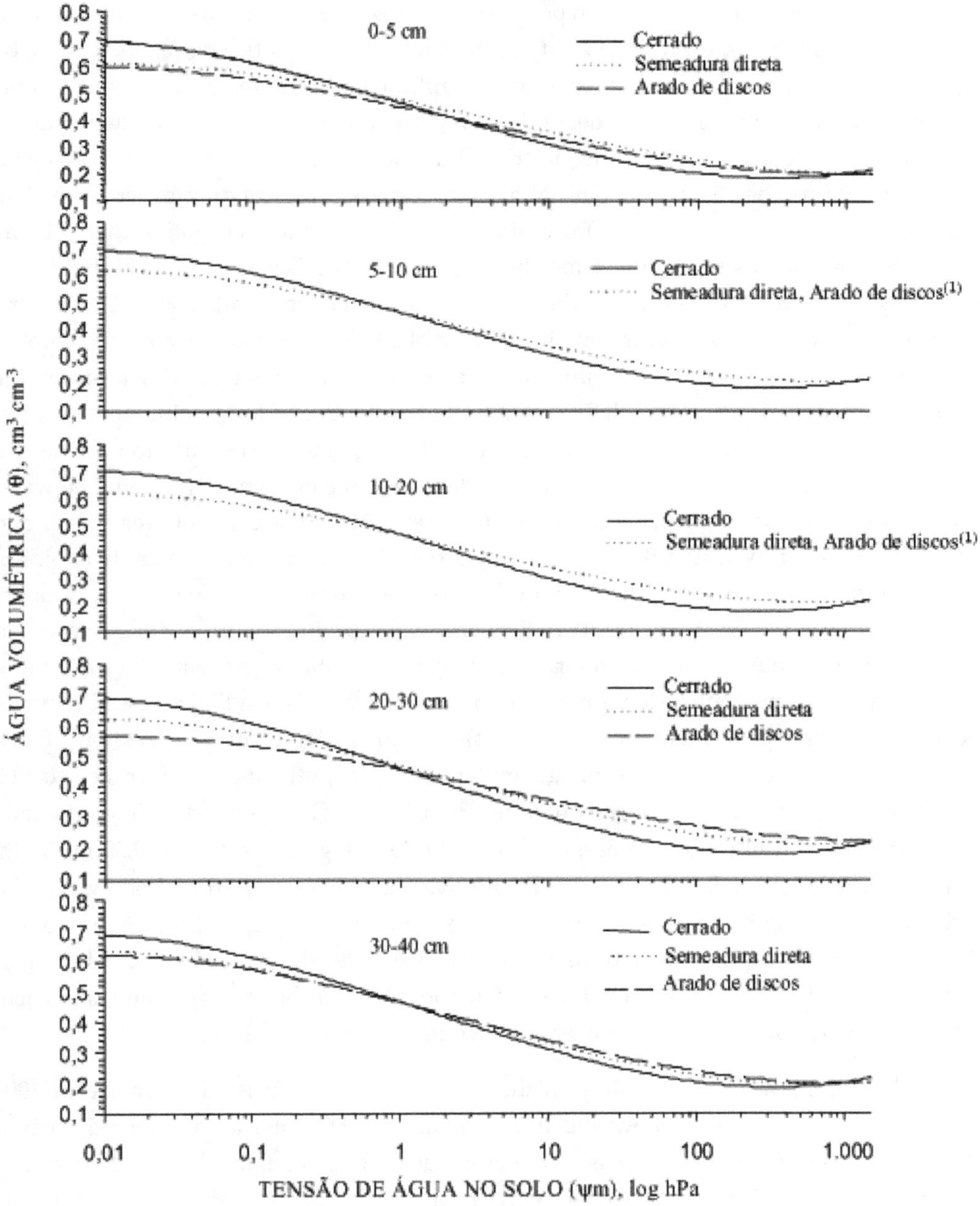

Figure 2. Retention curves of five different depths of Oxisol under cerrado, tillage and plowing discs preparation [31].

including the farmers. High concentrations of nitrate in the water for human consumption are worrying, as they cause methemoglobinemia, commonly known as blue baby syndrome. This is a problem that occurs only in children under six months old and pregnant women. Besides that, water contamination by nitrates and P has been connected with another abnormality denominated hypoxia (low oxygen levels) in water from Gulf of Mexico, which inhibits the production of shrimp and other acquatic species in this region [35]. For the USA, the maximum concentration of nitrate for water to be considered potable and not cause any harm to human health is 10 mg L^{-1} and 1mg L^{-1}, respectively, being the same values adopted by Brazilian Legislation, but for the European Union the accepted value of nitrate reaches 50 mg L^{-1} [36]. N leaching occurs in descending order for NO_3^- > NH_4^+ > Organic-N. Most leaching in form of nitrate occurs due to its negative charge being repelled by soil colloids where the same charge prevails [37]. The amount of N that is lost in leaching varies a lot according to the N dose, the method of application of fertilizers, the speed of mineralization and immobilization by plants and microorganisms, the amount of rain and the soil properties that influence its capacity to retain water, whichever are the texture, structure and porosity [38]. To quantify the elements due to leaching, it is necessary to know the water flow density in the soil and the element concentration in the soil solution, since their product generates the flow density of the element in the soil. The water flow density in the soil represents the water flow per unit cross-sectional area of soil and may be determined by Darcy-Buckingham equation [5]. Studies about water dynamics in field conditions emphasizing water flow in the crop root zone are not frequent and, often, incomplete, due to the great complexity of the necessary experimental procedures. However, as mentioned previously, knowing internal drainage is essential to estimate the leaching of chemicals down to the root system of the crop under analysis. When [39] assessed a nitrate leaching (total and provenient from fertilizers) to 0,80 m depth, in a succession of cultures corn-black oat-corn, under no-tillage implemmentation, utilizing fertilizer labeled ^{15}N, they obtained the results that are presented in Chart 3. It can be observed through the data in the chart, as concluded by the authors, that total leaching nitrate loss to 0,80 m depth, in the first corn crop, in 120 kg ha^{-1} of N dose, was approximately 96 and 68 kg ha^{-1}, for parcelling with 60 kg ha^{-1} and 30 kg ha^{-1} of N in sowing, respectively, of which only 3 and 1 kg ha^{-1} of N were provenient from the nitrogen-rich fertilizer. In the second year of cultivation, nitrate leaching was far lower than compared to the first year and the leachate nitrate provenient from the fertilizer was negligible (average 0,23%).

[40] continuing studies with leaching in the crop succession consisted of: maize in 2006 followed by brachiaria plus fallow and finally maize in 2007. Internal drainage decreased with the increase of applied N levels to the crop succession, changing from 31.5 to 73.4 % of the total rainfall (97 mm) in the first maize, from 26.1 to 58.1 % of the total rainfall (695 mm) in brachiaria + fallow, and from 56.6 to 87.4 % of the 419 mm of total rainfall in the second maize crop. The leaching of total nitrate (from fertilizer and other sources) was very low in the first maize crop for all applied N levels and significant for the rates of 120 and 180 kg ha-1 in the periods of brachiaria plus fallow (26.16 for 120 kg ha-1 and 39.8 for 180 kg ha-1) and of the second maize crop (approximately 23 kg ha-1 for both levels). There was no N leaching from fertilizer in the first maize crop and N leaching was very low in the brachiaria and second maize crop.

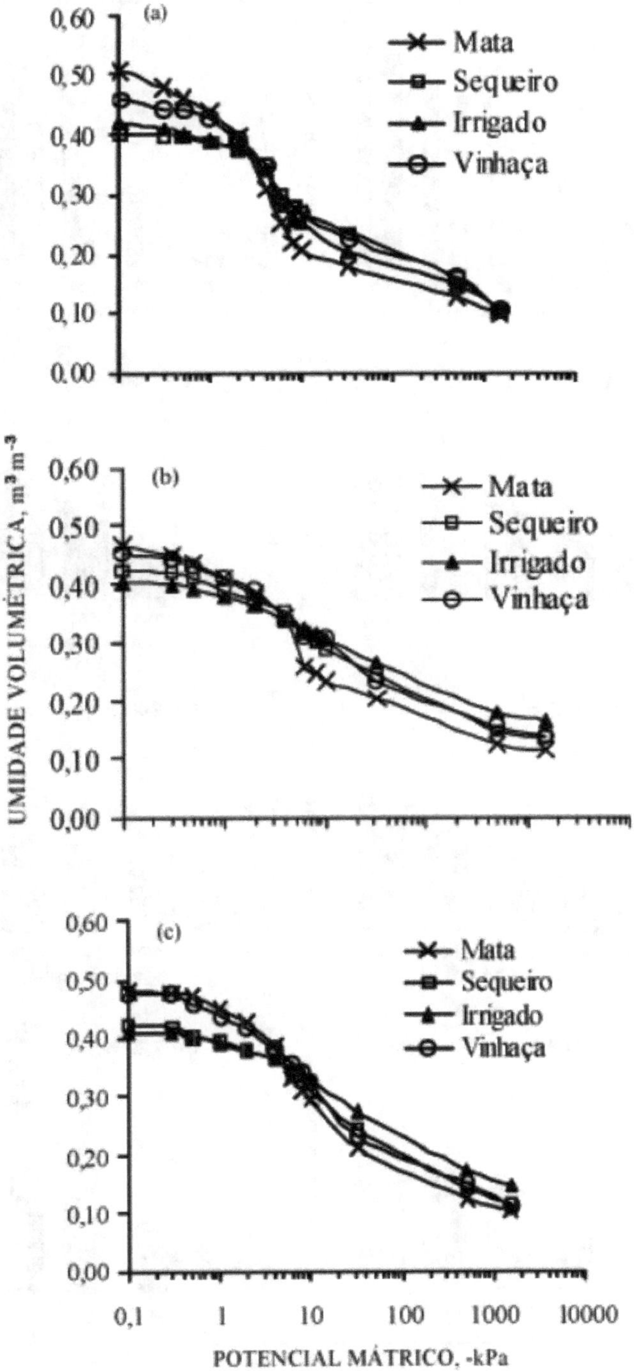

Figure 3. Retention curves in the layers (a) 0-0,2 m, (b) 0,2-0,4 m e (c) 0,4-0,6 m, for the use and management of the soils studied systems [32].

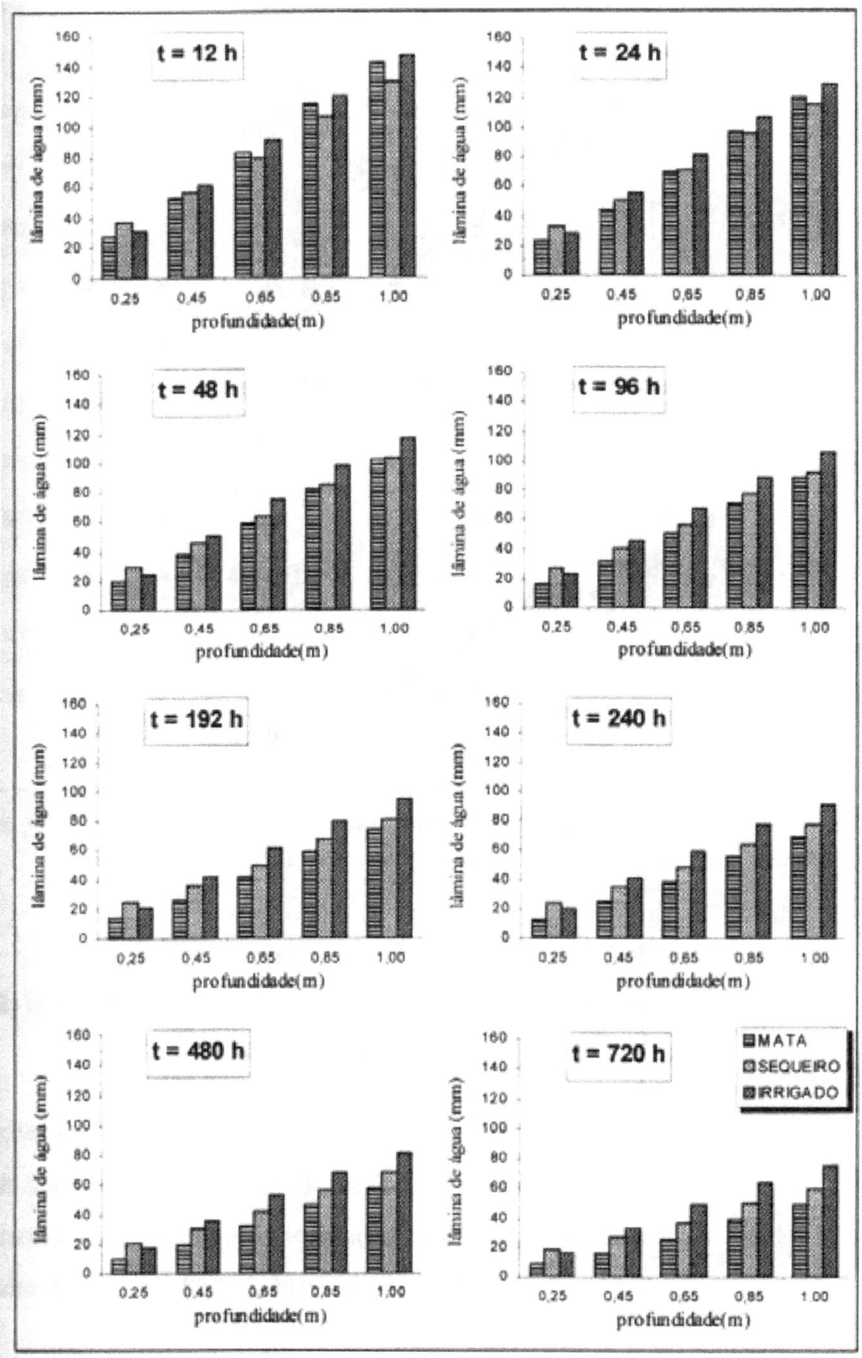

Figure 4. Variation of water storage in the soil for three managements, between the current humidity and PWP, for eight different drainage times and five depths [33]

Treatments 1	Treatments 2	DAE	Corn cultivation – harvest 2003/2004			
			Precipitation (mm)	Water drainage to 0,80 m (mm)	Leaching-NO^-_3 (kg ha^{-1}) Total	Leaching-NO^-_3 (kg ha^{-1}) PF
30	**90**	**0-30**	110,1	103,62 a	10,27 a	0,20 a
60	**60**			129,91 a	13,97 a	0,18 a
30	**90**	**30-60**	200,6	165,67 a	33,14 a	1,79 a
60	**60**			180,11 a	20,00 b	0,28 b
30	**90**	**60-90**	239,4	155,71 b	24,93 b	0,94 a
60	**60**			218,63 a	61,64 a	0,86 a
30	**90**	**90-120**	64,9	1,54 b	0,01 a	0,07 a
60	**60**			3,65 a	0,01 a	0,07 a
30	**90**	**Total**	**615,0**	**426,54 b**	**68,35 b**	**3,00 a**
60	**60**			**532,29 a**	**95,61 a**	**1,39 b**

Treatments 1	Treatments 2		Between the first corn cultivation and black oat			
			Precipitation (mm)	Water drainage to 0,80 m (mm)	Leaching -NO^-_3 (kg ha^{-1}) total	Leaching -NO^-_3 (kg ha^{-1}) PF
30	**90**	**Total**	296,0	151,18 b	2,81 b	-
60	**60**			243,24 a	4,77 a	-

Treatments 1	Treatments 2	DAE	Black oat – harvest 2004			
			Precipitation (mm)	Water drainage to 0,80 m (mm)	Leaching -NO^-_3 (kg ha^{-1}) total	Leaching -NO^-_3 (kg ha^{-1}) PF
30	**90**	**0-40**	0,0	6,68 b	0,00 a	-
60	**60**			11,88 a	0,00a	-
30	**90**	**40-80**	146,5	15,02 b	0,62 b	0,01 a
60	**60**			61,69 a	2,64 a	0,02 a
30	**90**	**Total**	**146,5**	**21,70 b**	**0,62 b**	**0,01 a**
60	**60**			**73,57 a**	**2,64 a**	**0,02 a**

Treatments 1	Treatments 2		Between black oat and the second corn cultivation			
			Precipitation (mm)	Water drainage to 0,80 m (mm)	Leaching -NO^-_3 (kg ha^{-1}) total	Leaching -NO^-_3 (kg ha^{-1}) PF
30	**90**	**Total**	136,3	56,78 b	0,07 a	-
60	**60**			142,14 a	0,03 a	-

Treatments 1	Treatments 2	DAE	Corn cultivation - harvest 2004/2005			
			Precipitation (mm)	Water drainage to 0,80 m (mm)	Leaching -NO^-_3 (kg ha^{-1}) total	Leaching -NO^-_3 (kg ha^{-1}) PF
30	**90**	**0-30**	181,9	113,16 b	4,80 a	-

60	**60**			180,17 a	6,45 a	-
30	**90**	**30-60**	231,7	132,50 b	11,79 a	-
60	**60**			211,40 a	9,80 a	-
30	**90**	**60-90**	78,4	0,61 b	0,00 a	-
60	**60**			4,74 a	0,00 a	-
30	**90**	**90-120**	164,4	133,37 a	6,43 b	0,13 a
60	**60**			134,59 a	12,12 a	0,13 a
30	**90**	**Total**	**656,4**	**379,64 b**	**23,02 a**	**0,13 a**
60	**60**			**530,90 a**	**28,37 a**	**0,13 a**

Obs. The treatments 30-90 and/or 60-60 refer to the application of N in the sowing in the 6-8 leaf stage. DAE stands for "days after emergency". Averages followed by the same letter do not differ from each other in significance level 5% by Tukey test. 1 and 2 refer to the nitrogen fertilization in corn, respectively in the sowing and coverage.

Table 3. Water drainage and total nitrate leaching provenient from the fertilizer (PF) to 0,80 m depth, in the succession corn-black oat-corn [39]

3. Conclusions

Conservative management systems, such as tillage, present an expressive improvement effect in the hydro-physical quality of tropical and subtropical soils. The soil in function of type has diverse physical, chemical and biological composition, making it difficult to have a pattern for the amount of hydro-physical indicators to evaluate their quantity. Therefore, there is a need for more detailed studies, particularly for tropical conditions that can define more precise values for the indicators of quality. Different types of soil have a great influence in the leaching process magnitude, however, other factors such as organic matter content, dose, type and time of fertilizer application, mainly the nitrogen-rich one, and also the weather, have great influence in this process. Thus, leaching studies aiming the use of fertilizers and their interaction with different soil managements are influential to Brazilian research, since there is a great lack of results for soils in tropical and subtropical weather countries.

Author details

Paulo Leonel Libardi[1] and Flávia Carvalho Silva[2*]

*Address all correspondence to: flcarvalhoblg@gmail.com

1 Exact Sciences Department– ESALQ-USP, Piracicaba, SP, Brazil

2 Agronomics Sciences Department – UEM, Umuarama, PR, Brazil

References

[1] Doran, J.W.; Parkin, T.B. *Defining and assessing soil quality*. In: DORAN, J. W. et al. (Eds.). Defining soil quality for a sustainable environment. Madisson: ASA/SSSA, 1994. p-3-21.

[2] Watson, K.K. An instantaneous profile method for determining the hydraulic conductivity of unsaturated porous materials. *Water Resources Research*, 1996; 2 709-715.

[3] Hillel, D.A; Krentos, V.K.; Stilianov, Y. Procedure and test of an internal drainage method for measuring soil hydraulic characteristics in situ. *Soil Science.*, 1972; 114, 395-400.

[4] Libardi, P.L.; Reichardt K.; Nielsen, D.P.; Biggar, J.W. Simple field methods for estimating soil hydraulic conductivity. *Soil Science Society of American journal*, 1980; 44, p. 3-7.

[5] Libardi, P.L. *Dinâmica da água no solo.* 2.ed. São Paulo: Editora da Universidade de São Paulo, 2005. 329p.

[6] Greacen, E.L. Soil water assessment by neutron method. CSIRO. Austrália, 140p, 1981.

[7] Carvalho, L.A. de. Condutividade *hidráulica do solo no campo: as simplificações do método do perfil instantâneo*, 2003. Tese (Mestrado) – "Escola Superior de Agricultura Luiz de Queiroz", Universidade de São Paulo, Piracicaba, 2003.

[8] Topp, G.C.; Davis, J.L.; Annan, A.P. Eletrotromagnetic determination of soil water content: measurement in coaxial transmission lines. *Water Resources Research*, 1980; 16(3): 574-582.

[9] Smith, K.A.; Mullins, C.E. Soil analysis-physical methods. New York: s.ed., 1991. 620p.

[10] Klein, V.A.; Libardi, P.L. Condutividade hidráulica de um Latossolo Roxo não saturado, sob diferentes sistemas e manejo. *Ciência Rural*: Santa Maria, 2002; 32 (6), 945-953.

[11] Angelotti Neto, A.; Fernandes, E.J. Condutividade hidráulica de um Latossolo Vermelho em pousio e cultivo intensivo. *Pesquisa Agropecuária Brasileira*, 2005; 40 (8), 797-802.

[12] Buckhingham, E. *Studies of the movement of soil moisture*. USDA Bur. Soil Bull. 38. US Government Printing Office. Washington D.C., 1907.

[13] Haines, W.B. Studies in the physical properties of soil: V. The hysteresis effect in capillary properties and the models of misture associated therewith. *Jounal Agronomy. Science*, 1930; 20, 97-116.

[14] Richards, L.A. A pressure membrane extraction apparatus for soil solution. *Soil Science*, 1941; 51, 377-386.

[15] Richards, L.A. Pressure membrane apparatus – construction and use. *Agronomy. Engineer*, 1947; 28, 451-454.

[16] Richards, L.A. Methods of measuring soil moisture tension. *Soil Science*. 1949; 68, 95 – 112.

[17] Kutílek, M; Nielsen, D.R. *Soil Hydrology*, Catena Verlag, Germany, 1994, 370p.

[18] Poulovassilis, A. The influence of the initial water content on the redistribution of soil water after infiltration. *Soil Science*, Baltimore, 1983; 135, p. 265-281.

[19] Klute, A. Water retention: laboratory methods. In BLACK, C.A., ed. Methods of Soil Analysis. I Physical and mineralogical methods. Madison, Winconsin, American Society of Agronomy, Inc., *Soil Science Society of America*, Inc. Publisher, 1986, 635-662.

[20] Moraes, S.O. *Heterogeneidade hidráulica de uma terra roxa estruturada*. Piracicaba, 1991, 141p. (Doutorado – Escola Superior de Agricultura "Luiz de Queiroz"/USP), 1991.

[21] Sharma, M.L; Uehara, G. Influence of soil structure on water relations in low humic latosols: J. Water retention. *Soil Science Society America Journal*, Madison, 1968; 32, 765-770.

[22] Reeve, M.J.; Smith, P.D.; Thomasson, A.J. The effect of density on water retention properties of field soils. *Journal of Soil Science*, Oxford, 173; 24, 355-367.

[23] Rawls. W.J. e Pachepsky, Y.A. Soil Consistence and Structure as predictors of water retention. *Soil Science Society. American. Journal*, 2002, 66, 1115-1126.

[24] Dexter, A.R. Soil physical quality – Part I. Theory, effects of soil texture, density and organic matter and effects on root growth. *Georderma*, 2004; 120, 201-214.

[25] Childs, E.C.; Colins George, N. The permeability of porous materials. Proc. Roy. Soc., *London, Ser. A*. 1950; 201, 392-405.

[26] Burdine, N.T. Relative permeability calculations from pore-size distribution data. *Petroleum Trans. Am. Inst. Mining Eng.*, 1953; 198, 71-77.

[27] Marshall, T.J. A relation between permeability and size distribution of pores. *Journal Soil Science*, v. 9, p. 1-9, 1958.

[28] Millington, R.J.; Quirk, J.P. Permeabilty of porous solids. Trans. *Faraday Society.*, 1961; 57, 1200-1207.

[29] Mualem, Y. A new model for predicting the hydraulic conductivity of unsaturated porous media. *Water Resources Research*, 1976; 12, 513-522.

[30] Van Genuchten, M.T. A closed-form equation for predicting the hydraulic conductivity of unsaturated soils. *Soil Science Society America Journal*, 1980; 44, 892-898.

[31] Oliveira, G.C.; Dias Júnior, M.C.; Resck, D.V.C.; Curi, N. Caracterização química e físico-hídrica de um Latossolo Vermelho após vinte anos de manejo e cultivo do solo. *Revista Brasileira de Ciência do Solo,* 2004; 28, 327-336.

[32] Silva, A.J.N.; Cabeda, M.S.V.; Lima, J.F.W.F. Efeito de sistemas de uso e manejo nas propriedades físico-hídricas de um argissolo amarelo de tabuleiro costeiro. *Revista Brasileira de Ciência do Solo,* 2005; 29,.833-842.

[33] Klein, V.A.; Libardi, P.L. *Propriedades físico-hídrico-mecânicas de um Latossolo Roxo, sob diferentes sistemas de uso e manejo.* 150 p., 1998. Tese (Doutorado em Solos e Nutrição de Plantas) – Escola Superior de Agricultura "Luiz de Queiroz", Universidade de São Paulo, Piracicaba, 1998.

[34] Stark, J.C.; Jarrel, W.M.; Letey, J. Evaluation of irrigation-nitrogen management practices for celery using continous-variable irrigation. *Soil Science Society American Journal,* Madison, 1980; 47, 95-98.

[35] Hoeft, R.G. *Desafios para a obtenção de altas produtividades de milho e de soja nos EUA.* Piracicaba: Potafos, p.1-4, 2003 (Informações Agronômicas, 104).

[36] Tundisi, J.G. *Água no Século XXI*: Enfrentando a escassez. São Carlos: RiMa IIE, 2003. 248p.

[37] Gonçalves, C.N.; Ceretta, C.A.; Basso, C.J. Sucessões de culturas com plantas de cobertura e milho em plantio direto e sua influência sobre o nitrogênio do solo. *Revista Brasileira de Ciência do Solo*: Viçosa, 2000; 24, 153-159.

[38] Kiehl, J.C. Nitrogênio: Dinâmica e disponibilidade no solo. In: ___. *Curso de atualização em fertilidade do solo.* Campinas: Fundação Cargill, 1987. p.139-157.

[39] Fernandes, F.C.S.; Libardi, P.L.; Carvalho, L.A. Internal Drainage and nitrate leaching in a corn-black oat-corn sucession with two split nitrogen applications. Scientia Agricola, Piracicaba, 2006; 63, 483-492.

[40] Fernandes, F.C.S.; Libardi, P.L. Drenagem interna e lixiviação de nitrato em um Latossolo sob sucessão milho-braquiária-milho, com diferentes doses de nitrogênio. *Revista Brasileira de Ciência do Solo,* Viçosa, 2009, 33, 1163-1173.

7

Calibration of a New Device to Measure Water Content of Rocks

Maria Clementina Caputo and Rita Masciale

1. Introduction

The vadose zone, which extends from the soil–atmospheric interface to the capillary fringe of the water table, is a fundamental part of the hydrologic cycle. It controls how the precipitation splits into infiltration, surface runoff, evapotranspiration, groundwater recharge; it regulates the storage, transfer, filtering, adsorption, retard and attenuation of solutes and contaminants before these reach the groundwater.

Until the last decades, the vadose zone was considered as a black box that merely connect surface water and groundwater. As consequence an incomplete understanding of the complex dynamics of the vadose zone existed.

Nowadays, instead, the monitoring of the hydrological processes that occur through the vadose zone, are receiving increased attention mainly with refer to contaminant transport processes. Fluxes of water and solutes strongly depend on water content thus its monitoring and estimation becomes an important issue.

A wide range of methods, sensors and technologies are available for the measure of soil water content, mainly used in management of precision farming.

The traditional standard method for direct measurement of soil water content is the gravimetric method [1]. This method implies the sampling of the soil to carry out laboratory measurements, modifying the natural condition. As gravimetric method is destructive, labour intensive, not timely and costly, many alternative non-destructive methods for measuring water content have been developed. However, none of all methods measure the water content directly but only some properties, named "surrogate measure", that changes as the soil water content changes. By measuring the value of the surrogate parameter, it is possible to estimated the value of water content, using a calibration curve that represents the relationship between the surrogate measure and the soil water content values [2].

Neutron Thermalization method [3, 4], Capacitance method [5-8], Time Domain Reflectometry (TDR) method [9-13], Frequency Domain Reflectometry (FDR) method [14, 15], Impedance method [16, 17], Electrical Restistance method [18, 19] and Tensiometer method [20-22] are among the well-known technologies utilized to develop different kind of devices and probes for measuring moisture content and soil water potential, respectively.

More details on these technologies and devices, advantages and limitations of each one, are given in several references [23, 24, 2].

Recently, remote sensing technologies, like passive or active radiometry, have been applied in order to provide wide-area indications of surface soil water content [25, 26]. However, the great influence of vegetation and surface structure on the quality of received signal, restricts the sampling depth to the uppermost part (2–5 cm) of soil [27], strongly limiting the usefulness of this technology in hydrological application.

The use of all these methods and devices in the soil is customary by now but when the vadose zone consists of rocks the monitoring of water content become more difficult for several aspects. The main difficulty regards the installation of the probes that are often very delicate and cannot be hammered or screwed into the rock. Furthermore, after their insertion, a good contact between the rock and the sensor must be ensured in order to minimize the gap effect which causes significant errors in the investigated properties, especially for dielectric ones [28, 29].

At present, very few applications of these techniques to rocks, by creating new devices or by adapting the existing ones, are documented in literature.

The first measurements in sandstone and tuff blocks by means of penetration type probes using TDR are reported in [30]. Topp's equation [9] was used to convert the dielectric constant (K) values into water content (θ) highlighting an overestimation of water content. In fact, especially for rocks with low effective porosity, an individual calibration was needed.

In [31] TDR was applied in granodiorite both in the laboratory and in the field. Employing surface probes, it was showed that K and θ were almost linearly related, revealing that TDR was capable to measure the volumetric water content changes.

Using brass rods, individual K - θ calibration functions were developed in [32] for nine different rock types with porosity values ranging from 1 to 54%. However, the experimentally determined K - θ relationships showed an abrupt jump near fully saturation of samples, suggesting that TDR would be accurate for applications in rocks only when the gap effect is negligible.

By applying TDR in seven different types of rocks [33], the authors demonstrated that conductive silicone fillings, carefully applied, can successfully eliminate gap effect in the case of penetration probes. Moreover, they compared the K-θ relationships obtained with surface probes against that obtained using penetration probes. The results showed systematic differences between two types of probes, explained on the basis of their geometry.

In this chapter, we intended primarily to give further contribute to the knowledge on the applicability on rocks of methods developed for measuring water content in the soils by focusing on the Electrical Impedance Spectrometry (EIS) method. In particular, this work presents the results obtained to calibrate a new device, called Z-meter 2, based on EIS method. Samples of calcarenite have been used for the calibration procedure carried out in laboratory under controlled condition.

The electrical impedance in complex form and its changes over time have been measured with the aim to:

a. verify the suitability of the device for water content estimation in rocks;

b. characterize the effects on the measured values of the electric frequency applied and of the electrical conductivity (EC) of the solution used to saturate samples;

c. determine specific calibration curves for the investigated lithotype.

2. Materials and methods

2.1. Electrical Impedance Spectrometry (EIS) method

The basic principle of the EIS method [34] is the measurement of the electrical impedance ($\hat{Z}$) which provides a measure of medium's opposition during the passage of a alternating current (AC). The impedance is described not only by the amplitudes of the voltage and the current, but also by their relative phase-displaced.

The impedance of a circuit element can be defined as the frequency domain ratio of the voltage phasor $\hat{U}$ to the current phasor $\hat{I}$ across the element:

$$\hat{Z} = \frac{\hat{U}}{\hat{I}} \tag{1}$$

Impedance is represented as a complex quantity. In the polar form, it is

$$\hat{Z} = |Z| e^{j\theta} \tag{2}$$

where $|Z|$ is the magnitude, the argument θ is the phase difference between voltage and current and j is the imaginary unit.

In cartesian form, it is

$$\hat{Z} = R + jX \tag{3}$$

where R is the resistance, that forms the real part of impedance, independent from the frequency, and X is the reactance, that forms the imaginary part of impedance, dependent from the frequency. The values of impedance are expressed, as those of resistance and reactance, in ohms [Ω].

The impedance depends on the characteristics of AC electric circuit which can be made by combination of resistor, inductor and capacitor, either in series or parallel mode.

For a real resistor-capacitor (RC) circuit and resistor-inductor (RL) circuit, the impedance can be expressed by the following relations

$$\hat{Z}_L = R + j \cdot X_L = R_L + j\omega L \tag{4}$$

and

$$\hat{Z}_C = R - j \cdot X_C = R_C - \frac{j}{\omega C} \tag{5}$$

where

R= resistance

$X_L = 2 \cdot \pi \cdot f \cdot L$ = inductive reactance

$X_C = -\frac{1}{2 \cdot \pi \cdot f \cdot C}$ = capacitive reactance

f= frequency (Hz)

$\omega = 2\pi f$ = angular frequency

C= capacitance or electrical capacity (F)

L = inductance (H)

The porous medium (e.g. soil and rock) can be described by the impedance of an equivalent electric circuit. It always has the character of a resistor, so that the real component of the total impedance $\hat{Z}$ of the circuit is always present.

In a dry porous medium the resistance is so high that it can be considered as a non-conductor (dielectric). Therefore it is possible to assume that dry porous medium is polarized by the electric field and thus it is charged as a capacitor characterized by its capacitance.

Using this assumption, the capacitance of medium is substantially higher than the inductance so it can be neglected and the dominant component of imaginary part of the impedance $\hat{Z}$ is the capacitive reactance (X_C).

As for mentioned above, the porous medium can be described as an equivalent RC circuit and its impedance, expressed by the equation (5), describes primarily electromagnetic properties of medium.

The solid part (grains) is formed by insulating materials, characterized by their dielectric constants, and hence it represents the imaginary part (X_C) of the measured impedance. If the material is in non-zero moisture condition, it can be considered as a conductor because water always contains a certain quantity of mineral salts (conductive material). In this case the resistance is lower and the degree of saturation of the material strongly influences the real part of the measured impedance.

2.2. Z-meter 2 device

The Z-meter 2 is a new device developed, constructed and realized within the European International Project E!3838 of EUREKA Program [35]. It's based on EIS method and it is able to measure the electrical impedance in complex form. Z-meter 2 was specifically designed to carry out long-term field experiments.

According to the equivalent diagram of the circuit in Figure 1, the resulting measured value of impedance is represented by a parallel combination of resistor and capacitor.

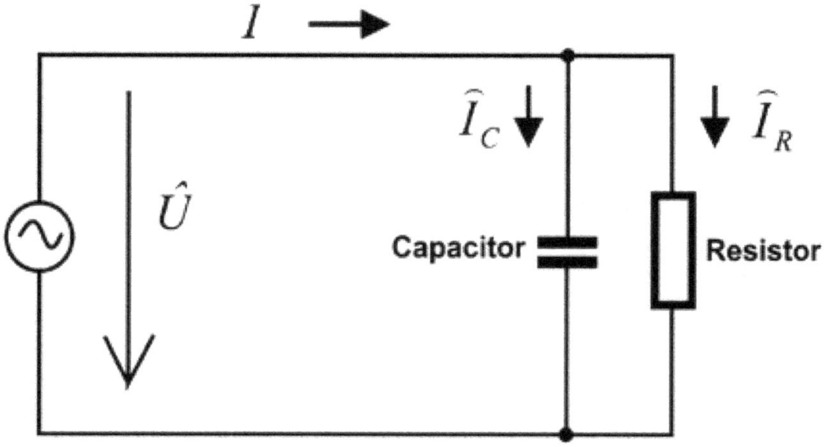

Figure 1. Equivalent diagram of measured impedance.

The electric current is divided into two orthogonal components, real $\frown{I}_R$ flowing through the real component of the equivalent resistor, and imaginary $\frown{I}_C$ flowing through the equivalent capacitor. In this way the Z-meter 2 allows measurement of both parts of the impedance: resistance and reactance. The basic parameters of the device are summarized in Table 1.

Many different kinds of probes can be used with Z-meter 2 for measuring the impedance of soils, either in the field or in laboratory: rod probes, modular probes with variable length constituted by stainless electrical conductive parts alternated with non conductive parts (polyamide), aluminium strip probes with a self-adhesive layer on one side that ensures perfect contact with the medium.

The kind of probe needs to be chosen in relation to the monitored medium, the environmental condition and the aim of the measurements. In this sperimentation, stainless steel road probes, called picket type probe, were used. These probes were chosen because they can be easily installed in the rock without be damaged.

PARAMETERS	Z-METER 2
Range of impedances	10 Ω – 1M Ω
Frequency range	100 Hz – 20 KHz
Accuracy of modulus of impedance	+/- 2% of the range
Accuracy of phase	+/- 2°
Level of measuring voltage	500mV – 5V with an interval of 500 mV
Communication with a PC bus	USB (COM port)
Integrated electric switcher	128 measuring points, each with 4 electrodes

Table 1. Basic parameters of a Z-meter 2 device.

2.3. Rock type and samples preparation

The rock used for the calibration tests was a marine sedimentary rock of Plio-Pleistocene age, known as "Calcarenite di Gravina" Formation. It is a porous rock with granular skeleton consisting of irregularly rounded lithoclasts of limestones and bioclasts, fragments of marine organisms, embedded in a fine, granular, carbonatic matrix, cemented to some degree. The matrix and the sparry cement, made up predominantly of mineral calcite, are entirely subordinate to the granular skeleton. The calcarenite's composition mostly consists of calcium carbonate (>95%) with an insoluble residue characterized mainly by clay minerals. Previous laboratory tests assessed that porosity ranges between 40% and 47% and the hydraulic conductivity (K) is about $6 \div 7x10^{-5}$ m/s for the coarse calcarenite and $3x10^{-5}$ m/s for the medium calcarenite [36-38]. The calcarenites are diffused in Apulia Region and frequently constitute part of the vadose zone or form significant porous aquifers in the region, playing an important role in drinking supply. Particularly, the samples of calcarenite used for the tests came from an active quarry, located in Canosa di Puglia, a city in the South of Italy (Figure 2).

Three blocks of calcarenite, hereafter called E1, E2 and E3, cut from one large block of rock, were used in order to verify the repeatability of measurements. The rock samples appeared relatively homogeneous at the sample scale. The samples were oven dried in order to obtain the dry weight. The dimensions of samples and their main physical properties are summarized in Table 2.

The lateral surface of the samples was sealed using a bi-component epoxy resin in order to ensure a one-dimensional water flow during the evaporation process.

The picket type probes, 10 mm diameter, 120 mm long with a conical bit 15 mm long, were used as electrodes in the tests (Figure 3). Two guide holes, 9 cm apart, with approximately 10 mm diameter, were drilled into rock samples for the installation of the probes. These ones were then slowly inserted into the sample with twisting motion and, alternatively, with the help of hammer, leaving 2 cm protrusion from the top edge of sample. In this way a good contact between the rock and the sensor was ensured.

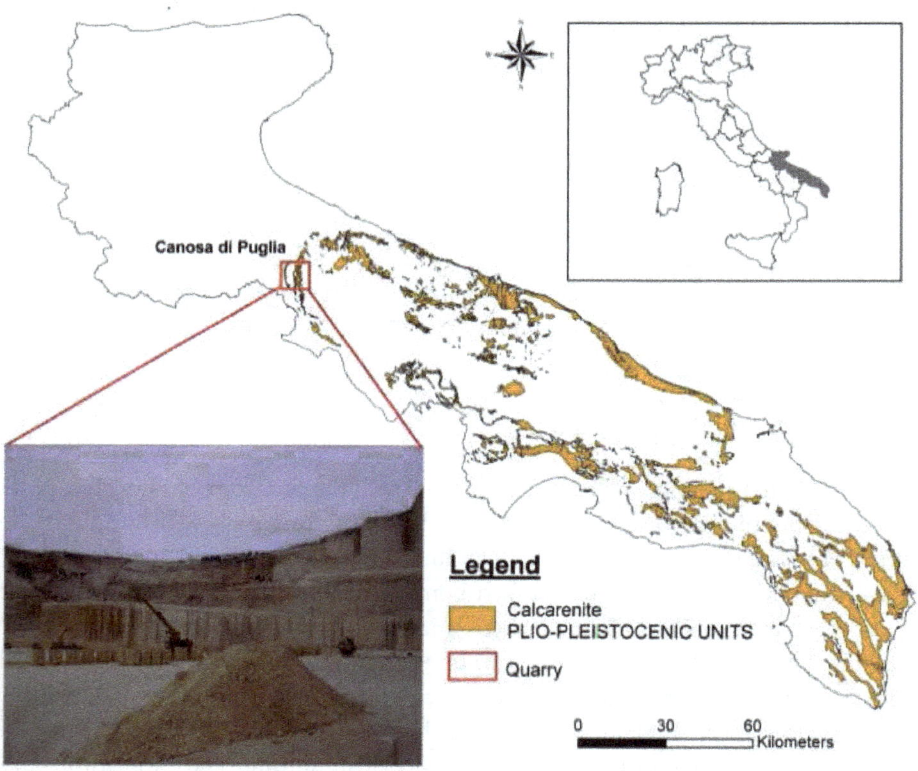

Figure 2. Regional distribution of calcarenites and location of quarry from which the rock samples have been collected.

	Sample E1	Sample E2	Sample E3
Dimensions (cm)	25x16x15	25x15x15	24.9x15x14.7
Bulk density (g/cm³)	1,525	1,515	1,500
Particle density (g/cm³)	2,71	2,71	2,71
Porosity (%)	43,73	44,11	44,66

Table 2. Physical characteristics of the calcarenite samples tested.

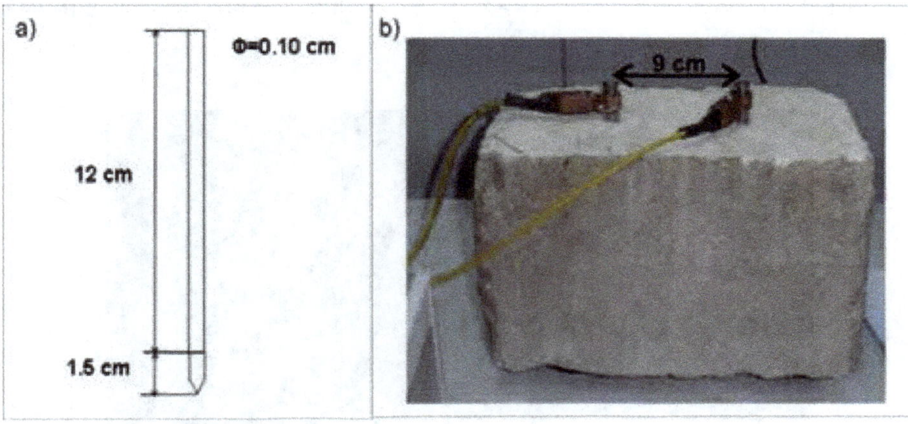

Figure 3. a) Scheme of picket type probe; b) probes position.

After the probes installation, the sample was placed in a water-filled vacuum bell jar, under confining pressure of 1 bar for at least 24 hours, in order to achieve almost full saturation. The saturation process took place from bottom to up by avoiding both air entrapment and uneven distribution of water. Three different solutions were used to saturate the samples in order to establish the effect of electrical conductivity (EC) on the output of the device: deionized water, 0,01 M and 0,03 M KCl solutions. The EC value as well as the KCl concentrations and the density of each water solution are reported in the Table 3.

KCl solution molarities	EC values (mS/cm)	Salt concentration g/L	Water solution density g/cm³(at 25°C)
deionized water	$3{,}4 \cdot 10^{-3}$	0	0.9971
0.01	1,75	0,745	0.9976
0.03	4,70	2,235	0.9988

Table 3. Physical properties of water solutions used to saturate the samples.

2.4. Measuring procedures

The entire cycle of measurements, comprising the measurements for all chosen frequencies, was repeated over time: 2-3 times per day in the first 15 days, 1 time in the following days, according to the expected rate of the evaporation process.

Contemporarily the weight loss of the rock sample, due to evaporation, was measured by electronic balance. The whole experimental set up is shown in Figure 4.

The tests have been performed in a thermostatic room at temperature of 25 ±1°C provided with a ventilation system. This is because temperature variations cause errors in all indirect moisture measurement methods due to the dependence of the dielectric properties from the temperature [39,40]. The tests were stopped when the sample's weight reached approximately it's dry weight.

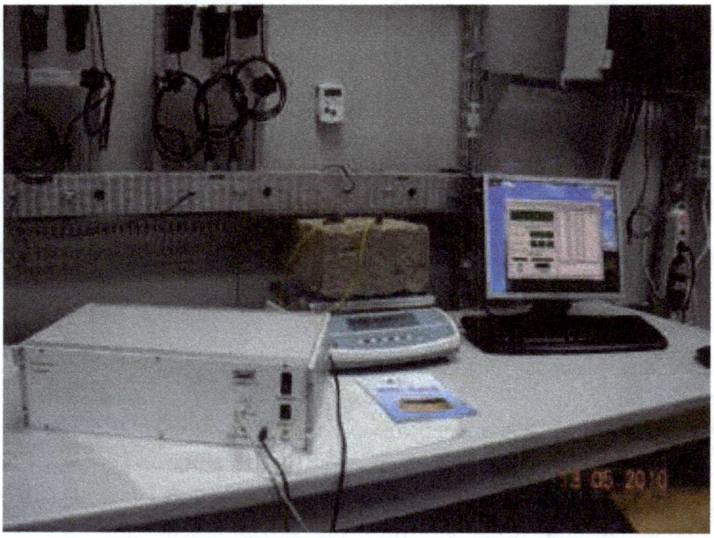

Figure 4. Experimental set up.

3. Results and discussion

3.1. Volumetric Water Content (VWC)

The first stage was to obtain the VWC of samples starting from the weight recorded at different times.

The rock VWC was determined, from the gravimetric one, in a standard way by the following equation (6) [41]:

$$\theta_v = \theta_g \left(\frac{\rho_b}{\rho_w} \right) \tag{6}$$

where

θ_v (cm^3 cm^{-3}) = VWC volumetric water content,

θ_g (g g^{-1}) = gravimetric water content,

ρ_b (g cm^{-3}) = rock bulk density,

ρ_w (g cm^{-3}) = water solution density.

The dynamics of the weight loss of the sample during the test is shown in Figure 5, in which the data for sample E2, saturated with 0.01 M KCl solution, are reported as an example.

The weight of the sample decreases over time in different way so that the evaporation from a soil or rock may be modelled as a two-stages process. In the first stage, corresponding to high water contents, the evaporation is governed by environmental conditions; in the second, when the water content decreases beyond a critical value, the evaporation is controlled by the ability of the medium to transmit water or, in other words, by its hydraulic properties [42, 43].

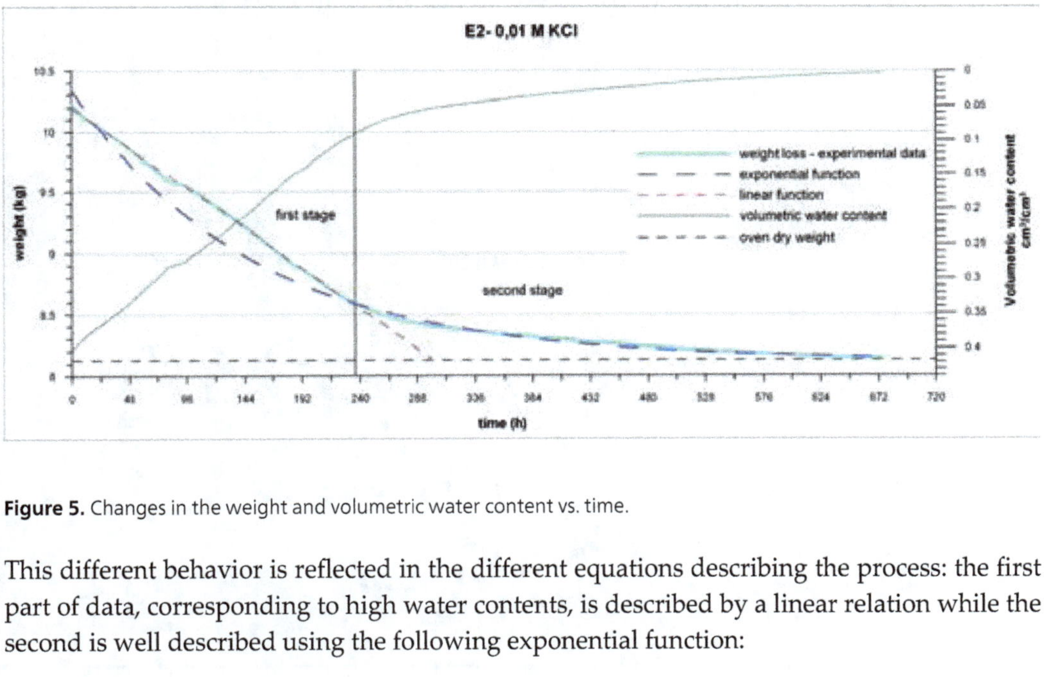

Figure 5. Changes in the weight and volumetric water content vs. time.

This different behavior is reflected in the different equations describing the process: the first part of data, corresponding to high water contents, is described by a linear relation while the second is well described using the following exponential function:

$$y = a + b\exp\left(-\frac{x}{c}\right) \tag{7}$$

where y (kg) is sample weight over time, a (kg) is its dry weight, b (kg) is the weight of water required to saturate the sample, x is the time (h) and c (h) is the only fitting parameter. The transition from the first to the second part occurs at a water content of about 6÷9%.

3.2. Resistance

The use of Z-meter 2 allows to investigate both the real (resistance) and the imaginary component (capacitive reactance) of impedance, being the two components recorded separately. In this study only the resistance was investigated because of its direct correlation with the water content.

The effective independence of the resistance (hereafter called Rx) from the frequency applied was checked and verified. Figure 6 shows the Rx output probe for all the tested measurement frequencies relative to sample E2 saturated with 0.01 M KCl solution.

The Rx values, measured for the same VWC, are very similar to each other for all tested frequencies, except that for VWC close to dry condition (less than 0.02). These means that, for VWC > 0.02, the device has a good accuracy and precision and that changing the frequencies has no effect on the Rx values that are therefore related with the VWC only. This results permits to use the complete dataset, consisting of medium Rx values recorded for each frequencies and the same VWC, in order to obtain the best fit curve, for each solution. For the reason mentioned above, all Rx data corresponding to VWC less than 0.02 are not included in the dataset.

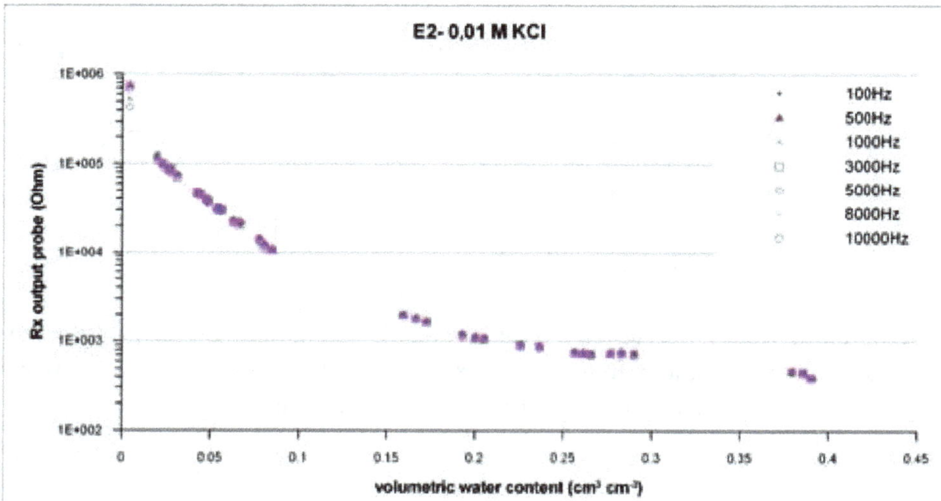

Figure 6. Rx values vs volumetric water content at all tested frequencies for sample E2.

The complete dataset of Rx values, recorded for the sample E2, are shown in Figure 7. Unlike what happens for the frequency, Rx values are strongly affected by changing the molarity, and hence EC value, of the solution used for the sample saturation. Moreover, as expected in a drying experiment, the VWC decreases over the course of the experiment and the corresponding Rx measurements increase significantly.

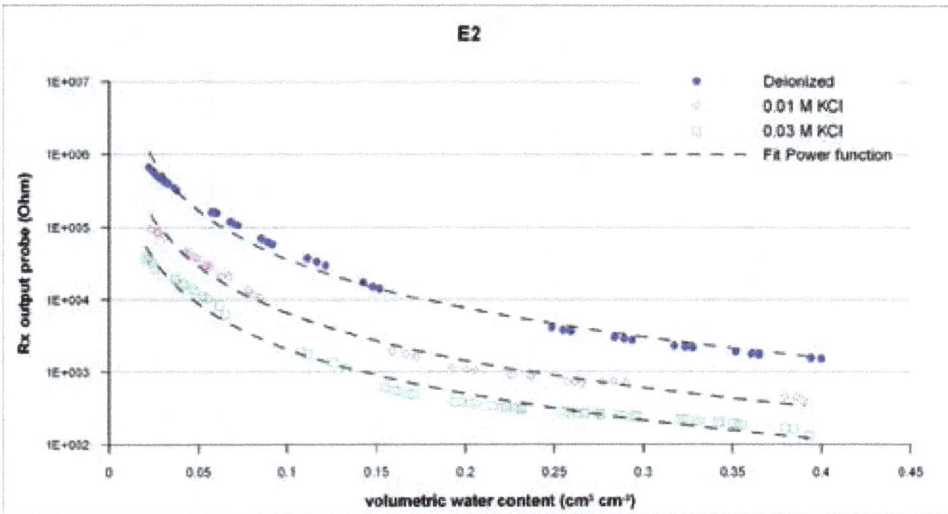

Figure 7. Rx complete dataset for the three solutions used to saturate sample E2.

The best interpolation function of dataset is represented by a power function of the type:

$$y = a \cdot x^{b} \tag{8}$$

This model can be linearized by taking logarithms:

$$\ln(y) = \ln(a) + b \cdot \ln(x) \tag{9}$$

The linearized experimental data, obtained for each solution used to saturate samples E1, E2 and E3, are shown in Figure 8.

Small differences are found in the three replicates obtained for the same solution that can be explained with the natural variability among different samples of the same rock type (see Table 2). The two replicates E1 and E2 have been used, as a unique dataset, to obtain the calibration function that describes the VWC-Rx relationships, for each solution, while the third replicate (E3) has been used for the validation test. In Figure 9, for each solution, the logarithms of calculated VWC are plotted against the logarithms of measured Rx coming from replicates E1 and E2. The regression curves are also reported in the graphs. The changes in Rx due to the changes in VWC are significant and the relationships show a negative slope. The information provided from the linear regression, as indicated by the value of R-squared coefficient, suggests good degree of correlation of the datasets to these functions.

The calibration function is $VWC = a \cdot Rx^{b}$, by converting the logarithmic function into power function. The values of parameters a and b, for each solution, are reported in Table 4. In order to check the performance of the calibration function obtained by regression of experimental data, a validation test was performed utilizing the data of the sample E3. The actual values of

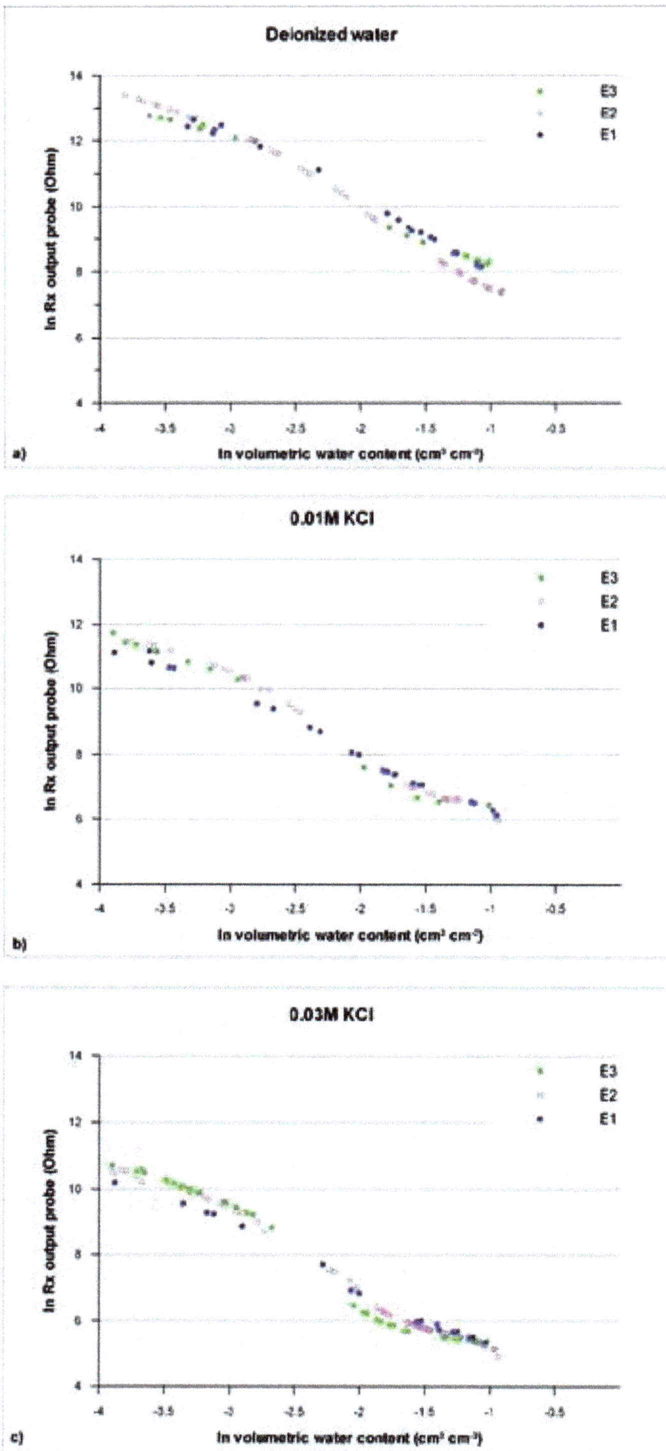

Figure 8. Three replicates (E1, E2, E3) for each solution (a, b, c).

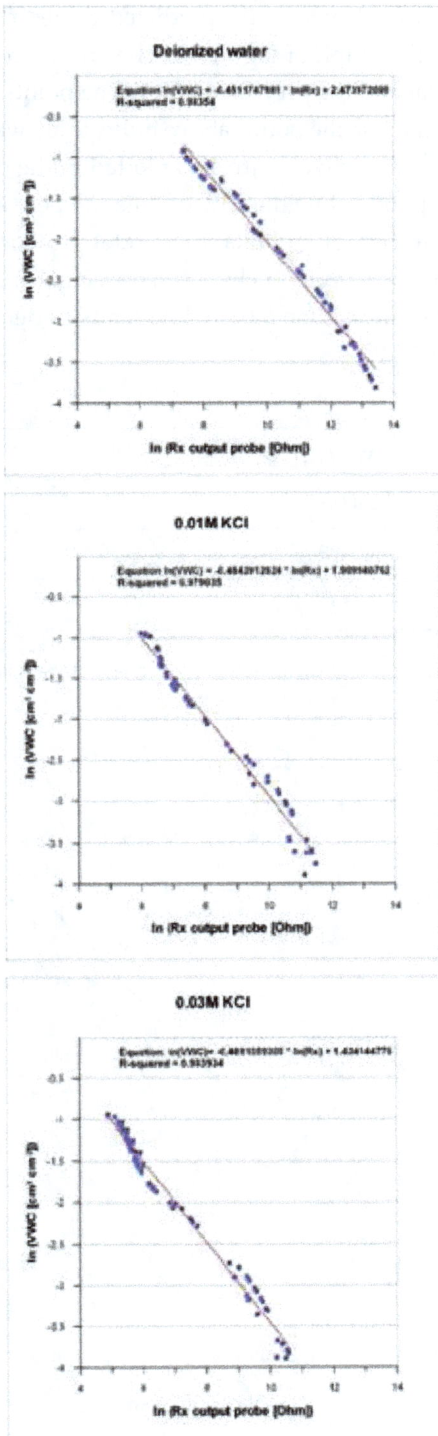

Figure 9. Calibration curves for each solution used to saturate samples.

VWC, measured during the test E3, against the predicted values, calculated using the calibration function, are plotted in the graph of Figure 10. The 1:1 line represents the perfect correspondence between the actual and the predicted value. The points below this line indicate an underestimation of the model while the points above indicate an overestimation. Two absolute error lines of ±0.03 and ±0.06 respectively, are also plotted in the graph. Only the 10% of the data fall outside prescribed limits while the 23% of data fall between the ±0.03 and the ±0.06 error lines. For the remaining 66% of the data, the model provides an estimate of the VWC with an error less than ±0.03. The correlation between the actual and predicted values is much higher for low values of VWC (less than 0.07) while tends to decrease for greater values of VWC.

KCl solution molarities	a	b
deionized water	2.4739	-0.4512
0.01	1.9099	-0.4843
0.03	1.4341	-0.4891

Table 4. Parameters of the calibration curve for each solution used to saturate samples.

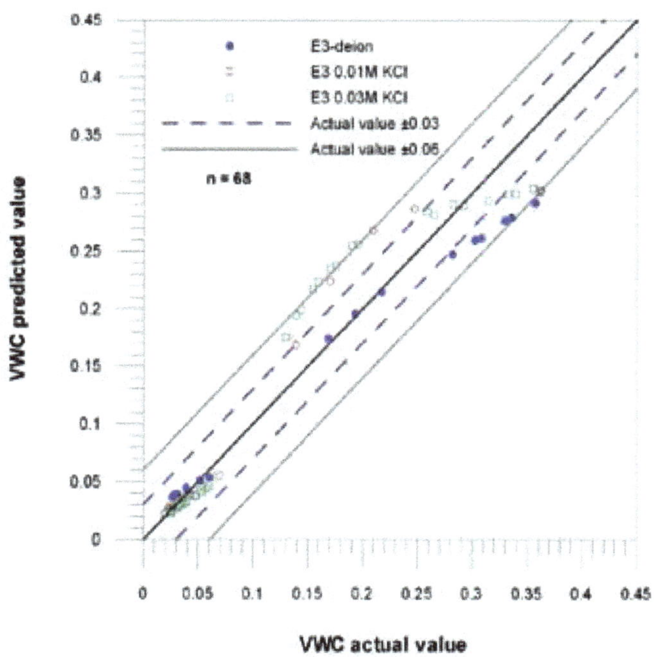

Figure 10. Validation test.

4. Conclusions

In this chapter has been described the experimental procedure used to obtain the calibration curves for a new device Z-meter 2, to measure water content in porous rock. Experimental tests have been carried out in laboratory, using samples of sedimentary rock, known as calcarenite, collected in a quarry located in Canosa, a city in South of Italy.

The device, based on the EIS method, gives as output both the resistance and the reactance values separately; this study considered the resistance because of its high correlation with the water content.

The tests were aimed at understanding the effects of the applied frequency and EC of KCl solution used to saturate the samples, on the device outputs and, consequently, on the calibration functions. For this reason a wide range of frequencies, from 100Hz to 10000Hz, was investigated and three KCl solutions with different EC values were used to saturate the rock samples.

The probes used were short stainless steel road probes, called picket type probes, chosen because they can be easily installed in the rock without be damaged. By analyzing the different data sets, none abrupt change in resistance values was observed during the evaporation process. This means that the probes work well in the calcarenite and that are able to record measurements of the impedance without any gap effects. The evaporation process monitored during the test showed how the weight of the sample decreases over time with different rates leading to model the process as a two-stages process. The first, corresponding to high water contents, was described by a linear relation while the second by an exponential function. The independence of the resistance from the frequency applied was observed. The resistance values obtained for the same water content at all tested frequencies were almost coincident, except close to dry condition (less than 2%). This means that, for water content higher than 2%, the device has a good accuracy and precision and that changing the frequencies has no effect on the resistence values.

Unlike the frequency, the resistance values were strongly affected by the molarity of the solution used for the sample saturation. For this reason the calibration curves have been obtained considering the data sets recorded for each solutions, separately. The calibration functions, that describe the relationships between the water content and resistance values, are of power type with a good degree of correlation as shown by value of R-squared of about 0.98.

Acknowledgements

We would like to thank the Project E!3838 "Research, design and realization of the automatized system of the observation of soil moisture changes" and the Project E!4981 "Computerised Measuring System for Analysis of Chosen Characteristics and Processes in Porous Environment by EIS Method" of the Europe International Program EUREKA. A particularly special thanks to Ing. Jana Pařílková professor at Brno University of Technology, Faculty of Civil

Engineering and co-applicant of the EUREKA program in the Czech Republic, in order to have permitted free use of the Z-meter 2 device for the laboratory tests. Thanks to Francesco De Benedictis also for his technical support during the experimentations in the laboratory.

Author details

Maria Clementina Caputo* and Rita Masciale

Water Research Institute (IRSA), National Research Council (CNR), Bari, Italy

References

[1] Gardner, W.H. Water Content. In: Klute A. (ed.) Methods of Soil Analysis, Part I, Physical and Mineralogical Methods. Madison, Wisconsin, USA: American Society of Agronomy and Soil Science Society of America, Inc; 1986. p493-544

[2] Evett, S.R., Heng, L.K., Moutonnet, P., Nguyen, M.L., editors. Field Estimation of Soil Water Content: A Practical Guide to Methods, Instrumentation, and Sensor Technology. IAEA-TCS-30. Vienna, Austria: International Atomic Energy Agency; 2008.http://www-pub.iaea.org/MTCD/Publications/PDF/TCS_30_web.pdf

[3] Greacen, E.L., editor. Soil Water Assessment by the Neutron Method, Melbourne, Australia: CSIRO; 1981.

[4] Evett, S.R. Measuring soil water by neutron thermalization. In: Stewart B.A. and Howell T. A. (eds.). Encyclopedia of Water Science. New York: Marcel Dekker; 2003. p889–893.

[5] Dean, T.J., Bell, J.P., Baty, A.J.B.,. Soil moisture measurement by a capacitance technique, part 1. Sensor design and performance. J. Hydrology 1987; 93, 67–68.

[6] Robinson, M. and T.J. Dean. Measurement of near surface soil water content using a capacitance probe. Hydrol. Proc. 1993;7, 77–86.

[7] Baumhardt, R.L., Lascano R.J. and Evett S.R. Soil Material, Temperature, and Electrical Conductivity Effects on Calibration of Multisensor Capacitance Probes. Soil Science Society of American Journal 2000; 64, 1940-1946.

[8] Kizito F., Campbell C.G., Cobos D.R., Teare B.L., Carter B. and Hopmans J. W. Frequency, Electrical Conductivity and Temperature Analysis of a Low-Cost Capacitance Soil Moisture Sensor. Journal of Hydrology 2008; 352, 367-378.

[9] Topp G.C., Davis J.L. and Annan A.P. Electromagnetic determination of soil water content: Measurement in coaxial transmission lines. Water Resour. Res. 1980; 16, 574–582.

[10] Topp G.C. and Davis J.L. Measurement of soil water content using time-domain reflectometry in field evaluation.Soil Science Society of America 1985; 49(1) 19–24

[11] Dalton F.N. and van Genuchten M.Th. The time-domain reflectometry method for measuring soil water content and salinity. Geoderma 1986; 38, 237-250.

[12] Jones S.B. and Or D. Frequency Domain Analysis for Extending Time Domain Reflectometry Water Content Measurements in Highly Saline Soils. Soil Science Society of American Journal 2004; 68, 1568–1577.

[13] Robinson D.A., Jones S.B., Wraith J.A., Or D. and Firedmena S.P. A Review of Advances in Dielectric and Electrical Conductivity Measurements in Soils Using Time Domain Reflectometry. Vadose Zone Journal 2003; 2, 444-475.

[14] Veldkamp E. and O'Brien J.J. Calibration of a frequency domain reflectometry sensor for humid tropical soils of volcanic origin, Soil Sci. Soc. Am. J. 2000; 64, 1549–1553.

[15] Bilskie J. Using dielectric properties to measure soil water content. Sensors Magazine 1997; 14, pp.26–32.

[16] Gaskin G.J. and Miller J.D. Measurement of Soil Water Content Using a Simplified Impedance Measuring Technique. J. Agric. Eng. Res. 1996;63, 153 – 160.

[17] Seyfried M.S. and Murdock M. D. Measurement of Soil Water Content with a 50-MHz Soil Dielectric Sensor. Soil Science Society of American Journal 2004; 68, 394-403.

[18] McCann I.R., Kincaid D.C. and Wang D. Operational characteristics of the watermark model 200 soil water potential sensor for irrigation management. Applied Engineering in Agriculture 1992;8 (5) 603–609.

[19] Shock C.C. Soil water potential measurement by granular matrix sensors. In Stewart B.A. and Howell T. A. (eds.). Encyclopedia of Water Science. New York, NY: Marcel Dekker; 2003. p899–903

[20] Richards L.A. The usefulness of capillary potential to soil moisture and plant investigators. J. Agric. Res. (Cambridge) 1928;37,719–742.

[21] Mullins C.E., Mandiringana O.T., Nisbet T.R. and Aitken M.N.,. The design limitations and use of a portable tensiometer. Journal of Soil Science 1986; 37,691–700

[22] Cassel D.K. and Klute A., 1986. Water potential: Tensiometry. In: Klute A. (ed.) Methods of Soil Analysis, Part I, Physical and Mineralogical Methods. Madison, Wisconsin, USA: American Society of Agronomy and Soil Science Society of America, Inc; 1986. p563–596.

[23] Dane, J.H., and G.C. Topp editors. Methods of Soil Analysis. Part 4-Physical Methods. Madison, Wisconsin, USA: American Society of Agronomy and Soil Science Society of America, Inc; 2002.

[24] Evett, S.R. Soil water and monitoring technology. In: Lascano R.J. and Sojka R.E. (eds.) Irrigation of Agricultural Crops. Madison, Wisconsin, USA: American Society of Agronomy and Soil Science Society of America, Inc; 2007. p25–84.

[25] Wigneron, J.-P., Calvet, J.-C., Pellarin, T., Van de Griend, A.A., Berger, M., Ferrazzoli, P. Retrieving near-surface soil moisture from microwave radiometric observations: current status and future plans. Rem. Sens. Environ. 2003;854, 489–506.

[26] Löw, A., Ludwig, R., Mauser, W.,. Derivation of surface soil moisture from ENVISAT ASAR wide swath and image mode data in agricultural areas. IEEE Trans.Geosci.Remote Sens. 2006; 44, 889–899

[27] Walker, J.P., Houser, P.R., Willgoose, G.R. Active microwave remote sensing for soil moisture measurement: a field evaluation using ERS-2. Hydrol. Proc. 2004;1811, 1975–1997

[28] Annan A. P. Time domain reflectometry - Air-gap problem for parallel wire transmission lines. In: Rep. Activ., Part B, Pap. 77–1B. Ottawa, Ont., Canada: Geol. Surv. of Canada; 1977. p. 59–62.

[29] Knight, J.H., Ferré P.A,. Rudolph D.L., and Kachanoski R.G. A numerical analysis of the effects of coatings and gaps upon relative dielectric permittivity measurement with time domain reflectometry, Water Resour. Res. 1997; 33,1455–1460.

[30] Hokett, S. L., Chapman J. B. and Russell C. E. Potential use of time domain reflectometry for measuring water content in rock, J. Hydrol. 1992;138, 89–96.

[31] [31]Schneebeli M., Fluhler H., Gimmi T., Wydler H., Laser H.P. and Bear T. Measurements of water potential and water content in unsatured crystalline rock. Water Resour. Res. 1995;31,1837-1843.

[32] Sakaki T., Sugihara K., Adachi T., Nishida K. and Lin W. R. Application of time domain reflectometry to determination of volumetric water content in rock, Water Resour. Res. 1998;34, 2623–2631.

[33] Sakaki T. and Rajaram H. Performance of different types of time domain reflectometry probes for water content measurement in partially saturated rocks. Water Resour. Res. 2006; 42, W07404, doi 10.1029/2005WR004643.

[34] Barsoukov E. and Macdonald R. J., Editors. Impedance Spectroscropy Theory, Experiment, and Applications. Hoboken, New Jersey, USA: John Wiley and Sons, Inc.; 2005.

[35] Parilkova J., Veselý J., PavlíkJ., Stoklásek R. Monitoring of the soil status using electrical impedance spectrometry method developed in Project E!3838 of the Europe International Program EUREKA. Proceedings of the XIX IMEKO World Congress

Fundamental and Applied Metrology, 6–11 September 2009, Lisbon, Portugal. http://www.imeko2009.it.pt/Papers/FP_443.pdf

[36] Iannone, A. and Pieri, P. Considerazioni critiche sui "Tufi Calcarei" delle Murge – Nuovi dati litostratigrafici e paleoambientali, Geografia Fisica e Dinamica Quaternaria 1979; 21,33–58.

[37] Andriani G.F. and Walsh N. Fabric, porosity and permeability of calcarenites from Apulia (SE Italy) used as building and ornamental stones. Bull. Eng. Geol. Env. 2003; 62, 77-84;

[38] Andriani G.F. and Walsh N. - Physical properties and textural parameters of calcarenitic rocks: qualitative and quantitative evaluation. Bull. Eng. Geol. Env. 2002; 67, 5-15.

[39] Ghretli M.,. Khalid K, Grozescu I. V., Sahri H. and Abbas Z. Dual frequency microstrip antenna sensor for water content measurements independent of temperature variation. Measurement Science and Technology 2007;18, 1054–1060.

[40] Evett, S.R., Tolk J.A., and Howell T.A. Soil profile water content determination: Sensor accuracy, axial response, calibration, temperature dependence and precision. Vadose Zone J. 2006; 5, 894–907.

[41] Black C.A., editor. Methods of Soil Analysis, Part I, Physical and Mineralogical Methods. Madison, Wisconsin, USA: American Society of Agronomy and Soil Science Society of America, Inc; 1965.

[42] Brutsaert W., editor. Evaporation into the Atmosphere: Theory, History and Applications, Hingham, Massachusetts: Kluwer Academic; 1982.

[43] Jury W., Gardner W. R. and Gardner W. H., editors. Soil Physics, 5th ed. New York: John Wiley and Sons, Inc.; 1991.

8

Time Stability of Soil Water Content

Wei Hu, Lindsay K. Tallon, Asim Biswas and
Bing Cheng Si

1. Introduction

Soil water is a key variable controlling water and energy fluxes in soils (Vereecken et al., 2007). It is necessary for plant and vegetation growth and development. Research has indicated that soil water content (*SWC*) varies both in space and time. Variations in both space and time present a substantial challenge for applications such as precision agriculture and soil water management.

Since the contribution of Vachaud et al. (Vachaud et al., 1985), a large body of research has indicated the presence of time stability of *SWC* (Biswas and Si, 2011c; Comegna and Basile, 1994; Grayson and Western, 1998; Hu et al., 2009; Martínez-Fernández and Ceballos; 2005; Mohanty and Skaggs, 2001), which means that the spatial pattern of *SWC* does not change with time at a certain probability. According to this concept, if a field is repeatedly surveyed for *SWC*, there is a high probability that a location with certain wetness characteristics (i. e., wet, dry, intermediate) will maintain those characteristics on subsequent occasions. Time stability has also been extended to describe the characteristics of *SWC* at point scales. A location will be regarded as time stable provided it can estimate the average *SWC* of an area.

Time stability of *SWC* has been observed at a large variety of scales ranging from plot (Pachepsky et al., 2005) to region (Martínez-Fernández and Ceballos, 2003) and related studies cover a range of investigated areas, sampling schemes, sampling depths, investigation periods, and land uses (Biswas and Si; 2011c; Brocca et al., 2009; Cosh et al., 2008; Hu et al., 2010a;Tallon and Si, 2004; Vachaud et al., 1985). As a result, a variety of methods have been developed to evaluate time stability of *SWC*, each with its own advantages and disadvantages. Time stability is usually used to characterize time persistence of the spatial pattern of *SWC* between measurement occasions, either at the measurement scale or at different scales (Biswas and Si, 2011c; Cosh et al., 2006; Kachanoski and De Jong, 1988; Vachaud et al., 1985).

On the other hand, *SWC* at time stable locations can be used to estimate the spatial average *SWC* of a given area (Grayson and Western, 1998). Therefore, quick and accurate estimation of soil water content at a field or large areas may be possible with only one representative location. Areal estimation of *SWC* from point source data has the potential to substantially reduce both the capital and labor costs involved in estimating average *SWC*, making the method appealing to a wide range of disciplines.

Time stability of *SWC* is controlled by various factors including soil, topography, vegetation, and climate (Brocca et al., 2009; Gómez-Plaza et al., 2001; Grayson and Western, 1998; Hu et al, 2010a; Tallon and Si, 2004). Information on the controls of time stability provides an essential insight into the mechanisms of soil water movement and storage. In addition, it is also important to identify the time stable location for average *SWC* estimation *a priori*.

The goal of this chapter is to provide a comprehensive review on the study of time stability of *SWC*. In doing so, we have introduced the concept, methodology, application, and the controlling factors step by step.

2. Concept

Soil water content varies with time and space. However, if a field is repeatedly surveyed for *SWC*, there is a high probability that a location with a certain relative wetness condition (i. e., wet, dry, and intermediate) at a given time will remain at the same relative wetness condition at other times. This phenomenon was first explained as time stability by Vachaud et al. (1985). This can be defined as the "time invariant association between spatial location and classical statistical parametric values" of *SWC*, most often the mean.

Time stability of *SWC* can be divided into two types, one is to describe the overall similarity of the spatial pattern between measurements, and the other is to describe the time-invariance of the relative *SWC* of a given location.

Repeated survey indicated that some locations are always wetter or drier than the average *SWC* of an area, resulting in the preservation of their ranks. Hence, the spatial pattern of *SWC* measured at one time will be similar to those measured on subsequent occasions. Statistically, this kind of time stability indicates that the orders, or ranks, of *SWC* at different locations do not change over time at some probability. If all locations were to maintain their rank on subsequent measurement occasions, then the spatial pattern of *SWC* will be identical. From this point, Chen (2006) augured that the term of "rank stability" or "order stability" should be better than "time stability" because "the stability is the order or rank of a soil property at different spatial points that does not change at some probability".

The most widely used method to describe time stability of *SWC* spatial patterns is Spearman's rank correlation analysis (Vachaud et al., 1985), while Pearson correlation analysis has also been widely used (Cosh et al., 2004). This is because Pearson correlation measures strictly linear relationships, while Spearman Rank correlation also measures nonlinear correlation between SWC values measured at two different occasions. Time stability of the spatial pattern of *SWC* is scale dependent because of the interaction between measurement locations (Kachanoski and De Jong, 1988). Scale dependence of the time stability was also investigated by different methods including spatial coherency analysis (Kachanoski and De Jong, 1988), wavelet coherency analysis (Biswas and Si, 2011c), and multivariate empirical mode decomposition (Rehman and Mandic, 2010). A detailed introduction of these methods is given in Section 3.

For time stability at point scales, *SWC* at each location is usually scaled in terms of the field average (Vachaud et al., 1985). This indicates that at a given location, the change in scaled *SWC* exhibits little change over time. Many indices are available to examine the degree of time stability at point scales. These indices include standard deviation of relative difference (Mohanty and Skaggs, 2001; Schneider et al., 2008; Vachaud et al., 1985), root mean square error (Bosch et al., 2006; Jacobs et al., 2004), width of the 90% empirical tolerance interval of relative water content (Guber et al., 2008), chi-squared statistic (Guber et al., 2008), root-mean-squared differences (Guber et al., 2008), temporal coefficient of variability (Starr, 2005), and mean absolute bias error (Hu et al., 2010a, 2010b). One of the applications of time stability at point scales is to identify catchment average soil moisture monitoring (CASMM) site (Grayson and Western, 1998) or benchmark location (Tallon and Si, 2004) for average *SWC* evaluation. Detailed introduction of these indices and applications is given in Section 3.

Generally, these two types of time stability are correlated. If the spatial pattern is time stable, there is a larger possibility of time stability at point scales. On the other hand, if more points are time stable, there is a larger possibility of time stability of spatial pattern. However, these two types of time stability are also distinguished for the following three aspects: (1) time stability of spatial pattern is used to evaluate the time stability of an area, while time stability of relative *SWC* is used to evaluate the time stability of a point. Therefore, time instability of spatial pattern does not mean the absence of time stable points (Grayson and Western, 1998; Schneider, 2008). On the other hand, if no points are time stable, it does not mean that spatial pattern is time unstable for any two measurement occasions; (2) they have different evaluation criteria as mentioned above, and (3) they have different applications. Time stability of spatial pattern is used to qualitatively describe the similarity of spatial distribution of *SWC* between different measurement occasions for the better understanding of soil water related process and influencing factors (Lin; 2006), while time stable locations can be selected to estimate the average *SWC* for an area (upscaling) by directly using the soil water content of the time stable location (Brocca et al., 2009; Cosh et al., 2008; Grayson and Western, 1998; Martínez-Fernández and Ceballos, 2005) or by considering the offset between the mean value and that of the time stable location (Grayson and Western, 1998; Hu et al., 2010b; Starks et al., 2008).

3. Methodology

There are different methods for analyzing time stability. We have documented the methods in two sections: (1) time stability of spatial patterns and (2) time stability of points.

3.1. Time stability of spatial patterns

3.1.1. Pearson correlation analysis

Pearson correlation coefficient is sensitive only to a linear relationship between two variables. A Pearson correlation coefficient, $r_{j,j'}$, between two spatial series of soil water content measured at time j and j' can be defined by

$$r_{j,j'} = \frac{\sum_i (SWC_j(i) - \langle SWC_j \rangle)(SWC_{j'}(i) - \langle SWC_{j'} \rangle)}{\sqrt{\sum_i (SWC_j(i) - \langle SWC_j \rangle)^2} \sqrt{\sum_i (SWC_{j'}(i) - \langle SWC_{j'} \rangle)^2}} \tag{1}$$

where $SWC_j(i)$ and $\langle SWC_j \rangle$ are soil water content at location i and spatial average at a given time j, respectively. $SWC_{j'}(i)$ and $\langle SWC_{j'} \rangle$ are soil water content at location i and spatial average at another time j', respectively. The resulting coefficients refer to the correlation of the spatial patterns of *SWC* from one time to another. It is expected that closely correlated patterns have a $r_{j,j'}$ near one, while uncorrelated patterns are indicated by $r_{j,j'}$ values near zero (Cosh et al., 2004). Student's t is generally used to test the significance of Pearson correlation coefficient. It can be implemented by some statistical software, such as Excel, SPSS, MathCad, Matlab, and SAS.

3.1.2. Spearman's rank correlation analysis

Spearman's rank correlation coefficient is a non-parametric measure of statistical dependence between two variables. It assesses how well the relationship between two variables can be described using a monotonic function. It is the Pearson correlation between the ranks of one series and the ranks of another series. Because ranking linearize some of the nonlinear relationships, it is sensitive to nonlinear relationships. Let R_{ij} be the rank of soil water content $SWC_j(i)$ and R'_{ij} the rank of $SWC_{j'}(i)$. The Spearman rank correlation coefficient, r_s, is calculated by

$$r_s = 1 - \frac{6\sum_{i=1}^{n}\left(R_{ij} - R'_{ij}\right)^2}{n(n^2-1)} \tag{2}$$

where n is the number of observations. A value of r_s=1 corresponds to identity of rank for any sites, or perfect time stability between time j and j'. The closer r_s is to 1, the more stable the spatial pattern will be. Student's t can be used to test the significance of r_s (Zar, 1972). Software such as Excel, SPSS, MathCad, Matlab, and SAS can implement this analysis.

3.1.3. *Average spatial coefficient of determination and temporal coefficient of variability*

If perfectly time stable soil water content pattern exists, the ratio of $SWC_j\ (i)$ to $\langle SWC_j \rangle$, can be a time independent scaling factor, $r\ (i)$, which is expressed as (Starr, 2005)

$$\frac{SWC_j(i)}{\langle SWC_j \rangle} = r(i) \tag{3}$$

Complete time stability is usually disturbed by a series of processes and therefore Eq. (3) needs an extra term to fit a more general situation (Starr, 2005)

$$\frac{SWC_j(i)}{\langle SWC_j \rangle} = r(i) + e_j(i) = s_j(i) \tag{4}$$

where $s_j\ (i)$ is a scaled factor depending on both time and location, and $e_j\ (i)$ is an additional term that accounts for random measurement errors, random sampling variability, and any true deviations from time stability. Assuming that $e_j\ (i)$ and $r\ (i)$ are independent, then the variance of $s_j\ (i)$ over space at any time, $\gamma_i^2[s_j\ (i)\ ]$, can be written as

$$\gamma_i^2\left[s_j(i)\right] = \gamma_i^2\left[r(i)\right] + \gamma_i^2\left[e_j(i)\right] \tag{5}$$

where $\gamma_i^2[r\ (i)\ ]$ and $\gamma_i^2[e_j\ (i)\ ]$ are the variances of $r\ (i)$ and $e_j\ (i)$, respectively over space. The $\gamma_i^2[s_j\ (i)\ ]$ can be calculated by averaging the calculated variances for each observation. Then, the average spatial coefficient of determination, R_s^2, which describes the proportion of the total variance explained by the time stability model of Eq. (3) can be written as

$$R_s^2 = \frac{\gamma_i^2\left[r(i)\right]}{\left\langle \gamma_i^2\left[s_j(i)\right]\right\rangle} \tag{6}$$

where $\langle \gamma_i^2[s_j\ (i)\ ] \rangle$ is the mean value of $\gamma_i^2[s_j\ (i)\ ]$ over time. Given $e_j\ (i)$ with an average value of zero over time, $r(i)$ can be approximated by time average of $s_j\ (i)$, $\langle s_j\ (i)\ \rangle$, i. e.,

$$r(i) = \langle s_j(i) \rangle \tag{7}$$

The value of R_s^2 approaches one for a complete spatial pattern described by Eq. (3) and zero for a situation where no variances can be explained by the time stability model. Therefore, greater R_s^2 means stronger time stability of spatial pattern. However, Starr (2005) pointed out it is not suitable for the situation where no spatial variability of *SWC* and completely time stability exists.

To deal with this issue, temporal coefficient of variability, CV_t, was developed to describe the degree of time stability (Starr, 2005) and can be written as

$$CV_t = 100\frac{1}{n}\sum_{i=1}^{n}\frac{\sigma_t\left[s_j(i)\right]}{\left\langle s_j(i)\right\rangle} \tag{8}$$

where $\sigma_t[s_j(i)]$ is the standard deviation of $s_j(i)$ over time, $\langle s_j(i)\rangle$ is the time average of $s_j(i)$. The smaller CV_t value indicates stronger time stability. The CV_t approaching zero indicates a perfect time stable pattern.

3.1.4. *Empirical orthogonal function method*

Empirical orthogonal function (EOF) analysis is a decomposition of a data set in terms of orthogonal basis functions which are determined from the data. It is the same as performing a principal components analysis on the data, except that the EOF method finds both time series and spatial patterns. The basic principle of EOF method is to partition a series into time-invariant spatial patterns (EOFs) of *SWC* and coefficients (ECs) which vary temporally but is constant spatially (Perry and Niemann, 2007). The original spatial series of *SWC* can be obtained by taking the sum of product of EOFs by ECs. A limited number of EOFs that present significant spatial variation of *SWC* can be selected and used to identify the dominant factors determining the spatial pattern.

First, the spatial anomalies of *SWC* are computed. The spatial anomaly at location i and time j, $z_j(i)$, can be calculated as

$$z_j(i) = SWC_j(i) - \frac{1}{n}\sum_{i=1}^{n} SWC_j(i) \tag{9}$$

Considering the spatial anomalies, the spatial covariance $v_{j,j'}$ at time j and j' can be calculated as

$$v_{j,j'} = \frac{1}{n}\sum_{i=1}^{n} z_j(i) z_{j'}(i) \tag{10}$$

where $z_{j'}(i)$ is the spatial anomaly at the same location i but time j'.

To consider all the measurement times, the matrix of spatial anomalies, Z, and its spatial covariance, V, can be written as

$$Z = \begin{pmatrix} z_{11} & \mathrm{K} & z_{1m} \\ \mathrm{M} & \mathrm{O} & \mathrm{M} \\ z_{n1} & \mathrm{L} & z_{nm} \end{pmatrix} \tag{11}$$

and

$$V = \frac{1}{n} Z^T Z \tag{12}$$

respectively, where m is the number of sampling times.

The next step is to diagonalize the spatial covariance matrix V by finding its eigenvectors, E, and eigenvalues, L. Mathematically, they should satisfy the equation

$$VE = LE \tag{13}$$

where E is an $m \times m$ matrix that contains the eigenvectors as columns

$$E = \begin{pmatrix} e_{11} & \mathrm{L} & e_{1m} \\ \mathrm{M} & \mathrm{O} & \mathrm{M} \\ e_{m1} & \mathrm{L} & e_{mm} \end{pmatrix} \tag{14}$$

and L is an $m \times m$ matrix that contains the associated eigenvalues along the diagonal and zeros at off-diagonals

$$L = \begin{pmatrix} l_{11} & \mathrm{L} & 0 \\ \mathrm{M} & \mathrm{O} & \mathrm{M} \\ 0 & \mathrm{L} & l_{mm} \end{pmatrix} \tag{15}$$

The eigenvectors in E represent the weights applied to each component in V to diagonalized V. This transformation is a rotation of the original axes in multi-dimensional space, where each dimension corresponds to a sampling time. Axes are orthogonal to each other and explain different amount of covariance in the spatial anomaly dataset. The eigenvalues contained in L represent the variance that occurs in the direction of each newly identified axis. After diagonalization of V, E and L are arranged accordingly to keep the eigenvalues in L sorted in a descending order. Therefore, the first axis explains the most covariance in the spatial anomaly dataset, and the second axis explains the second most covariance and so on. Each axis represents a direction in the multidimensional space. The total variance of the spatial anomaly data is the sum of the diagonal values in L and is equal to the total variance of the original spatial anomalies. Therefore, the portion of the variance, P_j, that the jth axis can explain is

$$P_j = \frac{l_{jj}}{\sum_{k=1}^{m} l_{kk}} \tag{16}$$

The relative importance of each axis in explaining the variability of spatial anomalies can be defined by the relative magnitude of the associated eigenvalue in L.

Each spatial anomaly can be described in terms of the new variable axis. A matrix F containing the coordinates of the spatial anomalies on the new axis can be obtained by projecting the spatial anomalies onto the rotated axis. Mathematically, this operation is

$$F = ZE \tag{17}$$

Each column in F is EOF, and corresponding column in E is expansion coefficient (EC). From the aspect of time stability application, usually a limited number of EOFs that explains significant amount of the spatial variability of the original SWC series is selected to present the underlying time-invariable spatial pattern. North et al. (1982) indicated a set of selection criterion. An EOF is considered significant provided the lower confidence limit of its eigenvalue is larger than the upper confidence limit of the eigenvalue of the next most important EOF (North et al., 1982).

3.1.5. Spectral coherency analysis

Spectral coherency is used to measure the similarity between two spatial series in a frequency domain, which can be converted to a spatial scale. This method assumes the spatial series to be stationary and linear. Therefore, it can be used to find the scale specific information on time stability of spatial pattern between different measurements.

Spectral coherency analysis involves calculation of the power spectrum of a variable $V_t^2(f_K)$ which estimates the observed variance as a function of spatial scale.

$$V_t^2(f_K) = (2h+1)^{-1} N^{-1} \sum_{l=-h}^{h} \left[\left(\sum_{j=1}^{N} SWC_t(j) \cos(2\pi f_{K+1} j) \right)^2 + \left(\sum_{j=1}^{N} SWC_t(j) \sin(2\pi f_{K+1} j) \right)^2 \right] \tag{18}$$

where $f_K = K/N$, K = 0, 1, 2, ..., N/2 cycles h^{-1}, and h is the smoothing coefficient determining the degree of averaging of adjacent independent frequencies and the degrees of freedom for the individual spectral variance estimates. The value of h cannot be too large because power spectrum of a variable may incorporate more bias from smoothing of the spectra, although larger value of h will result in smaller variance of the estimate (Gómez-Plaza et al., 2000).

The covariance between $SWC_j(i)$ and $SWC_{j'}(i)$ can be estimated as a function of spatial scale by calculating sample cross spectrum $V_{j,j'}(f_K)$

$$V_{j,j'}\left(f_K\right) = (2h+1)^{-1}N^{-1}\sum_{l=-h}^{h} F_j\left(f_{K+1}\right)^{*}F_{j'}\left(f_{K+1}\right) \tag{19}$$

where the asterisk (*) is a complex conjugate of

$$F_j\left(f_K\right) = \sum_{i=0}^{N-1} SWC_j(i)\left[\cos\left(2\pi f_K i\right) - i\sin\left(2\pi f_K i\right)\right] \tag{20}$$

$$F_{j'}\left(f_K\right) = \sum_{i=0}^{N-1} SWC_{j'}(i)\left[\cos\left(2\pi f_K i\right) - i\sin\left(2\pi f_K i\right)\right] \tag{21}$$

where $f_K = K/N$, K = 0, 1, 2, ..., N/2 cycles h^{-1}, and $i = (-1)^{0.5}$.

Spectral coherency function $R^2_{j,j'}(f_K)$ can then be calculated in a similar manner to calculate coefficient of determination (Kachanoski and De Jong, 1988)

$$R_{j,j'}^{2}\left(f_K\right) = \frac{\left|V_{j,j'}\left(f_K\right)\right|^2}{V_j^2\left(f_K\right)V_{j'}^2\left(f_K\right)} \tag{22}$$

The coherency function $R_{j,j'}(f_K)$ estimates the proportion of the spatial variance of $SWC_j(i)$ which can be explained by the spatial variance of $SWC_{j'}(i)$, as a function of spatial scale. Thus, spectral coherency is useful to measure time stability of soil water content as a function of spatial scale.

The significance test method for spectral coherency consist of parametric methods based on an assumed theoretical distribution and nonparametric method such as reshuffling, bootstrapping, and bagging method. We encourage readers to see Si (2008) for a more detailed account on significance test.

3.1.6. *Wavelet coherency analysis*

Wavelet analysis, differing with spectral analysis, can identify localized features of soil processes (Si, 2008). It is suitable to reveal the scale and location specific time-persistence of spatial pattern of *SWC* between sampling occasions assuming the soil water content system is linear (Biswas and Si, 2011c). Wavelet coherency of two spatial series can describe the time stability of spatial pattern. It requires the calculation of wavelet coefficients for each of the two data series and associated cross-wavelet spectrum. Many publications on the introduction of wavelet analysis (Farge, 1992; Kumar and Foufoula-Georgiou, 1993, 1997) and wavelet coherency (Biswas and Si, 2011c; Grinsted et al., 2004; Si, 2008; Si and Zeleke, 2005) can be found. Here, we will present the basic procedure to calculate wavelet coherency.

First, wavelet coefficient, $W_i^Y(s)$, is calculated with the continuous wavelet transform (CWT) for a *SWC* series of length n (Y_i, $i = 1, 2, \ldots, n$) with equal incremental distance δx. This can be defined as the convolution of Y_i with the scaled and normalized wavelet using fast Fourier transform (Si and Zeleke, 2005; Torrence and Compo, 1998):

$$W_i^Y(s) = \sqrt{\frac{\delta x}{s}} \sum_{j=1}^{n} Y_j \psi\left[(j-i)\frac{\delta x}{s}\right] \tag{23}$$

where $\psi[]$ is the mother wavelet function and s is the scale. The mother wavelet functions include Morlet, Mexican hat, Harr, and others (Si, 2008). Depending on the purpose, different mother wavelet functions can be selected. The Morlet wavelets allow us to detect both location dependent amplitude and phase for different frequencies in the spatial series (Torrence and Compo, 1998), which can be written as

$$\psi[\eta] = \pi^{-1/4} e^{i\bar{\omega}\eta - 0.5\eta^2} \tag{24}$$

where ω is dimensionless frequency and η is dimensionlessspace. The wavelet is stretched in space (x) by varying its scale (s), so that $\eta = s / x$. The Morlet wavelet (with ω =6) is a good choice for feature extraction purpose like identifying the scales and locations, since it provides a good balance between space and frequency localization.

The wavelet coefficients $W_i^Y(s)$ can be expressed as $a + ib$ where a and b are the real and imaginary components of $W_i^Y(s)$. For the polar form of complex numbers, $W_i^Y(s) = |W_i^Y(s)|(\cos\theta + i\sin\theta)$, where $\theta = \arctan\frac{b}{a}$ is called the phase or argument of $W_i^Y(s)$. The wavelet power spectrum is defined as $|W_i^Y(s)|^2$ and the local phase is defined as the complex argument of $W_i^Y(s)$.

After calculating the wavelet spectra $W_i^Y(s)$ and $W_i^Z(s)$ corresponding to two *SWC* spatial series Y and Z, respectively for two different times, cross wavelet power spectrum $W_i^{YZ}(s)$ at scale s and location i can be defined as

$$W_i^{YZ}(s) = \left[W_i^Y(s)\right]\left[\overline{W_i^Z(s)}\right] \tag{25}$$

and the wavelet coherency of two spatial series can be written as

$$\rho_i^2(s) = \frac{\left|\overleftrightarrow{W}_i^{YZ}(s)\right|^2}{\left|\overleftrightarrow{W}_i^{Y}(s)\right|^2\left|\overleftrightarrow{W}_i^{Z}(s)\right|^2} \tag{26}$$

where $\overleftrightarrow{(\cdot)}$ is a smoothing operator, which can be written as

$$\overleftrightarrow{W} = SM_{scale}\left[SM_{space}(W)\right] \tag{27}$$

where SM_{space} denotes smoothing along the wavelet scale axis and SM_{scale} smoothing in spatial distance. The following smoothing function is the normalized real Morlet wavelet and has a similar footprint as the Morlet wavelet

$$\frac{1}{s\sqrt{2\pi}}\exp\left(-\frac{\tau^2}{2s^2}\right) \tag{28}$$

where τ denotes location. Therefore, the smoothing along locations can be written as

$$SM_{space}(W,s,\tau) = \sum_{k=1}^{m}\left\{W(s,\tau)\frac{1}{s\sqrt{2\pi}}\exp\left[-\frac{(\tau - x_k)^2}{2s^2}\right]\right\}\Big|_s \tag{29}$$

The Fourier transform of Equation (28) is $\exp(-2s^2\omega^2)$, where ω is the frequency. Eq. (29) can be implemented using Fast Fourier Transform (FFT) and Inverse FFT (IFFT) based on convolution theorem and is written as

$$SM_{space}(W,s,x) = IFFT\left\{FFT\left[W(s,\tau)\right]\left[\exp\left(-2s^2\omega^2\right)\right]\right\} \tag{30}$$

The smoothing along scales can be written as:

$$SM_{scale}\left(W,s_k,x\right)=\frac{1}{2p+1}\sum_{j=k-p}^{k+p}\left[SM_{space}\left(W,s_j,x\right)\Pi\left(0.6s_j\right)\right]\Big|_x \tag{31}$$

where P is the number of terms on each symmetrical half of the window, and Π is the rectangle function. The factor of 0.6 is the empirically determined scale decorrelation length for the Morlet wavelet (Si and Zeleke, 2005; Torrence and Compo, 1998).

There are many methods available to test the statistical significance of wavelet coherency. The Monte Carlo simulation or reshuffling method is among the suggested ones. A detailed description of these methods can be found in Si (2008). Matlab codes for calculating wavelet coherency are available at URL: http://www.pol.ac.uk/home/research/waveletcoherence/ (Grinsted et al., 2004).

3.1.7. *Multivariate empirical mode decomposition*

Empirical mode decomposition (EMD), extracts oscillations from the soil water content series into a finite and often small number of intrinsic mode functions (IMFs) according to the energy associated with different space scales (Huang et al., 1998; Huang and Wu, 2008). Unlike spectral and wavelet methods, EMD does not call for any assumption of the data and works directly in the spatial domain with the basis completely derived from the data. Therefore, it is intuitive, direct, *a posteriori* and adaptive (Huang et al., 1998; Huang and Wu, 2008). The locality and adaptivity of EMD can deal with different types of spatial series including non-stationary and nonlinear (Huang et al., 1998). Therefore, EMD has a great potential to find the underlying scale of spatial series of soil moisture without imposing any mathematical assumption on the measured data (Biswas and Si, 2011b).

Multivariate empirical mode decomposition (MEMD) is the multivariate extensions of standard EMD (Huang et al., 1998). An important step for MEMD is the computation of the local mean because local extrema are not well defined directly for multivariate spatial data. Moreover, the notion of "oscillatory model" defining an IMF is rather confusing for multivariate spatial data. To deal with these problems, Rehman and Mandic (2010) produced a multiple m-dimensional envelopes by taking projections of multiple inputs along different directions in an m-dimensional space.

Assuming $\{v(s)\}_{s=1}^{S}=\{\upsilon_1(s),\ \upsilon_2(s),\ \ldots,\ \upsilon_m(s)\}$ being the m spatial data sets of soil water content as a function of space (s), and $x^{\theta_k}=\{x_1^k,\ x_2^k,\ \ldots,\ x_m^k\}$ denoting a set of direction vectors along the directions given by angles $\theta^k=\{\theta_1^k,\ \theta_2^k,\ \ldots,\ \theta_{m-1}^k\}$ on a m-1-dimensional sphere (k is the number of direction used to calculate the projections and envelope curves). Then, IMFs of the m spatial data sets can be obtained by MEMD as follows:

1. Choose a suitable point set for sampling on an m-1-dimensional sphere. This can be done by sampling unit hyperspheres (m-spheres) based on both uniform angular sampling methods and quasi-Monte Carlo-based low-discrepancy sequences.

2. Calculate a projection, denoted by $p^{\theta_k}(t)\}_{s=1}^{S}$, of the spatial data sets $\{v(s)\}_{s=1}^{S}$ (S is the number of sampling point for each time) along the direction vector x^{θ_k}, for all k (the whole set of direction vectors), giving $p^{\theta_k}(s)\}_{k=1}^{K}$ as the set of projections.

3. Find the spatial instants $s_i^{\theta_k}$ corresponding to the maxima of the set of projected data sets $p^{\theta_k}(s)\}_{k=1}^{K}$.

4. Interpolate $\left[s_i^{\theta_k}, v\left(s_i^{\theta_k}\right)\right]$ to obtain multivariate envelope curves $e^{\theta_k}(s)\}_{k=1}^{K}$ for all the data sets considered.

5. For a set of K direction vectors, the mean $m(s)$ of the envelope curves is calculated as:

$$m(s)=\frac{1}{K}\sum_{k=1}^{K} e^{\theta_k}(s) \tag{32}$$

6. Extract the "detail" $d(s)$ using $d(s)=v(s)-m(s)$. If the "detail" $d(s)$ fulfills the stoppage criterion for a multivariate IMF, apply the above procedure to $v(s)-d(s)$, otherwise apply it to $d(s)$. The stoppage criterion for multivariate IMFs is similar to that proposed by reference (Huang et al., 2003).

MEMD has the ability to align "common scales" present within multivariate data. Each common scale is manifested in the common oscillatory modes in all the variates within an m-variate IMF. After MEMD analysis, scale and location specific time stability of *SWC* can be easily identified by comparing and calculating spearman's rank correlation coefficient between the IMFs with the same numerical numbers for different measurements using Eq. (2). The exact scale for each IMF can be obtained from the instantaneous frequencies by Hilbert transformation with IMF (Huang et al., 1998). The instantaneous frequencies of soil water content can be converted to period (1/frequency), which was further converted to the spatial scale after multiplying the period with the sampling interval. MEMD can be completed using Matlab program such as that written by Rehman and Mandic (2009) (http://www.commsp.ee.ic.ac uk/~mandic/research/emd.htm). This method requires that soil water content measurement should be obtained in intervals with equidistance.

3.2. Time stability at points

Time stability at point scales are usually obtained in terms of average *SWC* estimation quality. Direct and indirect methods can be used to estimate average *SWC* by the measurement at time stable locations. The direct method estimates average *SWC* directly by measuring *SWC* at a time-stable location (Brocca et al., 2009; Cosh et al., 2008; Grayson and Western, 1998; Pachepsky et al., 2005). The indirect method estimates average *SWC* by considering the offset between the average *SWC* and the measurement value at a time-stable location (Grayson and Western, 1998; Starks et al., 2006). All the indices are listed in Table 1. Each index has its own advantages and disadvantages, and these are discussed in the application (Section 4.2). For all indices, smaller value of time stability index at a given location indicates stronger time stability and also higher quality of average *SWC* estimation by that location.

Index	Formula and explanation	Average SWC estimation Method†	Reference
Standard deviation of relative difference, $SDRD(i)$	where $SDRD(i)=\sqrt{\frac{1}{m-1}\sum_{j=1}^{m}(\delta_j(i)-\langle\delta_i\rangle)^2}$ and $\delta_j(i)=\frac{SWC_j(i)-\langle SWC_j\rangle}{\langle SWC_j\rangle}$	Direct and indirect	Vachaud et al., 1985
Root mean square error, $RMSE(i)$		Direct	Jacobs et al., 2004
Standard deviation of relative SWC, $\sigma(\beta_i)$ ‡	$\langle\delta_i\rangle=\frac{1}{m}\sum_{j=1}^{m}\delta_j(i)$ where $RMSE(i)=(\langle\delta_i\rangle^2+SDRD(i)^2)^{1/2}$ and $\sigma(\beta_i)=\sqrt{\frac{1}{m-1}\sum_{j=1}^{m}(\beta_j(i)-\langle\beta_i\rangle)^2}$	Indirect	Pachepsky et al., 2005
Width of the 90% empirical tolerance interval of relative water content, $T(i)$	where $\beta(i)_{P=0.95}$ and $\beta(i)_{P=0.05}$ are the relative SWC values at cumulative probability of 95% and 5%, respectively.	Indirect	Guber et al., 2008
Chi-squared statistic, $\chi^2(i)$	where σ_j is the standard deviation of SWC at observation time j.	Indirect	Guber et al., 2008
Root-mean-squared differences, $D(i)$		Indirect	Guber et al., 2008
Mean absolute bias error, $MABE(i)$		Indirect	Hu et al., 2010a, 2010b

†Direct refers to estimating average *SWC* directly by measuring *SWC* at a time-stable location and indirect refers to estimating average *SWC* by considering the offset between the mean value and the measurement value at a time-stable location.

‡ $\sigma(\beta_i)=SDRD\,(i)$.

Table 1. Time stability indices at point scales and their formulas

4. Application

4.1. Time stabilityof spatial pattern

Information on the spatial pattern of soil water content with time is important to understand the changes in hydrological processes and in making predictions with physical and statistical models. In this section, we have documented the use of each method introduced above to characterize the similarity of the spatial pattern of *SWC*.

Spearman's rank correlation analysis is the most widely used method to examine the similarity of the spatial patterns of *SWC*. Numerous studies have used this method and have confirmed the presence of time stability of the spatial pattern (Biswas and Si, 2011c; Hu et al., 2009, 2010b; Martínez-Fernández and Ceballos, 2003; Vachaud et al., 1985; Vivoni et al., 2008). In the first of this kind, Vachaud et al. (1985) measured soil water storage of 0 to 1.0 m depth at 17 neutron access tubes located at 10 m intervals for 24 occasions over a time span of two and a half years. According to the cumulative probability function for the two extreme conditions, i. e., the wettest on 7 September 1982, and the driest on 17 August 1983, many locations maintained their ranks (Figure 1). For example, location #1 always had the maximum soil water storage, location #4 had the minimum soil water storage irrespective of the environmental conditions (e. g., wet or dry). Spearman's rank correlation coefficients of the soil water storage between measurements were also calculated to examine the degree of time stability of the spatial pattern. As an example, the correlation between seven measurements representing the range of the water storages recorded over the entire observation period is presented in Table 2. All rank correlation coefficients were highly significant at the 0.1% two-tailed test, indicating strong time stability of the spatial pattern of soil water storage. On the other hand, time instability of spatial pattern between measurements was also reported (Comegna and Basile, 1994; Mohanty and Skaggs, 2001). Although it is very popular among researchers, Spearman's rank correlation analysis should be viewed only as a statistical tool measuring the degree of concordance between two rankings (Vachaud et al., 1985). It may be questionable if differences between measured values are smaller than experimental uncertainties. This may be the case in situations where either the probability density function is very uniform or the experimental determinations are very crude.

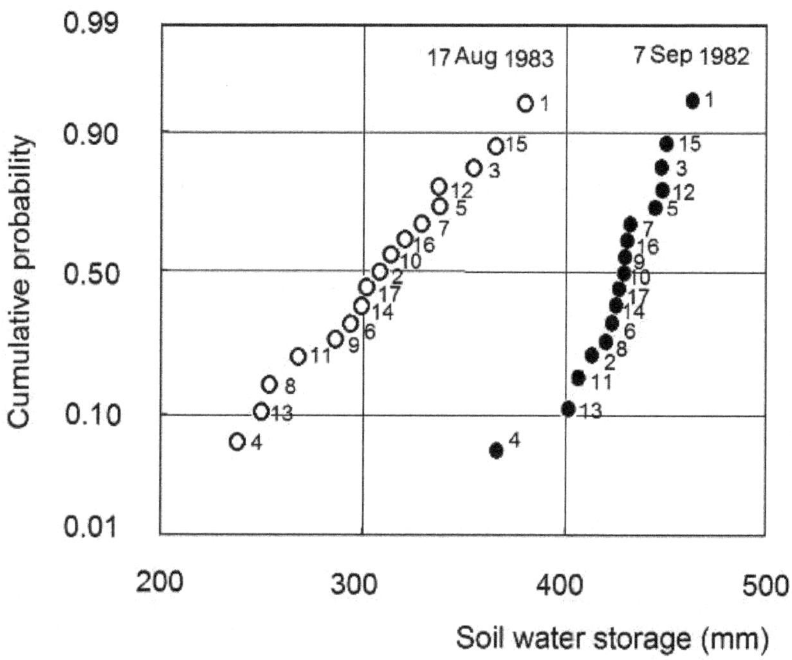

Figure 1. Cumulative probability function of soil water storage at 0-1.0 m measured in the most dry, and the most wet conditions. Numbers refer to measuring locations. (Reproduced from Vachaud et al. (1985)).

Dates	7 Sep 1982	28 Oct 1981	2 Dec 1982	29 Jul 1981	18 Aug 1981	16 Jul 1982	25 Aug 1981
Average storage (mm)	428.1	425.77	424.77	382.06	354.72	347.23	344.46
(1)	1						
(2)	0.953	1					
(3)	0.941	0.953	1				
(4)	0.988	0.961	0.946	1			
(5)	0.953	0.922	0.882	0.968	1		
(6)	0.863	0.789	0.843	0.836	0.882	1	
(7)	0.860	0.824	0.824	0.794	0.863	0.939	1

Table 2. Matrix of Spearman's rank correlation coefficient between series of soil water storage measurements obtained on the 17 locations on seven dates. (Reproduced from Vachaud et al. (1985))

As an alternative to Spearman's rank correlation analysis, Pearson correlation coefficients were also used to describe time stability of the spatial pattern (Cosh et al., 2004). It was observed that the Pearson correlation analysis revealed a pattern similar to that of the Spearman's rank correlation analysis (Cosh et al., 2004). Note that the tendency of soil moisture sensors to occasionally report erroneous measurements may negatively affect the value of Spearman's rank correlation coefficient, while the Pearson correlation coefficient is less sensitive to this problem. This is because the total number of measurements will not affect Pearson correlation coefficient as it affects the rank correlation coefficient (Cosh et al., 2006, 2008). However, both Spearman's rank correlation and Pearson correlation analysis are only suitable to describe how much spatial pattern at one measurement time can be preserved at another time (Rolston et al., 1991). If we want to get the information on the similarity of spatial pattern over multiple measurement occasions, mean value of correlation coefficients can be used (Hu et al., 2009).

With the aim to characterize the degree of time stability of spatial pattern among multiple data series, Starr (2005) suggested two indices, i. e., average spatial coefficient of determination (Rs^2) and temporal coefficient of variability (CV_t). The authors reported that 26 to 76% (with mean of 47%) of the observed variability was explained by the time stable pattern. The reported CV_t value ranged from 11% to 18% (with mean of 16%) for different sampling transects. On the other hand, the Rs^2 may not be a good measure of how time stable a *SWC* pattern is at a field that is uniform and completely time stable. This is because the Rs^2 will be zero in this case (Starr, 2005).

Perry and Niemann (2007) used the EOF method to characterize the time stable spatial pattern of *SWC*. They identified two dominant spatial patterns of soil water variability at Tarrawarra catchment (Figure 2). The first spatial pattern was more important for medium wet periods and was associated with lateral redistribution of soil water. The second spatial pattern served to modify the first spatial pattern at some days which was associated to evapotranspiration. The importance of each spatial pattern on each measurement occasion can be reflected by the

EC values (Figures 2c and 2d). There was a large decrease in EC1 weight during the dry season, which agrees well with the finding of Western et al. (1999) that soil moisture patterns on moderately wet occasions show a strong topographic influence. The EOF method can also be used to predict the *SWC* distribution for unobserved times if the average *SWC* and the temporal varying coefficient (ECs), which reflect the relative importance of different spatial patterns can be obtained. Average *SWC* can be estimated by many methods such as linear regression (Western et al., 1999), a dynamic multiple linear regression of soil moisture against topographic attributes (Wilson et al., 2005), and by measurement from representative location (Grayson and Western, 1998). Examples of EOF method can also be found in other study (Ibrahim and Huggins, 2011; Jawson and Niemann, 2007; Joshi and Mohanty, 2010; Korres et al., 2010; Yoo and Kim, 2004). Different numbers, usually one to four, of significant EOFs were obtained to explain most of *SWC* variations (usually more than 70%). The time-invariant spatial patterns were usually related to topography-related factors (Yoo and Kim, 2004), sand content (Jawson and Niemann, 2007), mixed influences of rainfall, topography, and soil texture (Joshi and Mohanty, 2010), elevation and wetness index under wet conditions and soil properties (ECa and bulk density) in dry conditions (Jawson and Niemann, 2007).

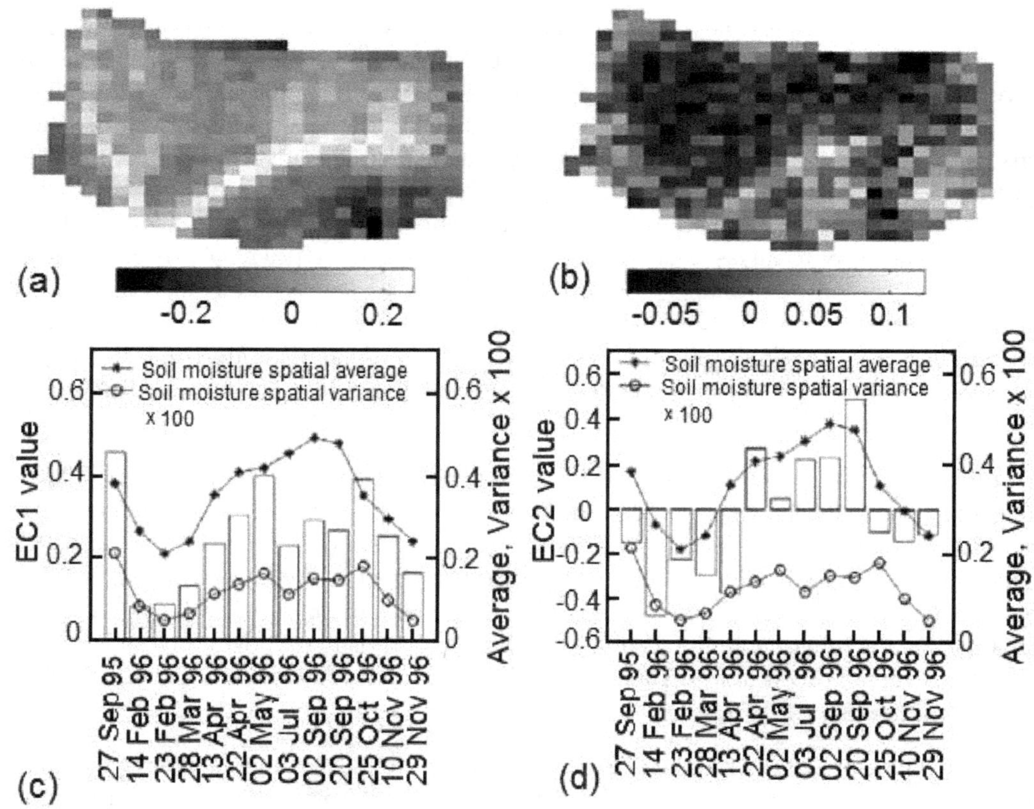

Figure 2. a) EOF1 and (b) EOF2 computed from the spatial anomalies of soil moisture. The bars in (c) and (d) are EC1 and EC2, respectively. The spatial average and variance of the soil moisture on each sampling date are also shown in (c) and (d) for comparison. (Reproduced from Perry and Niemann (2007))

Time stability of soil water content is usually scale dependent because of the interactions of soil moisture between measurement locations (Kachanoski and De Jong, 1988). The first effort to reveal the scale specific time stability was made by Kachanoski and De Jong (1988) who used spectral coherency to explore the time stability of soil water storage in a rolling landscape (Figure 3). They observed that for both wet and drying periods, the spectral coherency was high at large scales (>40 m). For smaller scales (<40 m), however, the spectral coherency was high for the drying period but greatly weakened during the alternating wetting and drying period. Therefore, their results clearly showed the loss of time stability at small scales (<40 m). Gómez-Plaza et al. (2000) also used the spectral coherency to explore the scale dependence of time stability in burned and unburned areas in a semi-arid environment. They clearly observed that soil moisture patterns were maintained in all cases at large scales (transect scale), while soil moisture tended to become time unstable, depending on the range of scale and section of transect considered at medium scales.

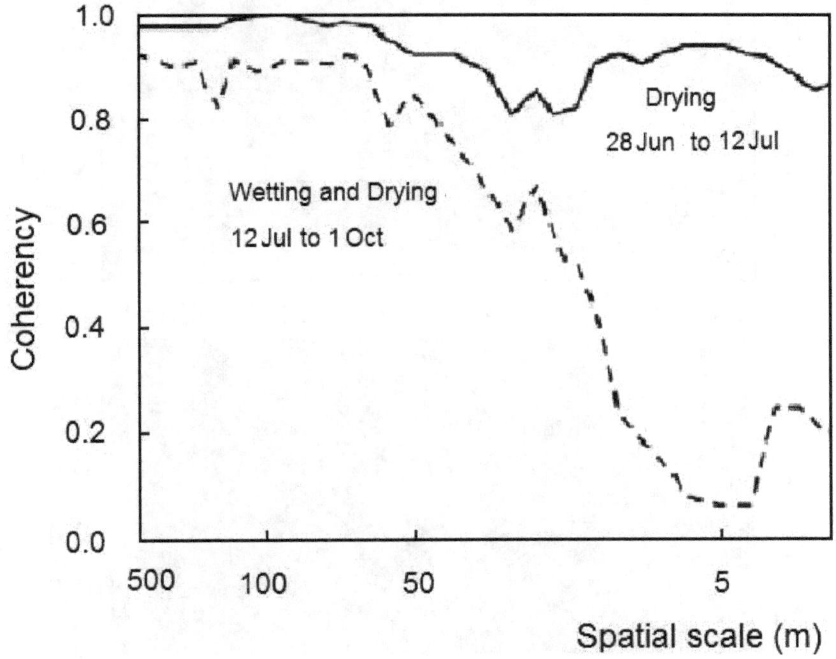

Figure 3. Coherency spectrum for soil water storage for the drying (June 28-July 12) and recharge (July 12-October 1) period. (Reproduced from Kachanoski and De Jong (1988))

Spectral coherency was verified to be robust to look for the scale dependency of time stability. Its multi-variate extension is also feasible (Si, 2008), although no publications have been found on this point. However, spatial coherency analysis assumes the stationarity in the measured data series over space, hence may not capture localized features of time stability (Biswas and Si, 2011a).

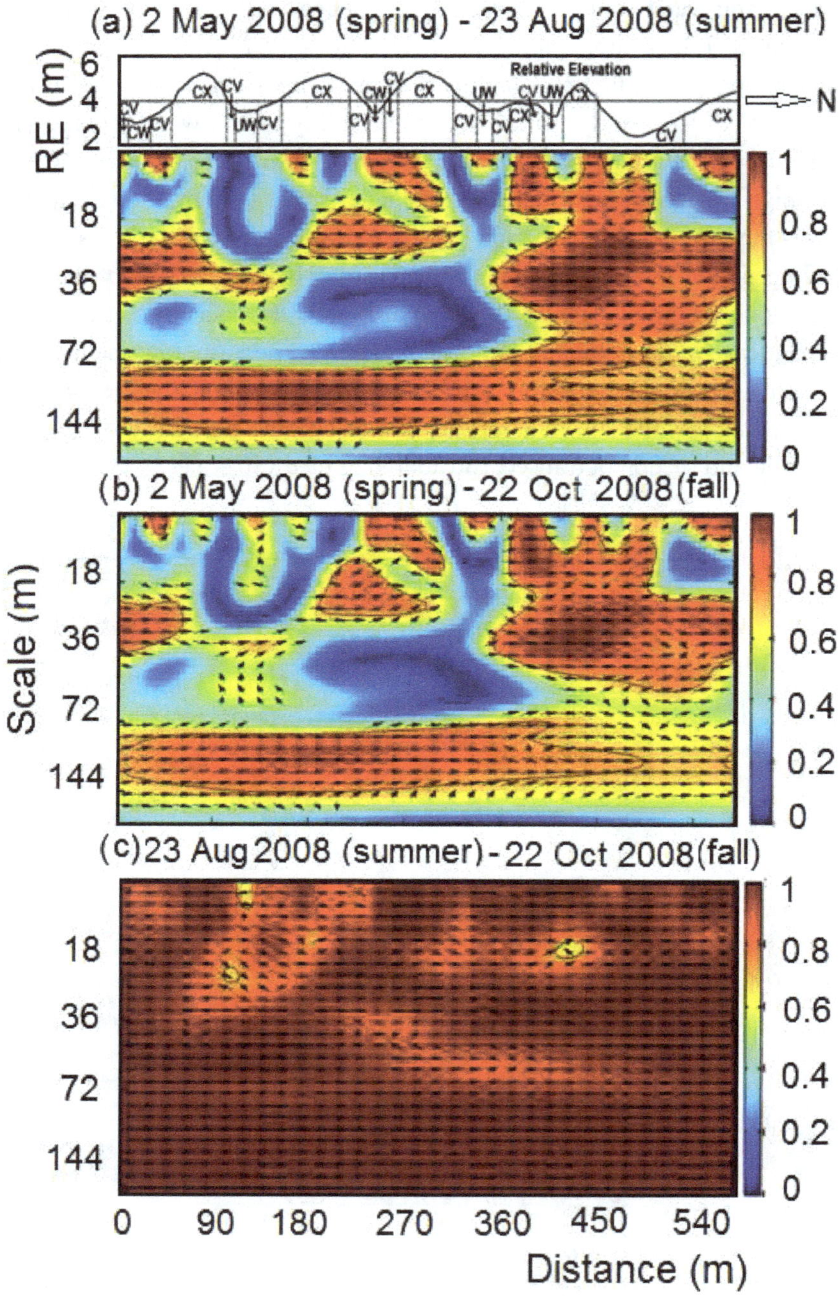

Figure 4. The inter-season wavelet coherencies between the soil water storage measured during (a) spring (2 May 2008) and summer (23 August 2008), (b) spring (2 May 2008) and fall (22 October 2008), and (c) summer (23 August 2008) and fall (22 October 2008) for the whole soil profile. Cross sectional view of the transect with landform elements is shown at the top. The X-axis indicates distance along the transect; the Y-axis indicates the scale in meter; the solid black line indicates 5% significance level; the color bar indicates strength of correlation, and the direction of arrow indicates the phase relationship or type of correlation. (Reproduced from Biswas and Si (2011c)).

Wavelet coherency can deal with the non-stationary series, it is therefore suitable to identify both scale and location specific time stability (Biswas and Si, 2011a). An excellent application of wavelet coherency in time stability analysis was made by Biswas and Si (2011a). They identified scales and locations of time stability between the spatial patterns of soil water storage in a hummocky landscape in central Saskatchewan, Canada. As Figure 4 shows, strong time stability at all scales and locations existed between summer and fall, when the vegetation was present in field. The spatial pattern in spring was different from that of summer or fall. The large-scale (> 72 m) time stability was present over the whole transect. The medium-scale spatial patterns were time stable in large depressions that had fewer landform elements. The small-scale time stability was mostly random.

Wavelet coherency can also be used to identify the correlation between different variables at different scales and positions (Lakshmi et al., 2004; Si and Zeleke, 2005). One limitation of wavelet coherency analysis is that it is based on the assumption of linearity of systems. Moreover, the application of wavelet coherency on time stability analysis was only confined to two spatial series. Extension of this method to multiple spatial series is expected to understand the scale and location specific time stability when soil water content from different seasons are considered together.

MEMD is a data-driven method that can deal with both non-stationary and nonlinear series. Recently, Hu et al. (unpublished data) (2012a) applied MEMD to time stability analysis using soil water storage measurements from 20 occasions in a hummocky landscape in central Saskatchewan, Canada. At first, multiple spatial series of soil water storage were separated into different IMFs. The IMFs for the 0-20 cm layer on a selection of measurement occasions is given in Figure 5. Each IMF corresponds to a specific scale common to all the spatial series considered. Spearman's rank correlation analysis was then conducted for each scale. According to their study, the dominant scale of variations of soil water storage was 104-128 m (IMF5), for 0-20 cm, 20-40 cm, 40-60 cm, and 60-80 cm, and 315-412 m (IMF7) for 80-100 cm, 100-120 cm, and 120-140 cm. Time stability generally increased with scale and it was the strongest at the dominant scale of variation. At different scales, the time stability generally ranked in the order of intra-season>inter-annual>inter-season. Therefore, the study of Hu et al. (2012a) verified that scale specific time stability of soil moisture can be identified using MEMD method.

According to the study reviewed above, time stability analysis is developing from two spatial series to multiple series, from one scale to multiple scales. With the advance of methodology, a deeper understanding of spatial patterns of soil moisture can be expected. Future work should make full use of various methods, especially those most recently developed, such as EOF, wavelet coherency, and MEMD to gain greater insight into the spatial pattern dynamics at different scales.

4.2. Identify time stable locations for average soil water estimation

One of the most important applications of time stability is to identify a representative location for average soil water content estimation for a given area. Identification of representative locations cover different land uses, different climate zones, and places with various topographical attributes (Gao et al., 2011).

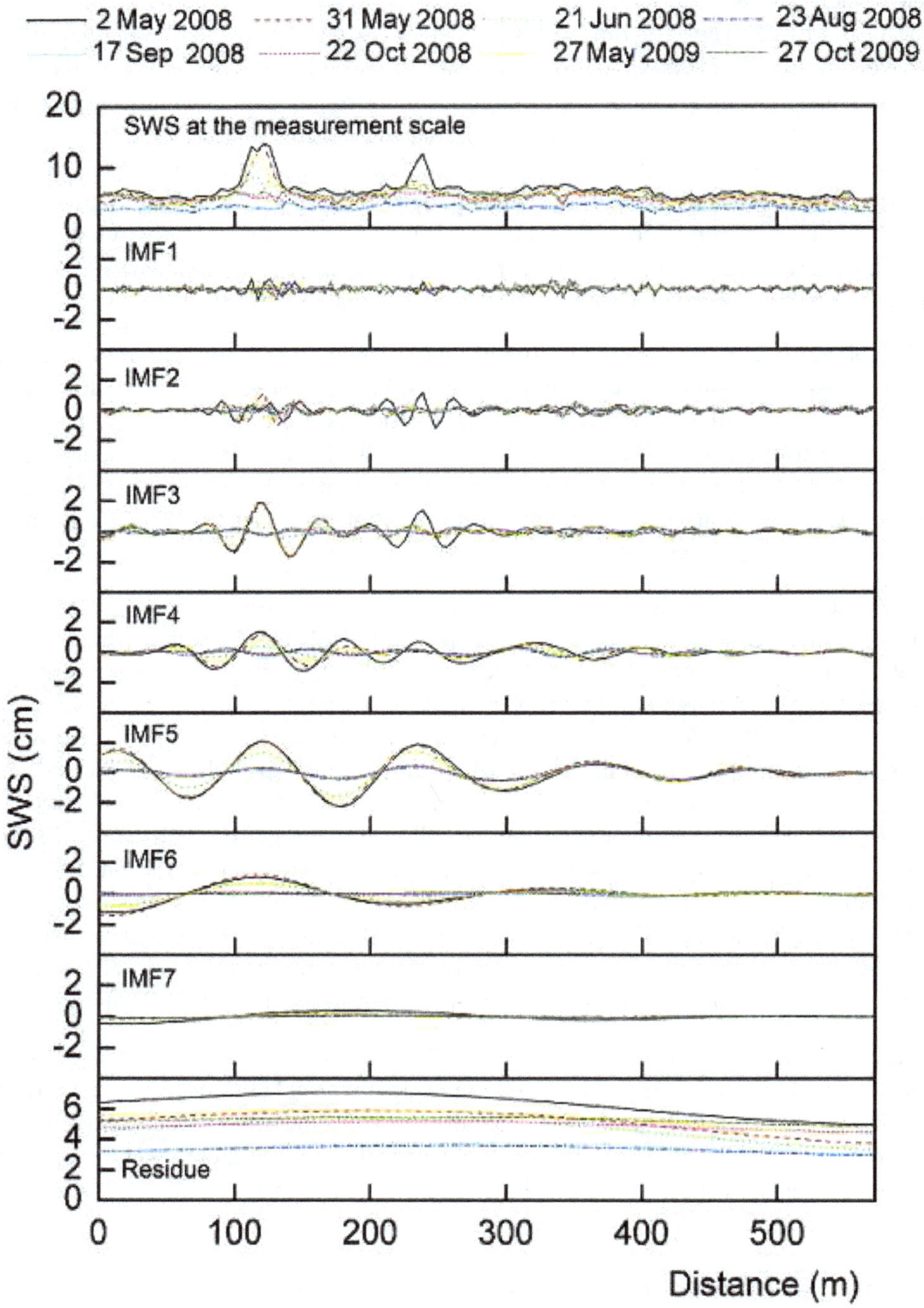

Figure 5. Intrinsic mode functions (IMFs) and residues based on multivariate empirical mode decomposition using soil water storage series (top) of the surface layer (0-20 cm). The horizontal axis is the distance between a sampling location and the origin of the transect. The vertical solid bar in Y-axis shows the scale of IMFs. (Reproduced from Hu et al. (2012a)).

Various time stability indices for use at the point scale have been used to identify representative locations. Most studies have been aimed at finding locations where soil water content can directly represent the mean of an area (Brocca et al., 2009; Cosh et al., 2008; Grayson and Western, 1998; Martínez-Fernández and Ceballos; 2005). In this way, *SWC* at the time stable location should approximate the average *SWC* of the study area. By extension, point scale time stability representing the areal average also means that the mean relative difference of soil water content, $\langle \delta_i \rangle$, at the stable location *i* should be close to zero at all times. Practically, it is difficult to find a location where the mean relative difference is the close to zero at all times. In this situation, the time stability index *SDRD* is widely used to judge the persistence of the stable location. Therefore, the mean relative difference and associated *SDRD* should be considered together. Successful application of this method can be found in many publications (Brocca et al., 2009; Cosh et al., 2006, 2008; Gómez-Plaza et al., 2000; Grayson and Western, 1998; Tallon and Si, 2004). In the study of Grayson and Western (1998), although the overall spatial pattern of soil moisture was time unstable, some locations were time stable and can represent the average *SWC* of a landscape. In the Tarrawarra watershed (Figure 6), the mean relative differences for locations 1, 4, 12, and 13 were close to zero, and their *SDRD* were also relatively small, so these locations were time stable and could be used as representative location for average *SWC* estimation. In the study of Tallon and Si (2004) which was conducted on a chernozemic soil on the Canadian prairies (75 km northeast of Saskatoon, SK, Canada) time stability of spatial patterns of soil moisture for different depths (30, 60, 90, 120 and 160 cm) were observed. With the exception of one location that occurred at two separate depths, the locations representative of average *SWC* differed with soil depths.

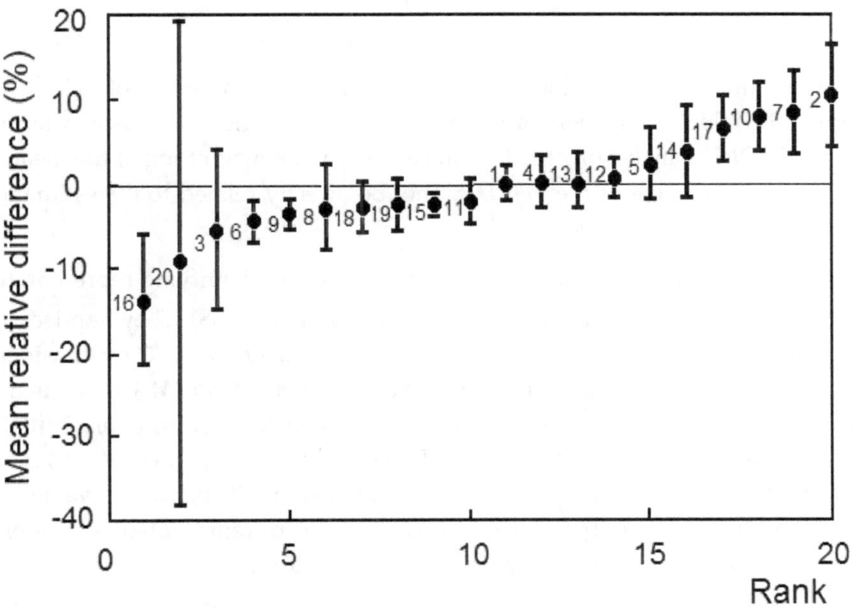

Figure 6. Ranked mean relative differences of soil water content for 0-60 cm measured by neutron probe in the Tarrawarra watershed. Also shown is one standard deviation error and site numbers (Reproduced from Grayson and Western (1998)).

Although mean relative difference and *SDRD* have been used successfully in the past in identifying representative locations for average *SWC*, the question becomes a matter of which index is more important when the location with mean relative difference closest to zero disagrees with that with the minimum *SDRD*. Most researchers believed that the *SDRD* was more important (Brocca et al., 2009; Cosh et al., 2006, 2008; Gómez-Plaza et al., 2000; Grayson and Western, 1998), while mean relative difference was also emphasized (Martínez-Fernández and Ceballos, 2003).

Instead of considering two indices simultaneously, one index, *RMSE*, combines both indices into one, thus making the identification of time stable location more objective. By this index, drier locations were usually found to have lower *RMSE* values than the wetter points (Cosh et al., 2004, 2006). This suggests that improved consistency will result from the selection of slightly drier sampling locations (Cosh et al., 2004). Operationally, it facilitates the selection of representative locations. However, it may confuse the concept of time stability, because for locations with large absolute values of relative difference (i. e. extremely dry or wet conditions) but very small *SDRD*, they may have large value of *RMSE* although they are completely time stable.

Another method for estimating average soil water content with time stable location involves a consideration of the constant offset (mean relative difference) for time stable location (Grayson and Western, 1998). Then $\langle SWC_j \rangle$ can be estimated by

$$\beta_j(i) = \frac{SWC_j(i)}{\langle SWC_j \rangle} \tag{33}$$

In this situation, time stable locations are usually defined by the value of *SDRD* (Grayson and Western, 1998). However, there are two limitations in using *SDRD* as a criterion. First, the value of the *SDRD* may be affected to some extent by the magnitude of the mean relative difference; second, the value of the *SDRD* cannot be directly related to the estimation error of the mean value (Hu et al., 2010b).

Indices that can consider the estimation error are needed for identifying representative time stable locations. One example was conducted by Guber et al. (2008). They applied $\chi^2(i)$ and $D(i)$ to rank the utility of sampling locations to estimate average *SWC* by considering the estimation error. Another alternative is the mean absolute bias error (*MABE*), which was developed directly to relate to the relative error of mean estimation (Hu et al., 2010a, 2010b). Although *MABE* was positively correlated to *SDRD* in some cases (Gao et al., 2011; Hu et al., 2010a, 2010b), the lowest value of *SDRD* did not coincide with the lowest value of *MABE* (Figure 7). Therefore, *MABE* outperformed *SDRD* in terms of representative location identification for indirect estimation of average *SWC* (Hu et al., 2010b).

The degree of representativeness of time stable locations may also change with season or year due to the change of the relative importance of different factors affecting time stability (Guber et al., 2008; Martínez-Fernández and Ceballos, 2003). Schneider et al. (2008) showed that the selected representative locations were appropriate to predict average soil water content of the sites for multiple years, although the time stability characteristics of some locations varied between years. An interesting study conducted by Guber et al. (2008) indicated that the utility of a location to estimate the estimation accuracy of field average *SWC* was not influenced by measurement frequency.

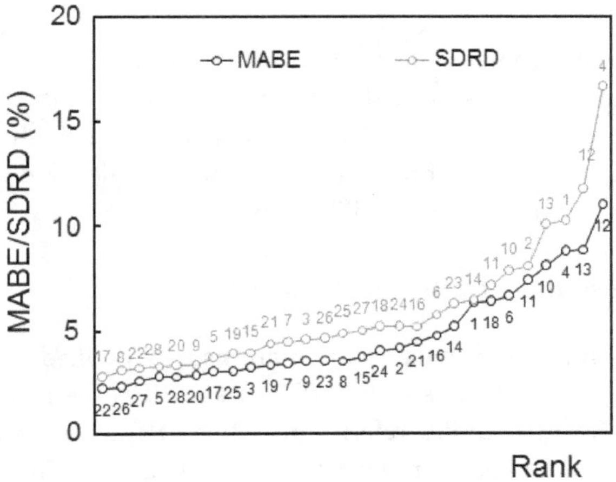

Figure 7. Rank ordered mean absolute bias error (*MABE*) and standard deviation of relative differences (*SDRD*) for the soil water storage of 0-1.0 m measured with neutron probe from October, 2004 to September, 2006. Also shown is location number (Reproduced from Hu et al. (2010b))

Many indices have been available to identify the representative location for average *SWC* estimation. However, few works have been conducted to compare their performances with a few exceptions (Gao et al., 2011; Hu et al., 2010a, 2010b, 2012b). *MABE* was reported to be a better index than *SDRD* in terms of average soil water storage estimation by the indirect method (Hu et al., 2010a, 2010b). More recently, with the soil water storage datasets from a transect in a Canadian prairie area and a watershed on the Chinese Loess Plateau, Hu et al. (2012b) evaluated the performance of different time stability indices in terms of the average soil water storage estimation quality judged by different goodness-of-fit indices. Their results showed that *MABE*, χ^2, *D*, and CV_t outperformed *SDRD* and *T*. If root mean squared deviations (*RMSD*) and Nash-Sutcliffe coefficient of efficiency (*NSCE*) were used as goodness-of-fit indices, *D* was the best time stability index. If absolute mean difference (*BIAS*) and absolute bias relative to mean (*RBIAS*) were adopted, *MABE* was the best time stability index. Their results also showed that average soil water storage estimation by the indirect method was more accurate than that by the direct method. It is necessary to compare their performances under other different hydrological backgrounds to obtain a more suitable index for identifying representative location. Furthermore, all these methods still need an intensive measurement of soil water content over time before a representative location can be identified. According to Martínez-Fernández and Ceballos (2005) about one year (a complete seasonal cycle) of measurements is needed to determine the representative location. Therefore, there is an urgent need to look for the possible definable features of the terrain and soils that could be used to determine the representative locations *a priori* (Martínez-FernándezandCeballos, 2005).

5. Controlling factors

Time stability of soil water content is controlled by factors that consistently influence soil moisture distribution in a similar way at all times. Most studies focus on factors influencing the underlying spatial pattern of soil moisture, and relatively few studies are associated with factors influencing the degree of time stability at point scales. We will review the controlling factors of the two types of time stability below.

5.1. Controlling factors of time stability of spatial pattern

Controlling factors of underlying spatial pattern are usually defined by comparing spatial patterns of time stability and various factors (Hu et al., 2010a; Perry and Niemann, 2007; Schneideret al., 2008). Related studies referred to different climate conditions, from semi-arid (Hu et al., 2010a) to humid environments (Perry and Niemann, 2007) and different land uses, such as rangeland (Mohanty and Skaggs, 2001), grass (Jacobs et al., 2004), forest (Lin, 2006), and agricultural land (Starr et al., 2005).

Soil properties such as soil texture are usually found to influence the spatial pattern of time stability. For example, Vachaud et al. (1985) found that the scaled factor of soil water content were significantly correlated to the silt plus clay contents on a field in St. Martin d'Heres, France. Mohanty and Skaggs (2001) observed that the sandy loam soil produced the best time-stable soil moisture pattern and silty loam soil produced an intermediate level of time stability in the Little Washita watershed of the southern part of the Great Plains of the US. Jacobs et al. (2004) pointed out that lower sand content may produce more time stable patterns in the Iowa. Grant et al. (2004) found a significant correlation between clay content, coarse fragment and the rank of mean relative difference of soil water storage in the Reynolds Mountain East catchment in the Owyhee Mountains. Starr (2005) found that coarser particle size classes were generally drier than adjacent finer particle size class soil on a potato farm located near Houlton, Maine. Hu et al. (2010a) recognized that sand and silt content can explain about 41.4% - 65.5 % of the spatial pattern of mean relative difference of soil water content. Except for soil texture, other soil properties such as organic matter content (Hu et al., 2009), bulk density (Jacobs et al., 2004), and soil thickness (Zhu and Lin, 2011) have also been recognized as the controlling factors for time stability of spatial pattern. However, soil properties may also have no influences on the spatial pattern of time stability. For example, Tallon and Si (2004) reported that clay content showed the least amount of control of spatial patterns on a chernozemic soil on the Canadian prairies (75 km northeast of Saskatoon, SK, Canada). On ungrazed sites, Schneider et al. (2008) found that neither bulk density, organic carbon nor sand and clay content could explain the time-stable characteristics of the sampling points. Comegna and Basile (1994) and Starr (2005) both observed the absence of the controlling influence of soil texture and they attribute it to the homogeneous distribution of soil texture in their study areas.

Topographic characteristics have also been recognized to influence the time stability of spatial patterns. Gómez-Plaza et al. (2000) showed that when the factors affecting soil moisture were limited to topographical position or local topography at the transect scale, spatial patterns of soil moisture presented time stability. Joshi et al. (2011) observed that fields with flat topography (LW21) showed the worst time stability features compared to the fields having gently rolling topography (LW03 and LW13) in Oklahoma. The influence of topography on the time stability of spatial pattern may be scale dependent. For example, although the Pearson correlation coefficient between soil water storage and elevation was very small, Biswas and Si (2011a) identified that strong scale-specific correlations existed between soil water storage and elevation at different scales using wavelet coherency, which contributed to the time stability of soil water storage at different scales. Kachanoski and De Jong (1988) attributed to the loss of time stability of soil water storage at scale smaller than 40 m to the role of spatial pattern of surface curvature.

Usually, soil properties and topographical properties jointly control the time stability of soil moisture spatial pattern. According to Joshi et al. (2011), soil properties (i. e., percent silt, percent sand, and soil texture) and topography (elevation and slope) jointly affected the spatiotemporal evolution and time stability of soil moisture at both point and footprint scales in the Little Washita watershed, Oklahoma and in the Walnut Creek watershed, Iowa. Zhu and Lin (2011) also observed that both soil and terrain controlled soil moisture variation at different seasons and soil depths at farm scale.

The relative role of soil and topographic properties in controlling the spatial variability and time stable pattern of soil moisture is usually related to the dominant hydrological process taking place for different soil water conditions between wet and dry periods. The *SWC* patterns existing between wet and dry periods are referred to as preferred states (Graysonet al., 1997; Western et al., 1999). When dry conditions dominate, evapotranspiration is greater than precipitation, and local controls dominate mainly by vertical movement of soil water with no connection between adjacent points. In this situation, differences in vegetation and soil properties such as texture would be responsible for the spatial pattern of soil moisture (Graysonet al., 1997). During wet states, evapotranspiration is less than precipitation and non-local controls govern soil moisture distribution. These controls are related to upslope topography and include catchment area, aspect, depth and soil profile curvature (Gómez-Plaza et al., 2001). Western at al. (1999) attributed these differences in local and non-local controls to the high degree of organization during wet periods that consist of connected bands of high soil water content in drainage lines. Soil water in these drainage lines is laterally redistributed by both overland and subsurface flow. Using EOF analysis, Perry and Niemann (2007) found that underlying spatial patterns of soil moisture were controlled by local soil properties in wet and dry conditions and topographic characteristics during intermediate conditions. Their results agreed with the findings of Western et al. (1999) that soil moisture patterns on moderately wet dates show a strong topographic influence, one that seems to diminish on both dry and very wet days. Note that soil porosity, rather than topography, will dominate the spatial distribution of soil moisture under very wet conditions (Western et al., 1999). Non-local control can also be found in the study of Biswas and Si (2012). In their study, the coefficient of determination between soil water storage and elevation increased more than eightfold after shifting the spatial series of soil water storage by a length equal to that of the existing slope. Although local and non-local control was widely recognized, it is still difficult to make a very clear limit between them, most likely due to the possible influences of topography on soil properties

Because of the different controls in different soil water conditions, time stability of soil water content tends to be stronger under the similar soil water conditions than those with quite different soil water conditions. For example, Martinez-Fernández and Ceballos (2003) observed that the periods with the lowest time stability coincided with situations involving the transition from dry to wet. Gao et al. (2011) revealed that the time stability of root zone soil water (0–60 cm) was higher in either dry or wet season than that including both, and soil water exhibited very low time stability during the transition period from dry to wet.

In addition to soil and topography, other factors such as vegetation (Pachepsky et al., 2005; Schneideret al., 2008), soil depth (Huet al., 2009) can also influence the time stability of soil moisture. For example, Gómez- Plaza et al. (2000) found that the existence of vegetation could destroy the time stability of spatial patterns of soil moisture. Mohanty and Skaggs (2001) found varying degrees of time stability depending on vegetation cover as well as topography. Generally, time stability of spatial pattern is strongest in deep soil layers than in shallow layers (Cassel et al., 2000). However, stronger time stability of spaital pattern at shallower depth (20 cm) than deeper soil depths was also observed (Hu et al., 2009, 2010a).

Although many factors have been revealed to influence the time stability of spatial pattern, no single, or even combined, factor can explain all time stable patterns. This suggests that some influencing factors may still remain undiscovered. In addition, random measurement errors also contribute to some variability of the time stable pattern (Starr, 2005). According to the study of Starr (2005), random measurement error can contribute 20% of the variability of underlying spatial pattern. Therefore, to better understand the spatio-temporal variability and time stability of spatial pattern of soil moisture, controlling factors should be made clearer when more accurate measurement of soil water content should also be expected.

5.2. Controlling factors of time stability at the point scale

Revealing the controlling factors of time stability at the point scale is crucial in identifying time stable locations for average *SWC* estimation. Soil and topography are usually found to influence the time stability of soil water content at point scales. Brocca et al. (2009) observed that the time stable positions were linked to topographic characteristics, primarily the upslope drainage area but also elevation and slope. Cosh et al. (2008) found that almost 60% of the *SDRD* variability can be credited to the soil type parameters. Soil parameters as major controlling factors is a finding that is in agreement with those of Hu et al. (2010a) who found that soil texture can significantly ($P<0.05$) affect the stability of soil water content. Gao et al. (2011) observed that time-stable locations corresponded to relatively high clay contents, mild slopes and planar surfaces, which agreed with the previous studies, where both soil type and topography influenced time stability at point scales (Jacobs et al., 2004; Joshi et al., 2011; Mohanty and Skaggs, 2011). However, poor relationships between time stable locations and soil and topographic properties were also found (Tallon and Si, 2004). Land use was observed to have no influences on time stability of soil water content at point scales (Hu et al., 2010a; Joshi et al., 2011).

Time stability at point scales was usually associated with the soil water conditions of that location. For example, Martinez-Fernández and Ceballos (2005) found that time stability as characterized by *SDRD* was usually stronger in dry locations than wet locations for all depths, and they attribute it to the predominance of the sandy fraction with the resultant weak ability to retain water in dry locations. Similar results were obtained in other studies (Bosch et al., 2006; Jacobs et al., 2004; Martínez-Fernández and Ceballos, 2003). However, the relationship of time stability and soil water condition is likely influenced by the time stability index used. In the LYMQ watershed on the Loess Plateau in China, Hu et al. (2010a) also found stronger time stability in drier locations using *SDRD*, while they observed stronger time stability in wetter locations when using *MABE*.

Time stability at point scales depends on different periods and soil depths. Generally, stronger time stability was found in wetter periods (Guber et al., 2008; Zhao et al., 2010; Zhou et al., 2007). Zhou et al. (2007) believed that the lateral redistribution of soil moisture in the wet/dry transition period and uneven evapotranspiration and root water uptake in drying process was the reasons for the relatively higher time stability during wetting periods. Guber et al. (2008) related the weaker time stability in the May–June period to the active vegetative growth of corn. Zhao et al. (2010) attributed the high temporal stability under wet conditions to an enhanced capillary movement of water from the subsoil to the topsoil. Time stability at point scale was usually found to be stronger at deeper depths observed by different time stability indices (Guberet al., 2008; Hu et al., 2010a, 2010b; Starks et al., 2006). This was generally in agreement with the relationship of time stability of spatial pattern with depth (Cassel et al., 2000).

Due to the complex influences of various factors on the time stability at the point scale, future work should be devoted to quantitatively define the contribution of various factors on time stability of soil water content. Models of the relationship of time stability and environmental factors should also be developed for the purpose of identification of time stable positions with known soil and topographic information. In addition, because of the different relationships between factors and different time stability indices (Hu et al., 2010a), attention should also be paid to the selection of time stability index for average *SWC* estimation in a field.

6. Conclusion

We reviewed the studies on time stability of soil water content, including its concept, methodology, application, and controlling factors. The following conclusions can be drawn: (1) Two types of time stability of soil water content can be classified; one is to describe the overall similarity of soil water content spatial patterns between different occasions, and the other is to describe the time invariance of the relative soil water content with time at point scales; (2) Time stability can be used to analyze the time persistence of spatial pattern of soil water content. Spearman's rank correlation, Pearson correlation analysis, average spatial coefficient of determination, and temporal coefficient of variability can be used to characterize the time stability of spatial pattern for two spatial series. Empirical orthogonal function can be used to extract the same spatial patterns of soil moisture for multiple spatial series. Spectral coherency, wavelet coherency, and multivariate empirical mode decomposition can be used to identify the scale specific time stability of a spatial pattern. However, the spectral coherency method assumes spatial series are stationary and linear, wavelet methods assume the spatial series to be linear, and both of them are only applicable to the case of two spatial series. For multivariate empirical mode decomposition, it can deal with non-stationary and nonlinear systems with multiple spatial series; (3) Time stability can be used to identify the most time-stable locations for average *SWC* evaluation for a given field. Standard deviation of relative difference, root mean square error, standard deviation of soil water content, width of the 90% empirical tolerance interval of relative water content, and mean absolute bias error can be used to identify the most time-stable location. However, the performance of these indices should be compared in different environmental backgrounds to look for the most suitable index; (4) Time stability of soil water content can be influenced by many factors, including soil, topography, vegetation, and climate. It also shows non-local control and local control, depending on the soil water conditions. Knowledge of the controlling factors for time stability of spatial pattern and that at point scale is important for understanding the soil water processes, soil water management, and identification of time-stable locations for average *SWC* evaluation. Future work should quantify the influences of different factors on time stability, and make deeper understanding of nonlocal and local control on time stability of soil water content; (5) Time stability of soil water content is also depth-wise and scale specific. Usually, time stability of soil moisture increased with depth and scale. Time stability analysis can not only facilitate our understanding to soil water related processes but also greatly reduce the sampling work for soil water content in fields.

Acknowledgements

The project was funded by Natural Science and Engineering council (NSERC) of Canada.

Author details

Wei Hu[1], Lindsay K. Tallon[1], Asim Biswas[2] and Bing Cheng Si[1]

*Address all correspondence to: bing.si@usask.ca

1 University of Saskatchewan, Department of Soil Science, Saskatoon, Canada

2 Department of Natural Resource Sciences, McGill University, Ste-Anne-de-Bellevue, Quebec, Canada

References

[1] Biswas, A., & Si, B. C. (2011a). Identifying Scale-Specific Controls of Soil Water Storage in a Hummocky Landscape Using Wavelet Coherency. Geoderma, 165, 50-59.

[2] Biswas, A., & Si, B. C. (2011b). Revealing the Controls of Soil Water Storage at Different Scales in a Hummocky Landscape. Soil sci. soc. am. j., 75, 1295-1306.

[3] Biswas, A., & Si, B. C. (2011c). Scales and Locations of Time Stability of Soil Water Storage in a Hummocky Landscape. J. hydrol., 408, 100-112.

[4] Biswas, A., & Si, B. C. (2012). Identifying Effects of Local and Nonlocal Factors of Soil Water Storage Using Cyclical Correlation Analysis. Hydrol. Process., doi: 10.1002/hyp.8459.

[5] Bosch, D. D., Lakshmi, V., Jackson, T. J., Choi. M., & Jacobs, J. M. (2006). Large Scale Measurements of Soil Moisture for Validation of Remotely Sensed Data: Georgia Soil Moisture Experiment of 2003. J. hydrol., 323, 120-137.

[6] Brocca, L., Melone, F., Moramarco, T., & Morbidelli, R. (2009). Soil Moisture Temporal Stability over Experimental Areas in Central Italy. Geoderma, 148, 364-374.

[7] Cassel, D. K., Wendroth, O., & Nielsen, D. R. (2000). Assessing Spatial Variability in an Agricultural Experiment Station Field: Opportunities Arising from Spatial Dependence. Agron. j., 92, 706-714.

[8] Chen, Y. (2006). Letter to the Editor on Rank Stability or Temporal Stability. Soil sci. soc. am. j., 70, 306.

[9] Comegna, V., & Basile, A. (1994). Temporal Stability of Spatial Patterns of Soil Water Storage in a Cultivated Vesuvian Soil. Geoderma, 62, 299-310.

[10] Cosh, M. H., Jackson, T. J., Bindlish, R., & Prueger, J. H. (2004). Watershed Scale Temporal and Spatial Stability of Soil Moisture and Its Role in Validating Satellite Estimates. Remote sens. environ., 92, 427-435.

[11] Cosh, M. H., Jackson, T. J., Moran, S., & Bindlish, R. (2008). Temporal Persistence and Stability of Surface Soil Moisture in a Semi-Arid Watershed. Remote sens. environ., 112, 304-313.

[12] Cosh, M. H., Jackson, T. J., Starks, P., & Heathman, G. (2006). Temporal Stability of Surface Soil Moisture in the Little Washita River Watershed and Its Applications in Satellite Soil Moisture Product Validation. J. hydrol., 323, 168-177.

[13] Farge, M. (1992). Wavelet transforms and their applications to turbulence. Annu. rev. of fluid mech., 24, 395-457.

[14] Gao, X. D., Wu, P. T., Zhao, X. N., Shi, Y. G., & Wang, J. W. (2011). Estimating Spatial Mean Soil Water Contents of Sloping Jujube Orchards Using Temporal Stability. Agr. water manage., 102, 66-73.

[15] Gómez- Plaza, A., Alvarez-Rogel, J., Albaladejo, J., & Castillo, V. M. (2000). Spatial Patterns and Temporal Stability of Soil Moisture across a Range of Scales in a Semi-Arid Environment. Hydrol. process., 14, 1261-1277.

[16] Gómez-Plaza, A., Martínez-Mena, M., Albaladejo, J., & Castillo, V. M. (2001). Factors Regulating Spatial Distribution of Soil Water Content in Small Semi-Arid Catchments. J. hydrol., 253, 211-226.

[17] Grant, L., Seyfried, M., & McNamara, J. (2004). Spatial Variation and Temporal Stability of Soil Water in a Snow-Dominated, Mountain Catchment. Hydrol. process., 18, 3493-3511.

[18] Grayson, R. B., & Western, A. W. (1998). Towards Areal Estimation of Soil Water Content from Point Measurements: Time and Space Stability of Mean Response. J. hydrol., 207, 68-82.

[19] Grayson, R. B., Western, A. W., Chiew, F. H. S., & Blöschl, G. (1997). Preferred States in Spatial Soil Moisture Patterns: Local and Nonlocal Controls. Water resour. res., 33, 2897-2908.

[20] Grinsted, A. J., Moore, C., & Jevrejeva, S. (2004). Application of the Cross Wavelet Transform and Wavelet Coherence to Geophysical Time Series. Nonlinear proc. geoph., 11, 561-566.

[21] Guber, A. K., Gish, T. J., Pachepsky, Y. A., van Genuchten, M. T., Daughtry, C. S. T., Nicholson, T. J., & Cady, R. E. (2008). Temporal Stability in Soil Water Content Patterns across Agricultural Fields. Catena, 73, 125-133.

[22] Hu, W., Biswas, A., & Si, B. C. (2012a). Application of Multivariate Empirical Mode Decomposition for Revealing Scale Specific Time Stability of Soil Water Storage. J. hydrol., submitted.

[23] Hu, W., Shao, M. A., Han, F. P., Reichardt, K., & Tan, J. (2010a). Watershed Scale Temporal Stability of Soil Water Content. Geoderma, 158, 181-198.

[24] Hu, W., Shao, M. A., & Reichardt, K. (2010b). Using a New Criterion to Identify Sites for Mean Soil Water Storage Evaluation. Soil sci. soc. am. j., 74, 762-773.

[25] Hu, W., Shao, M. A., Wang, Q. J., & Reichardt, K. (2009). Time Stability of Soil Water Storage Measured by Neutron Probe and the Effects of Calibration Procedures in a Small Watershed. Catena, 79, 72-82.

[26] Hu, W., Tallon, L. K., & Si, B. C. (2012b). Evaluation of Time Stability Indices for Soil Water Storage Upscaling. J. hydrol., accepted.

[27] Huang, N. E., Shen, Z., Long, S. R., Wu, M. C. L., Shih, H. H., Zheng, Q., Yen, N. C., Tung, C. C., & Liu, H. H. (1998). The Empirical Mode Decomposition and the Hilbert

Spectrum for Nonlinear and Non-Stationary Time Series Analysis. Proc. r. soc. a, 454, 903-995.

[28] Huang, N. E., Wu, M. C. L., Long, S. R., Shen, S. S. P., Qu, W., Gloersen, P., & Fan, K. L. (2003). A Confidence Limit for the Empirical Mode Decomposition and Hilbert Spectral Analysis. Proc. r. soc. a, 459, 2317-2345.

[29] Huang, N. E., & Wu, Z. (2008). A Review on Hilbert-Huang Transform: Method and Its Applications to Geophysical Studies. Rev. geophys, 46, RG2006, doi: 10.1029/2007RG000228.

[30] Ibrahim, H. M., & Huggins, D. R. (2011). Spatio-Temporal Patterns of Soil Water Storage under Dryland Agriculture at the Watershed Scale. J. Hydrol., 404, 186-197.

[31] Jacobs, J. M., Mohanty, B. P., Hsu, E. C., & Miller, D. (2004). SMEX02: Field Scale Variability Time Stability and Similarity of Soil Moisture. Remote sens. environ., 92, 436-446.

[32] Jawson, S. D., & Niemann, J. D. (2007). Spatial Patterns from EOF Analysis of Soil Moisture at a Large Scale and Their Dependence on Soil, Land-Use, and Topographic Properties. Adv. water resour., 30, 366-381.

[33] Joshi. C., & Mohanty, B. P. (2010). Physical Controls of Near-Surface Soil Moisture across Varying Spatial Scales in an Agricultural Landscape During SMEX02. Water resour. res., 46, W12503, doi: 10.1029/2010WR009152.

[34] Joshi. C., Mohanty, B. P., Jacobs, J. M., & Ines, A. V. M. (2011). Spatiotemporal Analyses of Soil Moisture from Point to Footprint Scale in Two Different Hydroclimatic Regions. Water resour. res., 47, W01508, doi:10.1029/2009WR009002.

[35] Kachanoski, R. G., & De Jong, E. (1988). Scale Dependence and the Temporal Persistence of Spatial Patterns of Soil Water Storage. Water resour. res., 24, 85-91.

[36] Korres, W., Koyama, C. N., Fiener, P., & Schneider, K. (2010). Analysis of Surface Soil Moisture Patterns in Agricultural Landscapes Using Empirical Orthogonal Functions. Hydrol. earth syst. sci., 14, 751-764.

[37] Kumar, P., & Foufoula-Georgiou, E. (1993). A Multicomponent Decomposition of Spatial Rainfall Fields: 1. Segregation of Large- and Small-Scale Features Using Wavelet Transforms. Water resour. res., 29, 2515-2532.

[38] Kumar, P., & Foufoula-Georgiou, E. (1997). Wavelet Analysis of Geophysical Applications. Rev. geoph., 35, 385-412.

[39] Lakshmi, V., Piechota, T., Narayan, U., & Tang, C. L. (2004). Soil Moisture as an Indicator of Weather Extremes. Geophys. res. lett., 31, L11401, doi:10.1029/2004GL019930.

[40] Lin, H. (2006).Temporal Stability of Soil Moisture Spatial Pattern and Subsurface Preferential Flow Pathways in the Shale Hills Catchment. Vadose zone j., 5, 317-340.

[41] Martínez-Fernández, J., & Ceballos, A. (2003). Temporal Stability of Soil Moisture in a Large-Field Experiment in Spain. Soil sci. soc. am. j., 67, 1647-1656.

[42] Martínez-Fernández, J., & Ceballos, A. (2005). Mean Soil Moisture Estimation Using Temporal Stability Analysis. J. hydrol., 312, 28-38.

[43] Mohanty, B. P., & Skaggs, T. H. (2001). Spatio-Temporal Evolution and Time-Stable Characteristics of Soil Moisture within Remote Sensing Footprints with Varying Soil Slope and Vegetation. Adv. water resour., 24, 1051-1067.

[44] North, G. R., Bell, T. L., Cahalan, R. F., & Moeng, F. J. (1982). Sampling errors in the estimation of empirical orthogonal functions. Mon. weather rev., 110, 699-706.

[45] Pachepsky, Y. A., Guber, A. K., & Jacques, D. (2005). Temporal Persistence in Vertical Distributions of Soil Moisture Contents. Soil sci. soc. am. j., 69, 347-352.

[46] Perry, M. A., & Niemann, J. D. (2007). Analysis and Estimation of Soil Moisture at the Catchment Scale Using Eofs. J. hydrol., 334, 388-404.

[47] Rehman, N., & Mandic, D. P. (2009). http://www.commsp.ee.ic.ac.uk/~mandic/research/emd.htm.

[48] Rehman, N., & Mandic, D. P. (2010). Multivariate Empirical Mode Decomposition. Proc. r. soc. a, 466, 1291-1302.

[49] Rolston, D. E., Biggar, J. W., & Nightingale, H. I. (1991). Temporal Persistence of Spatial Soil-Water Patterns under Trickle Irrigation. Irrig. sci., 12, 181-186.

[50] Schneider, K,, Huisman, J. A., Breuer, L., Zhao, Y., & Frede, H. G. (2008). Temporal Stability of Soil Moisture in Various Semi-Arid Steppe Ecosystems and Its Application in Remote Sensing. J. hydrol., 359, 16-29.

[51] Si, B. C. (2008). Spatial Scaling Analyses of Soil Physical Properties: A Review of Spectral and Wavelet Methods, Vadose zone j., 7, 547-562.

[52] Si, B. C., & Zeleke, T. B. (2005). Wavelet Coherency Analysis to Relate Saturated Hydraulic Properties to Soil Physical Properties. Water resour. res., 41, W11424, doi: 10.1029/2005WR004118.

[53] Starks, P. J., Heathman, G. C., Jackson, T. J., & Cosh, M. H. (2006). Temporal Stability of Soil Moisture Profile. J. hydrol., 324, 400-411.

[54] Starr, G. C. (2005). Assessing Temporal Stability and Spatial Variability of Soil Water Patterns with Implications for Precision Water Management. Agr. water manage., 72, 223-243.

[55] Tallon, L. K., & Si, B. C. (2004). Representative Soil Water Benchmarking for Environmental Monitoring. J. environ. inform., 4, 581-590.

[56] Torrence, C., & Compo, G. P. (1998). A Practical Guide to Wavelet Analysis. Bull. am. meteor. soc., 79, 61-78.

[57] Vachaud, G., Passerat De Silans, A., Balabanis, P., & Vauclin, M. (1985). Temporal Stability of Spatially Measured Soil Water Probability Density Function. Soil sci. soc. Am. J., 49, 822-828.

[58] Vereecken, H., Kamai, T., Harter, T., Kasteel, R., Hopmans, J., & Vanderborght, J. (2007). Explaining Soil Moisture Variability as a Function of Mean Soil Moisture: A Stochastic Unsaturated Flow Perspective. Geophys. res. lett., 34, DOI: 10.1029/2007GL031813.

[59] Vivoni, E. R., Gebremichael, M., Watts, C. J., Bindlish, R., & Jackson, T. J. (2008). Comparison of Ground-Based and Remotely-Sensed Surface Soil Moisture Estimates over Complex Terrain During SMEX04. Remote sens. environ., 112, 314-325.

[60] Western, A., Grayson, R., Blöschl, G., Willgoose, G., & McMahon, T. (1999). Observed Spatial Organization of Soil Moisture and Its Relation to Terrain Indices. Water resour. res., 35, 797-810.

[61] Wilson, D. J., Western, A. W., & Grayson, R. B. (2005). A Terrain and Data-Based Method for Generating the Spatial Distribution of Soil Moisture. Adv. water resour., 28, 43-54.

[62] Yoo, C., & Kim, S. (2004). EOF Analysis of Surface Soil Moisture Field Variability. Adv. water resour., 27, 831-842.

[63] Zar, J. H. (1972). Significance Testing of the Spearman Rank Correlation Coefficient. J. am. stat. assoc., 67, 578-580.

[64] Zhao, Y., Peth, S., Wang, X. Y., Lin, H., & Horn, R. (2010). Controls of Surface Soil Moisture Spatial Patterns and Their Temporal Stability in a Semi-Arid Steppe. Hydrol. process., 24, 2507-2519.

[65] Zhou, X., Lin, H., & Zhu, Q. (2007). Temporal Stability of Soil Moisture Spatial Variability at Two Scales and Its Implication for Optimal Field Monitoring. Hydrol. earth syst. sci. discuss, 4, 1185-1214.

[66] Zhu, Q., & Lin, H. (2011). Influences of Soil, Terrain, and Crop Growth on Soil Moisture Variation from Transect to Farm Scales. Geoderma, 163, 45-54.

Permissions

All chapters in this book were first published by InTech Open; hereby published with permission under the Creative Commons Attribution License or equivalent. Every chapter published in this book has been scrutinized by our experts. Their significance has been extensively debated. The topics covered herein carry significant findings which will fuel the growth of the discipline. They may even be implemented as practical applications or may be referred to as a beginning point for another development.

The contributors of this book come from diverse backgrounds, making this book a truly international effort. This book will bring forth new frontiers with its revolutionizing research information and detailed analysis of the nascent developments around the world.

We would like to thank all the contributing authors for lending their expertise to make the book truly unique. They have played a crucial role in the development of this book. Without their invaluable contributions this book wouldn't have been possible. They have made vital efforts to compile up to date information on the varied aspects of this subject to make this book a valuable addition to the collection of many professionals and students.

This book was conceptualized with the vision of imparting up-to-date information and advanced data in this field. To ensure the same, a matchless editorial board was set up. Every individual on the board went through rigorous rounds of assessment to prove their worth. After which they invested a large part of their time researching and compiling the most relevant data for our readers.

The editorial board has been involved in producing this book since its inception. They have spent rigorous hours researching and exploring the diverse topics which have resulted in the successful publishing of this book. They have passed on their knowledge of decades through this book. To expedite this challenging task, the publisher supported the team at every step. A small team of assistant editors was also appointed to further simplify the editing procedure and attain best results for the readers.

Apart from the editorial board, the designing team has also invested a significant amount of their time in understanding the subject and creating the most relevant covers. They scrutinized every image to scout for the most suitable representation of the subject and create an appropriate cover for the book.

The publishing team has been an ardent support to the editorial, designing and production team. Their endless efforts to recruit the best for this project, has resulted in the accomplishment of this book. They are a veteran in the field of academics and their pool of knowledge is as vast as their experience in printing. Their expertise and guidance has proved useful at every step. Their uncompromising quality standards have made this book an exceptional effort. Their encouragement from time to time has been an inspiration for everyone.

The publisher and the editorial board hope that this book will prove to be a valuable piece of knowledge for researchers, students, practitioners and scholars across the globe.

List of Contributors

Yongfu Xu
Department of Civil Engineering, Shanghai Jiao Tong University, Shanghai, China

Bill X. Hu
Collage of Water Resources and Environmental Sciences, China University of Geosciences, Beijing, China
Department of Earth, Ocean and Atmospheric Sciences, Florida State University, Tallahassee, USA

Juxiu Tong
College of Water Resources and Environmental Sciences, China University of Geosciences, Beijing, China

Katarzyna Marciszewska and Mirela Tulik
Department of Forest Botany, Warsaw University of Life Sciences - WULS-SGGW, Warsaw, Poland

Béatrice Yve
ANDRA, France

Marie-France Jutras and Paul A. Arp
Faculty of Forestry and Environmental Management, University of New Brunswick, Fredericton, Canada

Paulo Leonel Libardi
Exact Sciences Department - ESALQ-USP, Piracicaba, SP, Brazil

Flávia Carvalho Silva
Agronomics Sciences Department - UEM, Umuarama, PR, Brazil

Maria Clementina Caputo and Rita Masciale
Water Research Institute (IRSA), National Research Council (CNR), Bari, Italy

Wei Hu, Lindsay K. Tallon and Bing Cheng Si
University of Saskatchewan, Department of Soil Science, Saskatoon, Canada

Jean Luc Mari
IFPEN, France

Asim Biswas
Department of Natural Resource Sciences, McGill University, Ste-Anne-de-Bellevue, Quebec, Canada

Index

R

S

T

V

W

X

www.ingramcontent.com/pod-product-compliance
Lightning Source LLC
LaVergne TN
LVHW080806220926
848485LV00011B/276